SOLUTIONS FOR SUSTAINABLE DEVELOPMENT

PROCEEDINGS OF THE 1ST INTERNATIONAL CONFERENCE ON ENGINEERING SOLUTIONS FOR SUSTAINABLE DEVELOPMENT (ICES²D 2019), MISKOLC, HUNGARY, 3-4 OCTOBER 2019

Solutions for Sustainable Development

Editors

Klára Szita Tóthné, Károly Jármai & Katalin Voith

University of Miskolc, Hungary

CRC Press
Taylor & Francis Group
Boca Raton London New York

CRC Press is an imprint of the
Taylor & Francis Group, an **informa** business

A BALKEMA BOOK

Published by:
CRC Press/Balkema
P.O. Box 447, 2300 AK Leiden, The Netherlands
e-mail: Pub.NL@taylorandfrancis.com
www.crcpress.com – www.taylorandfrancis.com

First issued in paperback 2021

ISBN 13: 978-1-03-223778-7 (pbk)
ISBN 13: 978-0-367-42425-1 (hbk)

DOI: https://doi.org/10.1201/9780367824037

Typeset by Integra Software Services Pvt. Ltd., Pondicherry, India

Visit the Taylor & Francis Web site at
http://www.taylorandfrancis.com

and the CRC Press Web site at
http://www.crcpress.com

Library of Congress Cataloging-in-Publication Data

Table of contents

Preface   ix

Acknowledgements   xi

Committees   xiii

Part A: Process Engineering, Modelling and Optimisation

Analytical and numerical study for minimum weight sandwich structures   3
A. Al-Fatlawi, K. Jármai & G. Kovács

Estimation of compressive strength of current concrete materials: Effect of core diameter and maximum aggregate size   12
M.E. Belgacem, A. Brara & A. Ferdjani

The effect of the gear wear for the contact ratio   20
J. Bihari

Determination of GTN parameters using artificial neural network for ductile failure   25
Y. Chahboub, S. Szabolcs & H. Aguir

Investigating three learning algorithms of a neural networks during inverse kinematics of robots   33
H.M. Ghafil, L. Kovács & K. Jármai

Prediction of distillation plate's efficiency   41
V. Kállai, P. Mizsey & L.G. Szepesi

CAD tools for knowledge based part design and assembly versioning   49
K. Nehéz, P. Mileff & O. Hornyák

Heat transfer analysis for finned tube heat exchangers   56
M. Petrik, L.G. Szepesi & K. Jármai

Effect of soil reinforcement with stone columns on the behavior of a monopile foundation subjected to lateral cyclic loads   67
S.A. Rafa, I. Rouaz & A. Bouaicha

Performance of Cold Formed Steel Shear Wall Panel with OSB Sheathing under lateral load   73
I. Rouaz, S. Rafa & A. Bouaicha

The use of the linear sliding wear theory for open gear drives that works without lubrication   81
F. Sarka

Numerical simulation methods of stress corrosion cracking   86
B. Spisák, Z. Siménfalvi, Sz. Szávai & Z. Bézi

Implementation of a customized CAD extension to improve the calculation of center of gravity   94
M. Szabó, P. Mileff & K. Nehéz

Part B: Sustainable and Renewable Energy and Energy Engineering

Towards CHP system: Preliminary investigation and integration of an ORC cycle on a
simple gas boiler   101
M. Amara, H. Semmari & A. Filali

Examination the effect of environmental factors on a photovoltaic solar panel   108
I. Bodnár, L.T. Tóth, J. Somogyiné Molnár, N. Szabó, D. Erdősy & R.R. Boros

Thermal behaviour of PCM-gypsum panels using the experimental and the theoretical
thermal properties-Numerical simulation   115
F. Boudali Errebai, S. Chikh, L. Derradji, M. Amara & A. Terjék

Experimental and numerical study of the thermal behavior of a building in Algeria   123
L. Derradji, F. Boudali Errebai, M. Amara. A. Limam & A. Terjék

Comparison of Sakiadis and Blasius Flows using Computational Fluid Dynamic   129
M.M. Klazly & G. Bognár

Techniques for evaluation of mixing efficiency in an anaerobic digester   139
B. Singh, Z. Szamosi & Z. Siménfalvi

Part C: Waste Management and Reverse Logistics

Vehicle routing in drone-based package delivery services   151
A. Agárdi, l. Kovács & T. Bányai

Cyber-physical waste collection system: a logistics approach   160
M.Z. Akkad & T. Bányai

Efficiency improvement of reverse logistics in Industry 4.0 environment   169
I. Hardai, B. Illés & Á. Bányai

Application of sigmoid curves in environmental protection   178
F.J. Szabó

Part D: Environmental Management and Ecodesign

Comparative study of the influence of traditional walls with different typologies and
constructive techniques on the energy performance of the traditional dwellings of the Casbah
of Algiers   187
H.F. Arrar, A. Abdessemed Foufa & D. Kaoula

Rules of environmentally friendly packaging   196
Á. Takács

Part E: Circular Economy and Life Cycle Approaches

Aluminium infinite green circular economy – theoretical carbon free infinite loop,
combination of material and energy cycles   205
N. Babcsán

Life Cycle approach of a new Industrial Symbiosis alternative   211
B.S. Gál, R.S. Bodnárné & Zs. István

Possibilities for adopting the circular economy principles in the EU steel industry   218
Á. Kádárné Horváth, M. Kis-Orloczki & Á. Takácsné Papp

Influence of urban morphology on the environmental impactsof district. Applied LCA 227
D. Kaoula & A. Abdessemed Foufa

Life cycle extension of damaged pipelines using fiber reinforced polymer matrix composite
wraps 235
J. Lukács, Zs. Koncsik & P. Chován

Innovative solutions for the building industry to improve sustainability performance with
Life Cycle Assessment modelling 245
V. Mannheim, Zs. Fehér & Z. Siménfalvi

Investigation of different foam glasses with Life Cycle Assessment method 254
A. Simon, K. Voith & V. Mannheim

Circular economy solutions for industrial wastes 267
K.T. Szita, Zs. István, R.S. Bodnárné & A. Zajáros

New approach for characterizing the "naturalness" of building materials 275
A. Velősy

Part F: Smart Manufacturing and Smart Buildings

Grouping and analyzing PLC source code for smart manufacturing 287
O. Hornyák

Environmental assessment of buildings in Sudan 294
S.A.A. Ismail & Zs. Szalay

Part G: Innovation and Efficiency

The innovation shell, and barriers of disruptive innovation 307
Z. Bartha & A.S. Gubik

Lifetime analyses of S960M steel grade applying fatigue and fracture mechanical approaches 316
Zs. Koncsik

Efficient application of S690QL type high strength steel for cyclic loaded welded structures 325
H.F.H. Mobark & J. Lukács

Improvement of the modern house's energy efficiency in the regionof In Saleh 337
S. Oukaci, A. Hamid, D. Sammar, S. Sami & A. Naimi

Innovative and efficient production of welded body parts from 6082-T6 aluminium alloy 347
R.P.S. Sisodia & M. Gáspár

Redundancy Analysis of the Railway Network of Hungary 358
B.G. Tóth

Part H: Earth Science

A short discussion on some influencing factors of an artificial corrosion system and obtained
metallic pipe samples 371
H.D. Thien, B. Kovács, T. Madarász & I. Czinkota

Auothor index 379

Solutions for Sustainable Development – Szita, Jármai & Voith (eds.)
© 2020 Taylor & Francis Group, London, ISBN 978-0-367-42425-1

Preface

The International Conference on Engineering solutions and Sustainable Development is organized for the first time starting a – hopefully – long lasting tradition. The conference was held on 3–4 October, 2019 at the unique campus of University of Miskolc, Hungary.

The demands of this modern age brought up the idea in the head of the Organizing Committee members to put together this conference with the aim of creating an interdisciplinary platform for researchers and practitioners to present and discuss the most recent innovations, trends, and concerns as well as practical challenges encountered and solutions adopted in the fields of Technical and Environmental Science.

The conference covers the following topics:

- Process Engineering, Modelling and Optimisation
- Sustainable and Renewable Energy and Energy Engineering
- Waste Management and Reverse Logistics
- Environmental Management and Ecodesign
- Circular Economy and Life Cycle Approaches
- Smart Manufacturing and Smart Buildings
- Innovation and Efficiency
- Earth Science

Academics, scientists, researcher and professionals from different countries and continents have contributed to this book.

The editors would like to express their appreciation and gratefulness to the members of the Organizing, Scientific and Reviewing Committee. Special thanks to Dr. Viktória Mannheim (University of Miskolc) to do the first steps of starting this international conference. The editors would also like to express their thanks to all the reviewers, sponsors and conference participants for their time, help and cooperation to make this conference successful.

The editors are also thankful for all the help, patience and understanding of the Publisher during the preparation of the Proceedings.

K. Szita Tóthné
K. Jármai
K. Voith

Acknowledgements

This proceedings were prepared with the support of the following companies, institutions and organisations

MISKOLCI EGYETEM (UNIVERSITY OF MISKOLC)

GÉPÉSZMÉRNÖKI ÉS INFORMATIKAI KAR (FACULTY OF MECHANICAL ENGINEERING & INFORMATICS)

Felsőoktatási és Ipari Együttműködési Központ

Építésügyi Minőségellenőrző Innovációs Nonprofit Kft.

Bay Zoltán Alkalmazott Kutatási Nonprofit Kft.

Magyar Tudományos Akadémia (Hungarian Academy of Sciences), MISKOLCI TERÜLETI BIZOTTSÁGA Anyagtudományi- és Technológiai Szakbizottság Energia- és Környezetgazdálkodási Munkabizottság

LCA CENTER, Hungary

Solutions for Sustainable Development – Szita, Jármai & Voith (eds.)
© 2020 Taylor & Francis Group, London, ISBN 978-0-367-42425-1

Committees

ORGANIZING COMMITTEE

Prof. Dr. Klára SZITA Tóthné - LCA Center Hungary Association
Prof. Dr. Károly JÁRMAI - University of Miskolc
Adrienn BUDAY- MALIK - ÉMI Nonprofit Llc.
Zsolt ISTVÁN - Bay Zoltán Nonprofit Ltd.
Zsuzsanna KONCSIK - University of Miskolc
Tamás MADARÁSZ - University of Miskolc
Viktória MANNHEIM - University of Miskolc
Zoltán SIMÉNFALVI - University of Miskolc
Katalin VOITH – University of Miskolc

SCIENTIFIC COMMITTEE

Prof. Dr. Klára SZITA Tóthné - LCA Center Hungary Association
Prof. Dr. Károly JÁRMAI - University of Miskolc
Krisztián M. Baracza - University of Miskolc
Prof. Dr. Menouer BOUGHEDAOUI - University Blida
Adrienn BUDAY- MALIK - ÉMI Nonprofit Llc.
György GRÖLLER - LCA Center Hungary Association
Anthony HALOG - University of Queensland
Zsolt ISTVÁN - Bay Zoltán Nonprofit Ltd.
Prof. Dr. Zbigniew KŁOS - Poznan University of Technology
Zsuzsanna KONCSIK - University of Miskolc
Tamás MADARÁSZ - University of Miskolc
Viktória MANNHEIM - University of Miskolc
Paolo MASONI - Ecoinnovazione srl
Sándor NAGY - University of Miskolc
Zoltán SIMÉNFALVI - University of Miskolc
Prof. Dr. János SZÉPVÖLGYI - Hungarian Academy of Sciences
Petar VARBANOV - Brno University of Technology
Khalid M. YOUSRI - Housing and Building National Research Center, HBRC

LIST OF REVIEWERS

Árvai L., Bencs P., Berényi L., Boudali Errebai F., Czél Gy., Deme Bélafi Zs., Derradji L., Dobosy Á., Duarte Neves R., Dudra J., Dúl R., Gáspár M., Géber R., Gröller Gy., Gubik A., Hajtó J., Hegedűs Gy., Hégely L., István Zs., Jármai K., Józsa Zs., Király S., Kollányi T., Koritárné Fótos R., Kovács E., Kovács S., Kuzsella L., Láng P., Lévai Zs., Lukács J., Madarász T., Májlinger K., Mannheim V., Medgyasszay P., Mileff P., Mileffné Dudra J., Molnár V., Mosavi A., Nagy T., Nováky E., Sára B., Szabó J. F., Szalay Zs., Szamosi Z., Szaszák N., Szepesi G., Szépvölgyi J., Sziklai B., Szilágyi A., Szirbik S., Terjék A., Tervo H., Tímár I., Tóth L., Tóthné Szita K., Török T., Trampus P., Varga T.

Part A: Process Engineering, Modelling and Optimisation

Solutions for Sustainable Development – Szita, Jármai & Voith (eds.)
© 2020 Taylor & Francis Group, London, ISBN 978-0-367-42425-1

Analytical and numerical study for minimum weight sandwich structures

A. Al-Fatlawi, K. Jármai & G. Kovács
Faculty of Mechanical Engineering and Informatics, University of Miskolc, Miskolc, Hungary

ABSTRACT: Manufacturing a light weight structure with affordable cost without sacrificing strength has been a challenging task for designers. The 4-point bending test is required to obtain the static characteristics (deflection and stress with loading) for the sandwich structures consist of Aluminium honeycomb and foam core with composite material face sheets experimentally on the set of the specimens and numerically by using FEA (Digimat program) and compare between them. In this study, a minimum weight optimization for sandwich panels with two types of core (honeycomb and foam) is presented by using pure Matlab scripts (m. files) for honeycomb and foam core to compare between them. The weight of the sandwich panels considered is the objective function subject to required constraints based on the stiffness, face sheets failure, skin wrinkling and core shear as well as the unknown variables the thicknesses of the core t_c and the face sheet t_f.

1 INTRODUCTION

1.1 *Honeycomb core*

Bitzer (1997) described many methods to minimize the weight of the sandwich beams with isotropic symmetrical and unsymmetrical face sheets and subjected to bending and torsional stiffness requirement. 5052 expanded aluminum honeycomb materials are available in a wide selection of cell sizes and foil gauges. It expanded honeycomb is manufactured by bonding together sheets of aluminum foil, then expanding to form a cellular honeycomb configuration. Aluminum honeycomb materials specification grade honeycomb materials are predominantly used in sandwich structures to meet design requirements for highly engineered structural components. As a structural core material, it finds applications in all types of aerospace vehicles and containers as shown in Figure 1. This lightweight structure is attractive for utilizing in aerospace and automotive engineering applications due to its high performance like bending stiffness and strength to weight ratios. The minimum weight of sandwich structures is one of the important objectives in engineering applications. Many studies were emphasized on how to achieve the optimum design of sandwich structure with face sheets made of metal or composite include minimizing the weight and maximizing the stiffness, strength. Some studies were limited to a single objective optimization problem to minimize the weight of the sandwich structure. (Gibson 1997, Christos 1997 & Rodrigues 2015). A few studies included the weight and the cost of the sandwich structure as a design objective. Several different techniques and methods have been suggested to solve complexities in single or multi-objective design optimization problems. There are various design parameters to stratify an optimization technique for sandwich structure with composite face sheets like ply material, ply orientation angles, and core thickness. Rodrigues (2015) used Interior Point Algorithm with the finite element, ABAQUS to presented computational model combined topology and stacking sequence of composite laminated structures. Ju (2013) applied Multi-parameter approach to optimize the lightweight composite structure subjected to nonlinear constraints. Mamalis (2008) improved hybrid sandwich structures consist of metallic face sheets and wood layers

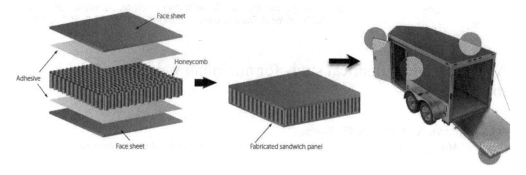

Figure 1. Honeycomb sandwich structures.

adhesively bonded between the face sheets and the core material under impact resistance. In this paper, we suggest an optimization approach to minimize the weight of honeycomb and foam sandwich panels made from composite face sheets.

1.2 *Foam core*

Cymat A35620SC 020SS stabilized aluminum foam (SAF) is closed cell aluminum foam formed from A356 aluminum alloy. The mechanical properties of SAF make it ideal for many varied applications like shipping and airplane container. These properties include high strength and stiffness to weight ratio and high mechanical energy absorption in all directions, not flammable or susceptible to environmental degradation, acoustic and thermal insulation properties, electromagnetic insulation properties and recyclable. There are many types of core as shown the Figure 2. (Gibson 1984) Described an analytical method to find the optimum thickness and density of the foamed core and the thickness of isotropic face sheets which minimize the weight of a sandwich beam for given stiffness and span length.

2 FOUR – POINT BENDING TEST

Hexcel Composites Publication. 2007 clarified the technical notes of the mechanical testing of the sandwich panels. There are many types of failure modes for the bending test as shown in Figure 3. Some of the more typical failure modes include tension and compression face sheet failure and wrinkling failure of face sheet. From the experimental and numerical 4-point

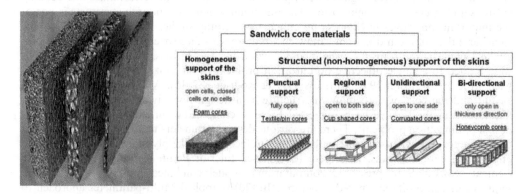

Figure 2. Aluminium foam core and sandwich core materials.

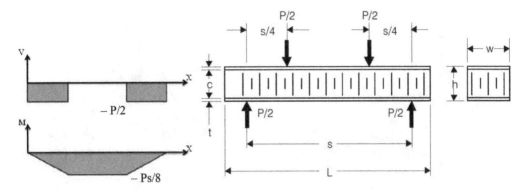

Figure 3. The norm of 4-point flexural test.

bending test for symmetrical sandwich panels, the stress and deflection with load can be calculated. These data (maximum deformation and maximum load define) can be used in Matlab software to define the load and the stiffness required. The standard of the 4-point bending test refers to MIL-STD-401B Sec.5.2.4 or ASTM C-393.

The average skin stress and modulus can be determined with the following equations:

$$\sigma = \frac{Ps}{8(h-t)wt} \tag{1}$$

$$E = \frac{11}{384} \frac{P}{d} \frac{s^3}{wt(h-t)^2} \tag{2}$$

Where: s = span, c = core thickness, P = total applied load, t = skin thickness, d = deflection at mid-span, w = panel width, h = panel thickness, P/d = load-deflection curve slope, L = specimen length, σ = skin stress, E = skin modulus.

Plascore Honeycomb Core Makes a sandwich structures lighter, stiffer and stronger than single sheet laminate. The core increases the flexural stiffness of a sandwich panel by effectively increasing the distance between the two stress skins. The honeycomb and foam cores also effectively provide shear resistance, a key component to overall flexural stiffness. The sandwich structure is stiffer and lighter than single sheet laminate. The stiffness of honeycomb and foam laminations allows container builders to use less material, reducing weight. The stiffness increases exponentially compared to single sheet material. The use of honeycomb and foam core creates a dramatic increase in stiffness with very little weight gain as shown in Figure 4.

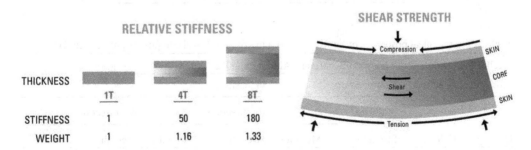

Figure 4. Relative stiffness and shear strength of sandwich structure.

5

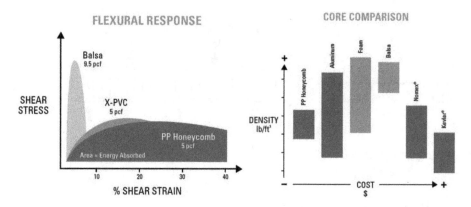

Figure 5.   Core comparison cost with density and flexural response.

To save the weight and money, the Aluminum honeycomb core is not only lightweight cores, but they are also more cost-effective than balsa and foam and do not absorb water, while Nomex and Kevlar, typically used in racing boats for their aerospace qualities, are extremely light, with high temperature stability for prepreg applications. The value of each core must be weighed in a strategy: which core delivers the best performance, is most compatible and is readily available at a competitive cost for the specific application. Most core materials respond similarly to stress under normal operating loads. As loading increases, the core begins to flex to accommodate the increase in shear stress on the core. Unlike other core materials that reach an ultimate yield stress and fail catastrophically as shown in Figure 5.

3   FINITE ELEMENT ANALYSIS (DIGIMAT–HC) PROGRAM

In this study, the sandwich panels made up of composite face sheets and Aluminium honeycomb and foam core. The face sheets were symmetrical with respect to the mid-plane of the sandwich panel. Every face sheet consisted of E-glass fabric/epoxy resin. The mechanical properties of the core and face sheet used in this study are given in Table 1, 2 and 3 respectively.

Table 1.   Foam core materials properties (Cymat A35620SC 020ss stabilized aluminum foam).

E	Compressive modulus	Mpa	600
ν	Poisson's ratio	—	0.33
G	Shear modulus	Mpa	200
σ	Compressive strength	Mpa	1.6
τ	Shear strength	Mpa	0.5
ρ	Density	kg/m^3	200

Table 2.   Mechanical properties of honeycomb core materials (Hexcel composites publication. 2007).

Mechanical properties of honeycomb materials Aluminum 5052							
Product construction		Compression		Plate shear			
		Stabilized		L - direction		W - direction	
Density	Cell size	Strength	Modulus	Strength	Modulus	strength	Modulus
kg/m3	mm	Mpa	Mpa	Mpa	Mpa	Mpa	Mpa
130	3	11	2414	5	930	3	372

Table 3.　Mechanical Properties of E-glass fiber composite materials/Epoxy resin from the website of performance composites ltd.

Loading axis 0/90°, dry, room temperature, Vf = 50%			
Property	Symbol	Units	Fiber glass
Young's Modulus 0°	E_1	Gpa	25
Young's Modulus 90°	E_2	Gpa	25
In-plane Shear Modulus	G_{12}	Gpa	4
Major Poisson's Ratio	v_{12}	-	0.2
Ultimate Tensile Strength 0°	X_t	Mpa	440
Ultimate Compression Strength 0°	X_c	Mpa	425
Ultimate Tensile Strength 90°	Y_t	Mpa	440
Ultimate Compression Strength 90°	Y_c	Mpa	425
Ultimate In-plane Shear Strength	S	Mpa	40
Density	ρ	kg/m^3	1900

4　OPTIMIZATION SOFTWARE

In this paper, the symmetric sandwich panel made up of composite laminate face sheets with foam and honeycomb core is investigated. Every face sheet consisted of E-glass/epoxy resin. The method which has been followed to calculate the optimal design variables were face sheet thickness t_f and core thickness t_c for both honeycomb and foam core to minimize the weight of sandwich panel (objective function). The mechanical properties of the foam core, honeycomb core and E-glass fabric/epoxy resin used in this study are given in Table 1, 2 and 3 respectively.

The equation of the weight of the sandwich panels was considered to minimize. Where the total weight of the sandwich panels is W_t, the weight of the face sheets is W_f, and the weight of the core is W_c for both honeycomb and foam core. Where l is the span length of the sandwich panel, b is the width of the sandwich panel, ρ_f is the density of the E-glass/epoxy, ρ_c is the density of the core (honeycomb and foam), t_f is the thickness of the E-glass/epoxy face sheet and t_f is the thickness of the core (honeycomb and foam). Therefore, to minimize the weight, equation (3) can be mathematically formulated subjected to design constraints. Kollár, L. & Springer, G. 2003 compute the equations of classical lamination theory and Tsai–Hill criteria of the first ply failure to calculate the mechanical properties of the laminates. The constraints of the optimization problem are stiffness, face-sheet failure, core shear and face-sheet wrinkling equation (4-7). The variable parameters of this problem are the core thickness and face sheet thickness. The panel stiffness, the face-sheet failure maximum load, the core shear maximum load and face sheet wrinkling maximum load for every core and face thickness were calculated. With all these data, for every step, it was calculated the minimum face thickness that accomplishes the load defined and the stiffness required equation (8). The objective function is the minimum weight of sandwich panels. The software calculates the minimum weight condition for sandwich structure which corresponds to the face sheet thickness and the core thickness. This program is a modified version of the program in the Composite Sandwich Optimizer 2017. GitHub, Inc. Where the program was developed to fit with the honeycomb and foam core sandwich panels and the mechanical characteristics of the materials used in addition to the type of test (equations of 4-point bending test). As well as the use of practical and numerical results in this program as inputs to achieve the desired results (maximum deformation δ_{max} and maximum load P_{max}).

Formulate the objective function of the weight of the honeycomb sandwich panels:

$$W_t = 2\,\rho_f\,b\,t_f\,l + \rho_c\,b\,t_c\,l \tag{3}$$

Calculate the stiffness of sandwich panel $(EI)_{min}$ by using given data from experimental and numerical results (maximum deformation δ_{max} and maximum load P_{max}):

$$(EI)_{min.} = \frac{P_{max.}l^3}{B_3 \delta_{max.}}, \text{ where } B_3 = \frac{384}{11} \text{ (4 – point bending)} \qquad (4)$$

This parameter has to be compared to the simplified stiffness expression for a sandwich panel in equation (5). The stiffness of the sandwich panel (D) in equation (5) must be greater than the stiffness of the sandwich panel (EI)min in equation (4).

Formulate the stiffness of the honeycomb sandwich panel constraint:

$$D = \frac{E_f \, t_f \, h^2 b}{2}, \text{ where } h = t_c + t_f \qquad (5)$$

Formulate the maximum load of the face-sheets failure constraint:

$$P_{ff} = \frac{\sigma_f \, b \, t_f h}{B_1 \, l}, \text{ where } B_1 = \frac{1}{8} \text{ (4 – point bending)} \qquad (6)$$

Formulate the maximum load of the core shear constraint:

$$P_{cs} = \frac{\tau_c \, b \, h}{B_2}, \text{ where } B_2 = \frac{1}{2} \text{ (4 – point bending)} \qquad (7)$$

Formulate the maximum load of the face-sheet Wrinkling constraint:

$$P_{wr} = \frac{b \, t_f \, h}{2 \, B_1 \, l \, [E_C \, G_c \, E_f]^{1/3}} \qquad (8)$$

The final results are optimum core thickness c_{opt}, optimum face thickness t_{opt} and Minimum weight W_{min}.

5 RESULTS, DISCUSSION AND CONCLUSIONS

According to the numerical results between honeycomb core and foam core of the sandwich panels, the deflection and the stress in the face sheet of honeycomb core sandwich panels was more than foam core of the sandwich panels whereas the shear stress in honeycomb core was less than foam core of the sandwich panels. Because of the honeycomb core are lightweight compared to foam core as well as the honeycomb core are more cost-effective than foam. Most core materials respond similarly to stress under normal operating loads. As loading increases, the core begins to flex to accommodate the increase in shear stress on the core. Unlike other core materials that reach an ultimate yield stress and fail catastrophically, honeycomb and foam continues to respond and perform. This continued response indicates the ability of the honeycomb and foam to absorb energy even after ultimate yield strength failure. In this comparison study, an optimization technique was applied on the sandwich structure with honeycomb and foam core, where the weight of the sandwich panel was minimized and the optimum values of design variables (core thickness t_c and face sheet thickness t_f) for every solution have been calculated. The design problem subjected to the constraint of bending stiffness, face sheets failure, core shear and face sheet wrinkling. The maximum deformation deflection δ_{max} and maximum load P_{max} which were given from experimental and numerical results. The optimal results of the design variables and the objective functions to

8

minimize the weight are shown in Table 4 & Table 5 and Figures 6–9 respectively. As can be seen, from the results of weight objective optimization the weight was reduced for the sandwich panels foam core less than honeycomb core for the same specimens (same maximum load). In this study, the designers can be select design variables including type of fiber, angle orientations of ply and the geometric parameters (face sheet thickness and core thickness) and provides to the designer the ability to meet the specified design requirements.

Table 4. Analytical results (MATLAB software), experimental results (4-point bending test) and numerical results (FEA by using DIGIMAT program) (4-point bending test) for honeycomb core composite sandwich panels.

Symbol	Dimension						Experimental* and Numerical data					Optimum results			
	l	s	b	h	t_c	t_f	W	P*	δ	σ_{skin}	τ_{core}	$t_{c\,opt.}$	$t_{f\,opt.}$	W_{min}	W_{red}
Unit	mm	mm	mm	mm	mm	mm	kg	KN	mm	Mpa	Mpa	mm	mm	kg	%
1	1000	840	120	22	20	1	0.615	1.869	57.36	77.3	0.887	29.285	0.423	0.545	11.382
2	1000	840	115	15	13	1	0.64	1.674	119.808	109	1.11	19.183	0.437	0.401	31.718
3	1130	840	54	20	18	1	0.37	0.733	61.097	74.8	0.742	27.265	0.394	0.228	38.378

Table 5. Analytical results (MATLAB software) and Numerical results (FEA by using DIGIMAT program) (4 - point bending test) for foam core composite sandwich panels.

Symbol	Dimension						Numerical data					Optimum results			
	l	s	b	h	t_c	t_f	W	P	δ	σ_{skin}	τ_{core}	$t_{c\,opt.}$	$t_{f\,opt.}$	W_{min}	W_{red}
Unit	mm	mm	mm	mm	mm	Mm	kg	KN	mm	Mpa	Mpa	mm	mm	kg	%
1	1000	840	120	22	20	1	0.768	1.869	47.977	66.7	3.37	25.245	0.661	0.762	0.781
2	1000	840	115	15	13	1	0.575	1.674	106.705	98.7	4.87	17.163	0.591	0.548	4.695
3	1130	840	54	20	18	1	0.366	0.733	52.121	65.4	3.35	23.224	0.633	0.319	12.841

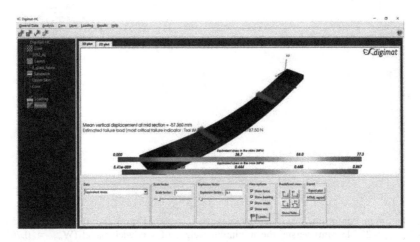

Figure 6. Numerical results, FEA by using Digimat – HC program, 4 - point bending test for honeycomb core composite sandwich panel for specimen No. 1.

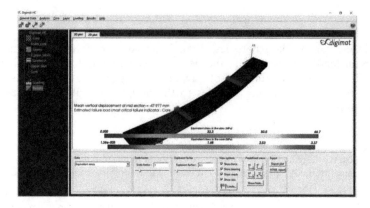

Figure 7.   Numerical results, FEA by using Digimat – HC program, 4 - point bending test for foam core composite sandwich panel for specimen No. 1.

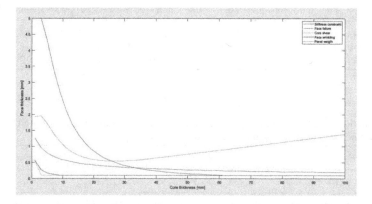

Figure 8.   Optimum composite face sheet thickness ($t_{f\ opt.}$ = 0.423 mm) and optimum honeycomb core thickness ($t_{c\ opt.}$ = 29.285 mm) versus the minimum weight of sandwich panel ($W_{min.}$ = 0.545 Kg for given Numerical results, FEA by using Digimat – HC program, 4 - point bending test (P_{max} = 1.869 KN and δ_{max} = 57.36 mm) for specimen No. 1.

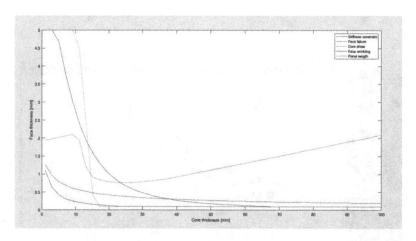

Figure 9.   Optimum composite face sheet thickness ($t_{f\ opt.}$ = 0.661 mm) and optimum honeycomb core thickness ($t_{c\ opt.}$ = 25.245 mm) versus the minimum weight of sandwich panel ($W_{min.}$ = 0.762 Kg for given Numerical results, FEA by using Digimat – HC program, 4 - point bending test (P_{max} = 1.869 KN and δ_{max} = 57.36 mm) for specimen No. 1.

10

REFERENCES

Allen, H. 1969. *Analysis and Design of Structural Sandwich Panel*. New York: Oxford Pergamon Press.

Bitzer, T. 1997. *Honeycomb Technology: Materials, Design, Manufacturing, Applications and Testing*. London: Chapman & Hall.

Christos, K. et al. 1997. Simultaneous cost and weight minimization of composite stiffened panels under compression and shear. *Composites Part A: Applied Science and Manufacturing*. 28(5): 419–435.

Composite Materials Engineering Specialists in Carbon Fibre. 2009. (Online), (Accessed March 2019). Available on: http://www.performance-composites.com/carbonfibre/mechanicalproperties_2.asp

Composite Sandwich Optimizer. 2017. GitHub, Inc. (Online), (Accessed March 2019). Available on: https://github.com/dinospiller/Composite-Sandwich-Optimizer.

Cymat A35620SC 020SS Stabilized Aluminum Foam. (Online), (Accessed March 2019). Available on: http://www.matweb.com/search/DataSheet.aspx?MatGUID=488c7071140a48ffb11a5cdb75cda ba9&ckck=1.

Jegorova, K. 2014. *Composite Honeycomb Cores*. Degree Thesis. Helsinki, Finland: Plastics Technology, Arcada University of Applied Sciences.

Ju, S. et al. 2013. *Multi-Parameter optimization of lightweight composite triangular truss structure based on response surface methodology*. Composite Structures. 97: 107–116.

Honeycomb Sandwich Design Technology. 2007. Hexcel Composites Publication. (Online), (Accessed March 2019). Available on: https://www.hexcel.com/user_area/content_media/raw/Honeycomb_Sand wich_Design_Technology.pdf.

Gibson, L.J. 1984. Optimization of stiffness in sandwich beams with rigid foam cores. *Materials Science and Engineering*. 67(2): 125–135.

Gibson, L. & Ashby M. 1997. *Cellular Solids*. Structure and Properties. London: Cambridge University.

Kollár, L. & Springer, G. 2003. *Mechanics of Composite Structures*. London: Cambridge University Press.

Mechanical Testing of Sandwich Panels: Technical Notes. 2007. Hexcel Composites Publication. (Online), (Accessed March 2019). Available on: https://www.hexcel.com/user_area/content_media/raw/Sand wichPanels_global.pdf.

Mamalis, A.G. et al. 2008. A new hybrid concept for sandwich structure. *Composite Structures*. 83(4):335–340.

Marine Core and Composite Panels for Recreational, Performance and Commercial Boat Building. (Online), (Accessed March 2019). Available on: https://www.plascore.com/download/datasheets/marine_documen tation/Plascore_Marine.pdf.

Rodrigues, G.P. 2015. *Combined topology and stacking sequence optimization of composite laminated structures for structural performance measures*. London: 4th Engineering Optimization Conference.

Solutions for Sustainable Development – Szita, Jármai & Voith (eds.)
© 2020 Taylor & Francis Group, London, ISBN 978-0-367-42425-1

Estimation of compressive strength of current concrete materials: Effect of core diameter and maximum aggregate size

M.E. Belgacem
National Center of Studies and Integrated Research on Building Engineering, CNERIB, Algiers, Algeria
University of Science and Technology HouariBoumedienne, USTHB, Algiers, Algeria

A. Brara
National Center of Studies and Integrated Research on Building Engineering, CNERIB, Algiers, Algeria

A Ferdjani
University of Blida (1), Blida, Algeria

ABSTRACT: Assessing the concrete compressive strength of the existing reinforced concrete (RC) structures by testing cores is the most reliable and effective method. However, the question that commonly arises concerning this method is the degree of representativeness of the test results. In this paper, a total number of 241 cores of current concrete material (C25/30 and C30/37 concrete class) was tested. The effect of core diameter (Ø =100 mm, 75 mm, 50 mm) for double length-to-diameter (l/d) ratio and maximum aggregate size (D_{max} = 25 mm, 16 mm, 8 mm, 4 mm) on the compressive strength of cores has been studied. For the two concrete classes studied, test results showed a decrease of core compressive strength with the increase of the maximum aggregate size. Conversely, the compressive strength of the core is increased as the diameter is increased. The effects of the two parameters are less pronounced for the C30/37 concrete class.

Keywords: Concrete compressive strength, core diameter, maximum aggregate size

1 INTRODUCTION

During the service life of RC structures, the concrete material faces several types of degradations that influence its quality. The main quality of concrete is compressive strength, which is usually monitored via testing concrete standard specimens that have been withdrawn in a construction site before or during the casting. These specimens are prepared, conserved and tested in accordance with well-regulated specifications and procedures Algerian standard (NA17004, 2008) and (NA16002, 2007). However, the real in site quality of concrete compressive strength remains unknown. Therefore in order to verify the conformity of concrete compressive strength that has been used during in structures or in the case having doubts about its quality, the common way of determining in situ compressive strength of concrete is to drill and test cores (Tuncan, 2008).

To obtain values which are correlated with the standard molded specimens, several standards such as (NA 17004), (NA 5071, 2005), (ASTM, 2001) and British standard (BS, 1976) recommend drilling 100 mm cores diameter with l/d ratio of 2. For cores with smaller diameters, the standards mentioned above are conservative and recommend increasing the number of cores to be tested. Nevertheless, it is often difficult to drill a core with such a diameter threshold of 100 mm and even with double length-to-diameter ratio. This is due to the limited thickness of the element, or the considerable concentration of steel bars in the structural components, particularly for the structures designed for seismic prone regions, such as Northen Algeria wherein the elements of the lateral load resisting system(beam, column and shear walls) the coating is usually less than 100 mm. In some countries, such as turkey, the TS EN 125 standard allows extracting of 50 mm even though there is a lack of reliability in the results of compressive tests (Tuncan, 2008).

Besides, using cores with smaller diameter yields to the advantages of reducing the cost of the equipment (coring dip, compression press and stoking space), Easy intervention and minimization of damages in the structure.

Generally, it is clear that the reduction of the specimen sizes can significantly affect the material strength. Moreover, in the case of concrete, which is a highly heterogeneous material, the reduction of the specimen sizes can dramatically change its properties (Tuncan, 2008). Moreover, there are several factors affecting cores compressive strength such as the slenderness, core diameter, maximum aggregate size D_{max}, strength class, casting and drilling directions, age of concrete and cure conditions. These factors should be taken into consideration for the interpretation of the results.

Concerning the effect of slenderness, (Tuncan 2006, 2008) and (Nikbin, 2009) have observed that for both molded specimen and drilled cores that the compressive strength increases when the slenderness decreases. According to these researchers for drilled cores, this increase offsets the loss of strength due to other factors as (diameter and D_{max}). Whereas, for slenderness (l/d) equal to 2 this effect is negligible. In this case, (Murdock & Kesler, 1957) proposed correction factors of slenderness that depend on the concrete class. For the effect of core diameter, which is the main factor of size effect, the compressive strength of molded specimens increases when the diameter of the specimen decreases. However, for drilled cores, compressive strength decreases when the core diameter decreases (Tuncan, 2006 & 2008) and (Nikbin, 2009). This reduction is due to the drilling operation, and the maximum aggregate size used that effects the concrete compressive strength negatively. Concerning the effect of D_{max} the literature reported the work of (Indelicato, 1993,1997) that detected no effect of maximum aggregate size in contrast with other investigations that have been done. The results of these investigations show that the increase of the maximum aggregate size effects negatively core compressive strength and this effect is more pronounced for small core diameter (Tuncan, 2006 & 2008).

This research paper aim to contribute at the enhancement of the field of assessing concrete compressive strength of existing RC structures through studying and examination of the effect of maximum aggregate sizes and lateral dimension of cores on the compressive strength for the strength levels commonly required in construction sites that is C25/30 and C30/37.

2 EXPERIMENTAL PROGRAM

Four classes of coarse aggregates were used (i.e. 0/4, 4/8, 8/16 and 16/25) with Portland cement CEMII42.5 as a binder. Various tests were carried out for the identification of the characteristics of these materials. The tests results showed that these materials are acceptable for the manufacture of hydraulic concrete.

In this study, two strength classes of concrete were adopted the C25/30 and C30/37. These classes correspond respectively to the characteristic strength of 25 and 30 MPa determined at 28 days on standard specimens 150×300 mm^2. These classes of concrete have been selected to reflect the strength levels commonly required in construction sites. The method of concrete formulation has been adopted is that of (Dreux & Festa, 1998). Mixture proportions of concrete mixtures are given in Tables 1 and 2.

Before casting of the concrete blocks, the class of these mixtures are confirmed by testing (06) standards specimens 150×300 mm^2, Figure 1. These are the reference specimens that give the average and characteristic concrete compressive strength F_{ck}. Therefore, blocks with different dimension to drill cores were casted and conserved in the open air, Figures 1 and 2.

Table 1. Proportions of concrete mixtures for the C25 concrete class.

Strength level (MPa)				C 25	
D_{max}(mm)	25	16	8	4	
Designation	C25/D25	C25/D16	C25/D8	C25/D4	
CEM II 42.5	315.00	358.33	383.33	CEM II 42.5	402.78
Sand 0/4	832.95	861.11	977.78	Sand 0/0.5	575.00
Gravel 4/8	170.35	238.89	752.78	Sand 0.5/2	291.67
Gravel 8/16	469.31	738.89	00.00	Sand 2/4	805.55
Gravel 16/25	433.27	00.00	00.00	Water (l)	244.44
Water(l)	210.55	203.89	227.78		
W/C	0.67	0.57	0.60	W/C	0.61
Density (kg/m³)	2431.43	2401.11	2341.67	Density (kg/m³)	2319.44

Constituents kg/m³

Table 2. Proportions of concrete mixtures for the C30 concrete class.

Strengthlevel (MPa)				C 30	
D_{max}(mm)	25	16	8	4	
Designation	C30/D25	C30/D16	C30/D8	C30/D4	
CEM II 42.5	355.88	355.25	395.52	CEM II 42.5	485.60
Sand 0/4	783.72	861.11	958.02	Sand 0/0.5	505.03
Gravel 4/8	145.77	238.49	739.03	Sand 0.5/2	258.32
Gravel 8/16	470.92	739.17	0.00	Sand 2/4	714.43
Gravel 16/25	436.39	0.00	0.00	Water(l)	277.05
Water (l)	214.06	217.98	225.66		
W/C	0.60	0.61	0.57	W/C	0.57
Density (kg/m³)	2406.73	2412.00	2 318.23	Density (kg/m³)	2240.43

Constituents kg/m³

Figure 1. Casting of standard specimens and block.

The properties of the different concrete mixtures are reported in Tables 3 and 4. The concrete class is determined based on the value of f_{ck}, which specifies the characteristics of compressive strength at 28 days.

3 RESULTS AND DISCUSSION

Core compressive strength results for respectively concrete classes C25 and C30 as a function of core diameter and maximum aggregate sizes D_{max} are presented in Tables 5 and 6. Noting

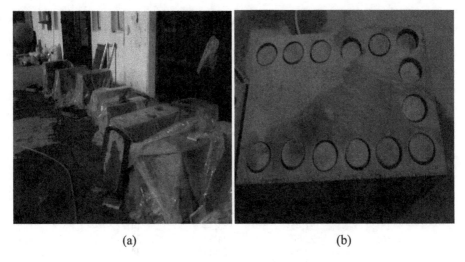

(a)   (b)

Figure 2.   Concrete blocks conservation (a) from which cores are extracted (b).

Table 3.   Properties of concrete (concrete class C25).

D_{max} (mm)	25	16	8	4
Slump (cm)	4.5	5	5.5	6
Fcm (MPa)	28.19	31.03	29.98	27.33
Fck[a] (MPa)	26.36	29.23	28.15	26.34
W/C	0.67	0.57	0.60	0.61
Standard deviation (MPa)	1.74	1.71	1.75	0.95
CoV (%)	6.18	5.52	5.83	3.46

CoV (%) = coefficient of variation

Table 4.   Properties of concrete (concrete class C30).

D_{max} (mm)	25	16	8	4
Slump (cm)	5.3	5.9	6.6	6.4
Fcm (MPa)	30.65	31.62	35.12	35.01
Fck[a] (MPa)	30.21	31.15	33.35	33.39
W/C	0.60	0.61	0.57	0.57
Standard deviation (MPa)	0.38	0.45	1.69	1.55
CoV (%)	1.25	1.43	4.8	4.41

a $F_{ck} = F_{cm} - k * \frac{S}{\sqrt{n}}$ With $k = 2,57$ (α(risk) = 5% (Neville,2000) v (degree of freedom) = n-1, v = 5)

that these cores were drilled, prepared and tested in accordance with (ASTM C42, 2001) and (NA 5071, 2005).

Besides, a post-processing of the results is needed to exclude the abnormal values which are referenced (Viviane, 2005), (Frédéric, 2006) and (ASTM, 2016).

3.1   *Effect of maximum aggregate size on the compressive strength*

Figures 3 and 4 illustrate the compressive strength results for the following core diameter: 50 mm, 75 mm and 100 mm, for C25 and C30 concrete classes, respectively.

Table 5.   Core compressive strength (concrete class C25).

D_{max} (mm)	Core compressive strength mean value (MPa)		
	Φ100 mm	Φ75 mm	Φ50 mm
25	20,50	19,97	14,14
16	17,30	13,48	15,58
8	19,96	17,06	21,01
4	18,34	18,51	19,16

Table 6.   Core compressive strength (concrete class C30).

D_{max} (mm)	Core compressive strength mean value (MPa)		
	Φ100 mm	Φ75 mm	Φ50 mm
25	21,45	23,34	14,72
16	25,30	24,87	24,44
8	30,28	28,86	34,76
4	28,53	29,53	30,75

The results depicted in Figures 3 and 4 show a global decrease of the compressive strength with an increasing maximal aggregate size for the different diameters, for the two concrete classes, as already observed by (Tuncan, 2006 & 2008).

This decrease is more pronounced for the C30 class. Some values of strength are somehow dispersed, notably for the smallest core diameter oφ 50 mm with $D_{max} = 8$ mm. This could be due to the drilling operation which affects more the smallest core diameter values, combined with the fact that the ratio of core diameter to D_{max} which is smaller compared to those of 75 mm and 100 mm core diameters. It is worth noting that NA 17004, ASTM and BS standards recommend that core diameter canto be at least three times the value of D_{max}. A ratio of 2.0 represents the minimum value in the standards mentioned above. However, it is observed that 50 mm core diameter with the value of D_{max} of 25 mm represent the minimal ratio recommended by these standards. Moreover, the 50 mm cores with the most significant value of D_{max} (i.e., 25 mm) do not represent the concrete compressive strength reliably.

As can be seen in Figure 3, the effect of D_{max} on the core compressive strength of core diameter 75 and 100 mm is not uniform for the concrete class C25. As for the cores of 50 mm diameter, there is a decrease of the compressive strength when D_{max} increases.

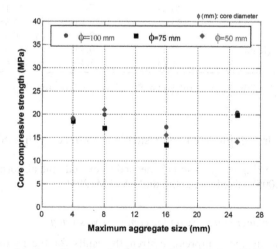

Figure 3.   Core compressive strength as function of D_{max} (concrete class C25).

16

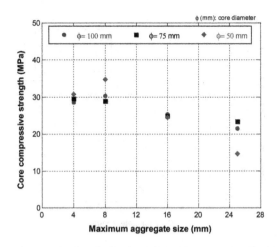

Figure 4.   Core compressive strength as a function of the D_{max} (concrete class C30).

For the concrete class C30, D_{max} affects the core compressive strength more significantly. When D_{max} increases cores, compressive strength decreases. For 75 and 100 mm cores, diameters the compressive strength gradually decreased as the D_{max} increases. On the other hand, for the 50 mm diameter core, it is observed that the effect of D_{max} is more pronounced and the lowest compressive strength is recorded for D_{max} of 25 mm. These findings are following those found by (Tuncan, 2006 & 2008).

3.2   *Effect of core diameter on the compressive strength*

Figures 5 and 6 depict the results of compressive strength evolution as a function of core diameter, for C25 and C30, respectively.

The evolution of the core compressive strength values as a function of core diameter is depicted in Figures 5 and 6. As shown by the figures, for the value of D_{max} equal to 16 and 25 mm, for both strength classes, the compressive strength decreases as the core diameter decreases, as observed by (Tuncan, 2006,2008) and (Nikbin, 2009). For cores with the value of D_{max} equal to 4 and 8 mm, the core compressive strength remains rather constant or slightly dispersed.

The fact that as the core diameter decreases the compressive strength decreases is due to the micro-cracks that appear as a result of the drilling operation. This latter affects negatively the strength of the concrete cores and especially the peripheral zone as observed during the

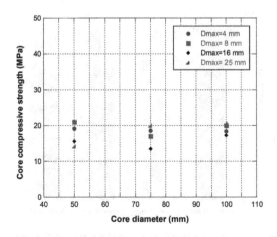

Figure 5.   Compressive strength vs core diameter for the different D_{max} (concrete class C25).

17

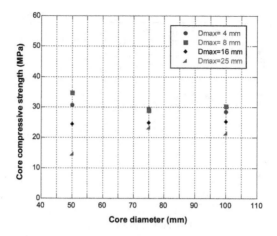

Figure 6. Influence of core diameter on compressive strength for the different D_{max} (concrete class C30).

compressive strength testing. This phenomenon is more pronounced for the smaller core diameter since the area affected by drilling operation compared to the volume of the core increases. This result holds for cores having the value of D_{max} equal to or greater than 16 mm. The explanation that can be given is that the effect of degradation during the drilling operation is less significant for the smallest aggregate size. This discrepancy seems to be due to the different ratio of D_{max}/D_{core} in, which the lowest ratio was used by Tuncan was around 0.15, while herein the ratio is around 0.04.

3.3 *Comparison between compressive strength of the core and standard specimen*

The histogram of Figure 7 presents the relative compressive strengths of 100 mm diameter cores, for both strength classes C25 and C30.

The histogram of Figure 7 shows that the relative strengths are comprised in the range of 60 to 90% for the two concrete classes. The ratios are slightly higher for the C30 class. These results are in contradiction with those observed by (Yip, 1988) where the relative strengths vary from 0.63 to 1.53.

Globally, test results show that generally as the D_{max} decreases the relative compressive strength increases. Core with smaller D_{max} has typically more quantity of cement, which

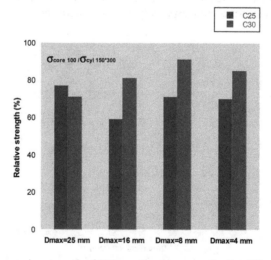

Figure 7. Relative compressive strength of 100 mm diameter cores for the different maximum aggregate sizes.

guarantees the link of the different aggregate with each other and lowers the probability of developing micro-cracks during the drilling operation.

4 CONCLUSIONS

From this study of the effects of maximum aggregate sizes and lateral dimensions on the core strength of current concrete materials, the following conclusions may be drawn:

- Core compressive strength decreases as the maximum aggregate size increases. This decrease is more pronounced in the case of 50 mm diameter cores, notably for the highest aggregate sizes.
- Generally, the compressive strength of cores decreases when the core diameter decreases.
- Dimension and aggregate sizes effects are slightly mitigated when enhancing concrete quality.

These results need to be confirmed by further experimental researches involving other core diameters and concrete classes for more broad generalization.

ACKNOWLEDGEMENTS

The financial support of the ministry of higher education MESRS in Algeria (Grant PNR size effect in the National center of studies and researches integrated of building) for conducting this study is greatly acknowledged. Many thanks go to Ms.Souhila BEDJOU for her contribution and valuable remarks that helped the authors to achieve this manuscript.

REFERENCES

Adam M. Neville. 2000. Properties of concrete translated by the CRIB research center Interuniversity on the concrete. Sherbrooke-Laval.

ASTM C 42/C. 2001. Standard method of obtaining and testing drilled cores and sawed beams of concrete, PA: American Society for Testing and Material. Philadelphia.

ASTM E178–16. 2016. Standard practice for dealing with outlying observations, ASTM International: PA. West Conshohocken.

BS, British Standard.1976. CSTR 11.Concrete core testing for strength. Technical report No. 11. The concrete society. London.

Frédéric et Myriam Bertrand. 2006. Valeurs non representatives.

Indelicato, F. 1993. A statistical method for the assessment of concrete strength through micro cores, Materials and Structures, 26, 261–267.

Indelicato, F. 1997. Estimate of concrete cube strength by means of different diameter cores: A statistical approach, Materials and Structures, Vol. 30, 131–138.

NA 17004 Algerien Standard. 2007. Assessment of the compressive strength on site of structures and prefabricated concrete elements.

NA 16002 Algerien Standard. 2008. Specifications, performances, production and conformity.

NA 5071: tests for concrete in structures. 2005. Core sampling, Examination and testing compression, Algiers: Algerian Institute of Standardization.

Georges Dreux and Jean Festa. 1998. New Guide of Concrete and its Constituents, Eighth Edition.

Nikbin, I.M.Eslami, M. and Rezvani, S.M. 2009. An Experimental Comparative Survey on the Interpretation of Concrete Core Strength Results, European, Journal of Scientific Research, Vol.N°(3), pp 445–445456.

Tuncan et al., 2006. A comparative study on the interpretation of core strength concrete results", Magazine of Concrete Research, 58(2), 117–122.

Tuncan et al. 2008. Assessing concrete strength by means clustering of small diameter cores, Construction and Building Materials, 22, 981–988.

Viviane Planchon. 2005. Treatment of aberrant values: Current concepts and general trends, Biotechnol. Agron. Soc. Environ., 9(1), 19–34.

Yip W.K. & Tam C.T. 1988. Concrete strength evaluation through the use of small diameter cores, Magazine of Concrete Research., Vol. 40(143), pp 99–99105.

Solutions for Sustainable Development – Szita, Jármai & Voith (eds.)
© 2020 Taylor & Francis Group, London, ISBN 978-0-367-42425-1

The effect of the gear wear for the contact ratio

J. Bihari
Institute for Machine and Product Design, University of Miskolc, Miskolc Hungary,

ABSTRACT: The contact ratio is such a parameter of a gear, that is very important because of the noise of the drive and the quality of the mesh. In general condition, the contact ratio can be considered constant during the whole lifetime. Although there are applications where the gears are working without any lubrication and the only criteria of the lifetime. Among these conditions the measure of the wear can be so extensive, that has a significant influence on the contact ratio in a quite long period of the lifetime. In these cases, the contact ratio decreases, which leads to the increase of the dynamic loads. The pressure line can be move, which results the modification of the tooth bending force's arm. It is easy to understand that this phenomenon depends on geometrical parameters, the initial conditions, relative movements and the loads. This means, not only the measure of the wear can be restricted by the strength, but the new geometry. In these conditions, the operational characteristics can be significantly different from the original, which was determined by calculation. Because of that, it is expedient to suppose more conditions as a basis in case of non-lubricated gear drives. It is recommended to calculate in every condition the safety factors. If we know the measure of the wear, we can determine when will be the gear pair inoperable with the use of known calculations. This paper shows an iterative process, in which the suspected measure of the wear is used as tolerance, the possible consequences were determined.

1 INTRODUCTION, EVALUATION OF CURRENTLY USED DESIGN METHODS

Open gear drives with spur gears are the simplest gear drives. These typically include a pair of gearwheels or a gear and a rack. These drives do not have such sealed housing that allows submerged oiling or oil mist lubrication. The grease lubrication is the only one opportunity to lubricate the gears. This is not a problem in many cases because these drives operate at low speed. The high consistency grease for such drives is not dispersed by the centrifugal force from the tooth surfaces. At high gear ratios such as actuators, or drum furnace drives grease scattered from the rapidly rotating pinion, it is constantly replaced by the lubricant applied to the other gear. For these drives, the service life can be calculated safely. Dimensioning can be performed according to DIN 3960 or DIN 3990, the proper lubricant material can be selected to lubricating factor. In most cases, the open design provides good cooling. In such drives, the primary cause of failure is the tooth fatigue or tooth surface fatigue, or in very harsh conditions when a significant amount of dust is added to the lubricant, the seizure of the tooth surfaces.

The design process is considerably more complicated if there is no possibility of lubrication of the gears. These are typical for food applications or for starter motors in motor vehicles. Polymers with good friction properties are often a good solution in the food industry and in the case of starter motors such materials cannot be used due to their strength characteristics. In these drives, the primary cause of failure is the wear of the tooth surfaces. In the absence of a better alternative, such drives are usually sized according to the standards mentioned above. It is easy to see that the calculation of the lifetime will be wrong because the standards do not know the state without lubrication, there are no values of lubrication factor in such cases. In the case of plastic gears, the failure caused by wear has become an accepted cause of failure.

Table 1. Damages and the opportunities to increase the safety against damage.

| | | How safety can be affected | | |
		Material	Construction, working conditions	Lubricant
Safety against damage	Brake of the tooth (the strength of the tooth)	yes	yes	-
	Pitting (the strength of the tooth surface)	yes	yes	moderately
	Seizure	moderately	yes	yes
	Excessive warming	-	yes	yes

In this case, the tooth will wear so thin that it will not be able to withstand the load on it, so it breaks down (Kissling 2013). It is suspected that this phenomenon may also occur in steel gear pairs without lubrication, especially at high gear ratios.

In vehicles, the gear drive between the starter and the freewheel gear rim cannot be lubricated because the scattered lubricant would render the clutch inoperative. However, these drives operate in dusty environments exposed to significant temperature fluctuations. It hasn't been a problem for a long time because the starter motor of an average car has only been running for a short time and rarely. Nowadays as the Start-Stop systems are widespread the starter motors of cars run one order of magnitude several times during the lifetime of the vehicle as before. Table 1 shows the possibilities of the designer in the traditional design process to avoid potential damage.

The table shows that lubrication is not influenced by lubricant in conventional theories. If we interpret the tooth breaking always as a consequence of inadequate root strength this is indeed the case. However, if the tooth fracture is due to thinning of the tooth, it is easy to see that the lubricant plays a significant role in avoiding it (Sarka 2019). It is also evident that during the design of a non-lubricated teeth, the lubrication cannot be influenced by pitting, excessive heating and the formation of seizure. In the case of food processing machines besides using polymers it is possible to solve the problem by the multiplication of the pinion gears or to insert the replacement of the pinon gear into the regular maintenance plan. But in case of cars, these are unacceptable requirements for customer service.

2 SPECIAL DESIGN REQUIREMENTS FOR VEHICLE STARTER

With the exception of mild hybrids, it is still the case that the internal combustion engines of cars are started with the flywheel gear rim of the engine, or in the case of automatic transmissions by the gear rim that mounted on the circumference of the switch plate. The drive is powered by an electric motor on which a small gear that fits to the gear rim, ensures the transmission of torque during the start-up process. This little gear is not in constant contact with the gear rim, but it moves in axial direction during the start-up process (Sarka 2019). At the end of the process, it also extends axially from the connection. Typical ratios are between 1:9 and 1:12 depending on the manufacturer. Typical modules range from 2 mm to 2.4 mm and small commercial vehicles to 3 mm. Heavy trucks also have larger modules.

Due to the axial coupling, the designer is already facing at the beginning of the design with the values of the design tolerances that are recommended in the standards are not proper, because a larger tooth gap is needed than usual to ensure axial coupling. The clearance of the tooth is basically influenced by the tolerance of the wheelbase and the tolerance of the base tangent length.

The wheelbase and its tolerance are typically not defined by the engineer they are derived from engine characteristics. On the castings of the engine, the fixing points of the starter motor are not customary or economical to be machined with the accuracy that the recommended tolerances of the wheelbase. The engine crankshaft runs on hydrostatic sliding

21

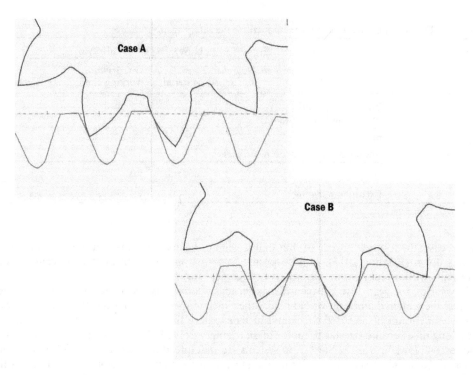

Figure 1. The base tangent length is on its minimum value, the centre distance is maximal (case A), and. minimal (case B).

bearings, resulting in significant radial runout. The crankshaft relative eccentricity also depends on the speed and current operating conditions. The gear rim is attached to the crank-shaft along with the flywheel. The crankshaft, during the start-up process and the increase of the speed and the build-up of lubricating film layer, continuously changes its position. The attaching of the gear rim on the flywheel brings additional uncertainty to the system. Together, these factors result in tolerances of up to 2 orders of magnitude greater than the tolerances that are optimal for the particular gear. These tolerances have a significant impact on the operation of the teeth. Another problem is that the value of the centre distance may also differ from the optimum for structural reasons. This can be corrected by profile shift within certain limits (Ungár 1996). With the data so recorded, the designer is no longer able to clearly determine the functionality of the teeth, since the operating characteristics may differ greatly from the tolerance limits. Therefore, it is not enough to count the nominal values. In such cases, the designer uses the dimensions for the tolerance limits as the nominal size for control. Control can be greatly accelerated by computer modelling (Bihari 2016).

Figure 1 also shows that the common tooth height and logically the contact ratio also vary significantly within tolerance limits. The length of the action can be increased by adjusting the pressure angle, but this is often not possible because the vehicle manufacturer limits that. To reduce noise, the pressure angle of less than 14 degrees is also common. Figure 2 shows the con-nections of the gear pair with the same parameters as shown in Figure 1, but with a 12-degree pressure angle, at the maximum and minimum values of the centre distance tolerance.

3   THE EFFECT OF WEAR TO THE CONTACT RATIO, THE EFFECT OF THE
    CONTACT RATIO

Based on many years of experience, the standards recommend that the value of the contact ratio must be maintained above 1.15 for the correct meshing of the teeth (DIN39601987)

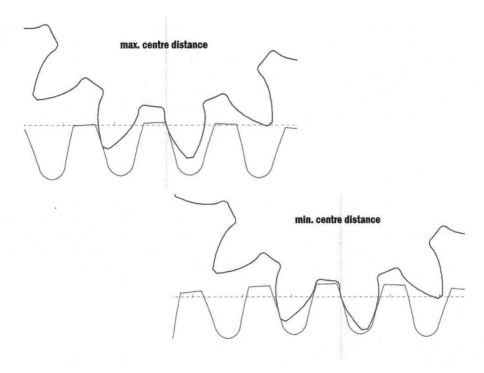

Figure 2. The connection of the teeth to the maximum and minimum values of the centre distance.

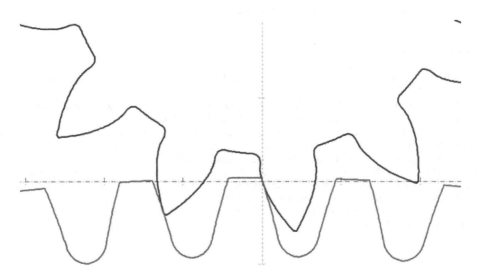

Figure 3. The change of the contact ratio by the influence of wear.

(DIN39901987). In Figure 2. the contact ratio of the gear pair is 1.6 and 1.08 in the two extreme positions. A gear with contact ratio 1.08 is still a functional gear, but the connection is significantly less favourable. In the case of starter motors due to the high gear ratios, there is always more wear on the pinon. These teeth are functional if they can transmit the correct torque, so they can be suitable when extreme wear occurs. However, due to the non-lubricating operation, wear is also greater than in the case of lubricated drives. Therefore, to determine the real life of the gears, it is not enough to take into account the tolerance limits.

With the wear of the pinion, the contact ratio decreases. Figure 3 takes into account a 0.2 mm wear by changing the base tangent length of the gear wheel shown in Figure 2. In this case, the contact ratio is only 0.98 the drive is still working, but safety factors, especially the safety factor of tooth surface fatigue, are only fragments of the original condition.

In this case, the tooth is greatly thinned, the topland approximate to zero, so there is a significant chance that the tip of the tooth will break. The contact ratio below 1, means that the teeth are not permanently connected, this means in every connection a shock to the tooth surfaces. These loads cause micro- and macro pitting.

4 CONCLUSIONS

The lifetime of drives without lubrication depends on several factors and the installation and operating characteristics of the starter motors as well. Scaling should always be done for the worst case. However, it should be considered that the contact ratio decreases due to wear, and because of it, wear is accelerated due to micro- and macro pitting. The designer can simulate these processes by creating different models, but the most expedient is to consciously monitor lifetime tests by continuously measuring wear and returning the results to the simulation.

ACKNOWLEDGEMENT

The described article/presentation/study was carried out as part of the EFOP-3.6.1-16-2016-00011 "Younger and Renewing University – Innovative Knowledge City – institutional development of the University of Miskolc aiming at intelligent specialization" project implemented in the framework of the Szechenyi 2020 program. The realization of this project is supported by the European Union, co-financed by the European Social Fund.

REFERENCES

Bihari, J. 2016. A kisméretű műanyag fogaskerekekkel szerelt hajtóművek tűrésezési problémái, GÉP 67(5-6), pp 14–17.
Kissling, U., Beerman, S., 2013 Verschleiß und Zahnbruch, Neueste Erkenntnisse zur Lebensdauer-Auslegung von Zahnrädern aus Kunststoff – Teil 1, Antriebstechnik 3/13.
Sarka, F. 2019. The use of the linear sliding wear theory for open gear drives that works without lubrication, Solutions for the Sustainable Development, Proceedings of the 1st International Conference for Solutions for the Sustainable Development, ICES²D.
Ungár, T., Vida, A., 1996. Segédlet a Gépelemek I – II. kötetéhez, 6. kiadás, Nemzeti Tankönyvkiadó.
DIN 3960, 1987-03 Begriffe und Bestimmungsgrößen für Stirnräder (Zylinderräder) und Stirnradpaare (Zylinderradpaare) mit Evolventenverzahnung, Beuth, 1987.
DIN 3990–1, 1987–12, Tragfähigkeitsberechnung von Stirnrädern; Einführung und allgemeine Einflußfaktoren, Beuth, 1987.

Solutions for Sustainable Development – Szita, Jármai & Voith (eds.)
© *2020 Taylor & Francis Group, London, ISBN 978-0-367-42425-1*

Determination of GTN parameters using artificial neural network for ductile failure

Y. Chahboub
University of Miskolc, Miskolc, Hungary

S. Szabolcs
Bay Zoltan kft, Miskolc, Hungary

H. Aguir
Mechanical Engineering Laboratory of Monastir- Tunisia

ABSTRACT: The Gurson–Tvergaard–Needleman (GTN) model, is widely used to predict the failure of materials based on lab specimens, The direct identification of the GTN parameters is not easy and its time consuming. The most used method to determine the GTN parameters is the combination between the experimental and Finite Element Modeling results and we have to repeat the simulations for many times until the simulation data fits the experimental data in the specimen level. In this paper, we determine the GTN parameters for the SENT specimen based on the fracture toughness test, and we are going to present how the artificial neural network (ANN) method could help us to determine the GTN parameters in a short time. The results obtained from this work show that the ANN is a great tool to determine the GTN parameters in addition to this the determined parameters respect very well the literature.

1 INTRODUCTION

Ensuring the Nuclear Safety of the Nuclear power plants, that's mean keep all the parts working properly and with high performance, therefore the pipeline is one this parts, the leakage problem in the pipelines is very critical issue that's might affect the performance of the Nuclear Power Plant if we didn't detect it from the beginning.

Ductile fracture is the main mode of fracture in the case of pipelines, the physical process in ductile fracture involves the nucleation, growth and coalescence of micro voids.

Comparing the GTN model to other damage models like Rice and Tracey Model, Johnson–Cook, Damage Model, the GTN model is a powerful tool that's used in the industry and in the research area, to predict the initiation and propagation of the crack, in addition to this the GTN model is implemented in FEM software products as MSC Marc Mentat.

The most used method to determine the GTN parameters is the combination between the experimental and Finite Element Modeling results and we have to repeat the simulations for many times until the simulation data fits the experimental data in the specimen level (axisymmetric tensile bar and CT specimens) but its time consuming, so for this reason, we are going to determine the GTN parameters by using the combination between the ANN and finite Element Modeling and experimental data in order to get the GTN parameters in a short time and with good accuracy. (Hamdi & Haykel 2010).

1.1 Gurson model

Gurson Tvergaard Needleman (GTN) model, it's a very known damage model that's widely used in engineering application to predict the failure of materials such as steel cast iron, copper, and aluminum and there are some studies which prove the usability of the model in the case of polymer also (Alpay Oral et al 2012) Gurson, Tvergaard and Needleman's damage model (Gurson 1975)is an analytical model that predicts ductile fracture on the basis of nucleation, growth and coalescence of voids in materials.

The model is defined as:

$$\phi = \frac{\sigma_e^2}{\sigma_M^2} + 2q_1 f^* \cosh\left[\frac{tr\sigma}{2\sigma_M}\right] - (1 + q_1^2 f^{*2}) \tag{1}$$

In which q_1 is the material constant, $tr\sigma$ is the sum of principal stresses, σ_M is the equivalent flow stress and f^* is the ratio of voids effective volume to the material volume ratio defined as follows:

$$f^*(f) = f_c \text{ If } f \leq f_c \tag{2}$$

$$f^*(f) = f_c + \frac{(1/q_1) - f_c}{f_f - f_c}(f - f_c) \text{ If } f \geq f_c \tag{3}$$

Where f is the voids' volume ratio, f_c is the voids' volume ratio at the beginning of nucleation and f_f is the voids' volume ratio when fracture occurs.

σ_M is the equivalent flow stress and it is obtained from the following work hardening relation:

$$\sigma_M\left(\varepsilon_M^{pl}\right) = \sigma_y\left(\frac{\varepsilon_M^{pl}}{\varepsilon_y} + 1\right)^n \tag{4}$$

In which n is the strain-hardening exponent and ε_M is the equivalent plastic strain. The voids' growth rate is the sum of existing voids growth f_g and the new voids' nucleation f_n.

$$\dot{f} = \dot{f_n} + \dot{f_g} \tag{5}$$

Where the components are further formulated as follows:

$$\dot{f_g} = (1 - f)tr\dot{\varepsilon}^{pl} \tag{6}$$

$$\dot{f_n} = A\dot{\varepsilon}_M^{pl} \tag{7}$$

$$A = \frac{f_n}{S_n\sqrt{2\pi}}\exp\left[-1/2\left(\frac{\varepsilon_M^{pl} - \varepsilon_N}{S_N}\right)\right] \tag{8}$$

In which $tr\varepsilon^{pl} = (\varepsilon_x + \varepsilon_x + \varepsilon_x)$ is the volume plastic strain rate, S_N is the voids' nucleation mean quantity, fn is volume ratio of the second phase particles (responsible for the voids' nucleation) and εN is mean strain at the time of voids' nucleation.

So parameters which can be defined in a vector form by:

$$\phi = \phi(q_1, q_2, f_0, f_c, f_n, f_f, \varepsilon_N, S_N) \tag{9}$$

In this work we are going to determine the values of f_0, fc and f_f as Marc MSC Mentat software give us the possibility to run the simulations without taking in consideration the nucleation parameters εn, fn and SN, as it will require a lot of time for computation and especially to train the Neural Network.

Table 1. Gurson parameters according to literature.

References	q_1	q_2	E_N	S_N	f_0	f_c	f_n	f_f	Material
Bauvineau et al. (1996)	1.5	1	-	-	0.002	0.004	-	-	CMn Steel
Decamp et al. (1997)	1.5	1	-	-	0.0023	0.004	-	0.225	CMn Steel
Schmitt et al. (1997)	1.5	1	0.3	0.1	0	0.06	0.002	0.212	Ferritic base Steel
Skallerud and Zhang. (1997)	1.25	1	0.3	0.1	0.0003	0.026	0.006	0.15	CMn Steel
Benseddiq and Imad. (2008)	1.5	1	0.3	0.1	0	0.004-0.06	0.002-0.02	~0.2	

According to the literature (Table 1) the values of q_1 and q_2, are almost fixed values $q_1 = 1.5$, $q_2 = 1$.

1.2 *Artificial neural network method*

Classically, the inverse identification procedure consists in finding the parameters which minimize the difference between a predicted response by the finite element method and the experimental one. The main drawback of this approach is its prohibitive time consuming especially when more than one test is used. In order to avoid this restriction, we substitute the FEM simulations by the ANN model during the optimization loop. The finite element simulations of the considered experimental tests are used only in the training of the ANN models. As the predicted response using the ANN model is almost instantaneous, more than one test can be done simultaneously to identify the material parameters. Therefore, the identification problem is converted into a multiobjective optimization using the genetic algorithm (GA) ANN model.

The architecture of a neural network depends on its network topology, transfer function, and learning algorithm. A multilayer perception trained by back propagation is among the most popular and versatile forms of neural networks and can deal with non-linear models with high accuracy. (Figure 1) presents the architecture of the ANN model. It is composed of three layers, namely, input layer, hidden layer, and output layer. For the hidden layer, each neuron receives total outputs from all of the neurons in the input layer as follows:

$$V_j = \sum_{i}^{n} W_{ij}x_i + b_j \tag{10}$$

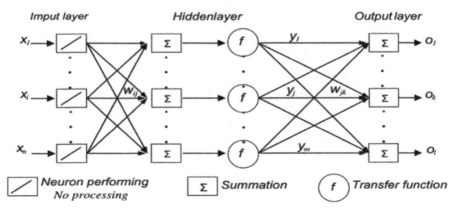

Figure 1. Multilayer perceptron.

where Vj is the potential input to the jth neuron in the hidden layer, n is the number of neurons in the input layer, Wij is the connection weight from the ith neuron in the input layer to the jth neuron in the hidden layer, xi is the input to the ith neuron in the input layer, and bj is the threshold value of the jth neuron in the hidden layer.

The output of a neuron in the hidden layer is calculated by applying the potential input to a transfer function. The sigmoid function shown in (11) is used as the transfer function between the input layer and the hidden layer.

$$y_i = F(V_j) = \frac{1}{1 + e^{-V_j}} \tag{11}$$

For the output layer, the output σ_k of the neuron k is given by the relation below:

$$\sigma_k = \sum_j^m W_{jk} y_j + b_k \tag{12}$$

where m is the number of neurons in the hidden layer, Wjk the connection weight from the ith neuron in the hidden layer to the kth neuron in the output layer, yj is the output from the jth neuron in the hidden layer, and hk is the threshold value of the kth neuron in the output layer.

2 METHODOLOGY AND RESULTS

In order to predict the ductile failure of the SENT specimen we need to determine the GTN parameters, which will be done by following these steps:

- Perform the small scale tests (CT, SENT) In order to provide the Experimental data
- Make the Finite Element Simulations 3D investigation and make the database for the neural network
- Determination of the GTN parameters by the combination between the experimental and FEM results and Artificial Neural Network

2.1 Fracture toughness test modeling

In order to determine the GTN parameters for the SENT specimen, we use the fracture toughness test data. These tests were performed on compact tension (CT) specimens, as shown in (Figure 2); its dimensions are also shown in the same figure.

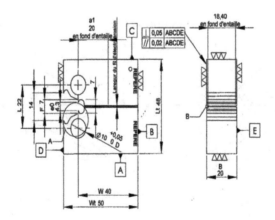

Figure 2. CT specimen.

28

Figure 3.　FEM of CT Specimen.

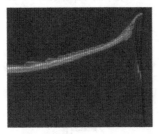

Figure 4.　Contour plot of the void volume fraction of the deformed specimen.

According to the literature (Bauvineau et al. 1996; Decamp et al. 1997; Schmitt et al. 1997; Skallerud & Zhang 1997; Benseddiq & Imad 2008), we were able to have initial values of GTN parameters as listed in Table 1 for steels.

We took advantage of the symmetry and we make the 3D FEM model just for the half of the CT specimen as shown in the (Figure 3) the FEM model contains a total of 58,103 nodes and 51,512 elements.

We shall proceed to the mesh refining near the crack tip because in this zone the crack propagates and the stress is very high, unlike the upper part of the specimen, which saves a little more of computing time the mesh size in the front of the pre-crack tip is 0.125 mm × 0.0625 mm, and the mesh is composed of quadratic axisymmetric elements with 8 nodes.

The contour plot of the void volume fraction of the deformed specimen is shown in (Figure 4), in which the crack has propagated into the specimen.

2.2　ANN and Data base creation

In order to provide the database necessary to train the ANN, 20 simulations were done for CT specimen with different settings of GTN parameters.

For the CT test, the trained ANN model consists of forty-eight neurons in the input layer, sixty neurons in the hidden layer and three neurons in the output layer (48-60-3). The neurons of the input layer represent the displacements of a reference point chosen in the boundary of the specimen corresponding to the given values of the reaction force F, and the neurons of the output layer are the GTN parameters to be identified (f_0, f_c and f_f).

After training the Neural Network, we could determine the three parameters and it didn't take a lot of time as if we did it by using the direct method which is the combination between the Experimental data and finite element data.

The GTN parameters determined by using the ANN are $f_0 = 0.001$, $f_c = 0.0052$, and $f_f = 0.2027$.

2.3 *Prediction of crack propagation for SENT SPECIMEN*

In order to check the validity of the GTN parameters that we found from CT Simulations, We are going to deal with the SENT specimen.

The dimensions of the SENT specimen are shown in the (Figure 5), and as we did for the CT simulation, we are going to use the axisymmetry, and make the 3D model just for the ¼ of the specimen, the FEM model contains a total of 75,461 nodes and 68,160 elements. The mesh size in the front of the pre-crack tip is the same as the CT specimen in order to avoid the effect of the sensibility of the mesh on the results (0.125 mm × 0.0625 mm) and the mesh is composed of quadratic axisymmetric elements with 8 nodes. The contour plot of the void volume fraction of the deformed specimen is shown in (Figure 6a), in which the crack has propagated into the specimen.

We run the FEM simulation based on the GTN parameters that we got from the ANN, the results show that the simulation curve fits the experimental curve and they are in a good agreement (Figure 7).

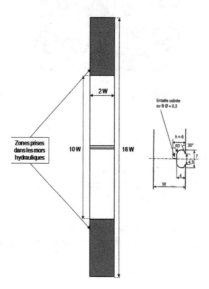

Figure 5.   Dimensions of SENT specimen.

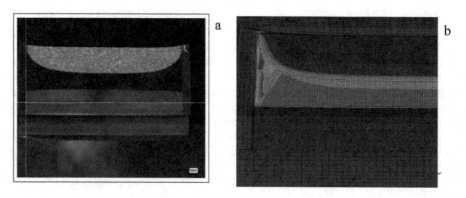

Figure 6.   (a) The contour plot of the void volume fraction of the deformed specimen; (b) the crack propagation in the SENT specimen.

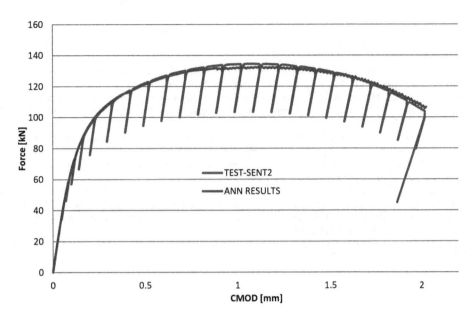

Figure 7.   Force-COD curve.

In addition to this and as we can see in the (Figure 6b), the GTN model predicts well the crack initiation and propagation which prove again the validity of this model.

3   CONCLUSION

As a conclusion of this work, we have performed 3D FEM analysis to study ductile fracture of the SENT specimen because it is a good representative of the PIPELINE.

To describe the crack propagation we did use the GTN model because it's a powerful and applicable tool comparing.

The GTN parameters are determined using the ANN, The GTN parameters found during the CT simulation predict well ductile fracture in SENT specimen.

More investigations are needed to determine the six Gurson parameters by using ANN, in addition to this a few experiments will be done in order to measure physically Gurson Parameters.

ACKNOWLEDGEMENTS

This work was carried out as part of the Advanced Structural Integrity Assessment Tools for Safe Long Term Operation project. The realization of this project is supported by the European Union.

The described work was partially supported by the EU as part of the EFOP3.6.1-16-2016-00011 "Younger and Renewing University Innovative Knowledge City – institutional development of the University of Miskolc aiming at intelligent specialisation" project implemented in the framework of the Szechenyi 2020 program.

REFERENCES

Clement Soret, Yazid Madi, Jacques Besson, Vincent Gaffard, 2015. *Use of the sent specimen in pipelinedesign.* 20th JTM - EPRG European pipeline research group, Paris, France. 34 p.

Alpay Oral, Gunay Anlas and John Lambros,2012 *Determination of Gurson−Tvergaard−Needleman Model Parameters for Failure of a Polymeric Material.* International Journal of Damage Mechanics.

Gurson, A.L., 1975 *Plastic Flow and Fracture Behavior of Ductile Materials Incorporating Void Nucleation, Growth and Interaction.* Ph.D. Thesis, Brown University, Providence, RI, USA.

Hamdi Aguir, Haykel Marouani (2010) *Gurson-Tvergaard-Needleman parameters identification using artificial neural networks in sheet metal blanking* International. Journal of Material Forming 3:113–116 *April.*

Bauvineau L., Burlet H., Eripret C., Pineau A. (1996) *Modelling ductile stable crack growth in a C-Mn steel with local approaches.* J Phys IV France 06(C6): C6-33–C36-42.

Decamp K., Bauvineau L., Besson J., Pineau A. (1997) *Size and geometry effects on ductile rupture of notched bars in a CMn steel: experiments and modelling.* Int J Fract 88(1): 1–18.

Schmitt W., Sun DZ., Blauel JG. (1997) *Damage mechanics analysis (Gurson model) and experimental verification of the behaviour of a crack in a weld-cladded component.* Nucl Eng Des 174(3): 237–246.

Skallerud B., Zhang ZL. (1997) *A 3D numerical study of ductile tearing and fatigue crack growth under nominal cyclic plasticity.* Int J Solids Struct 34(24): 3141–3161.

Benseddiq N., Imad A. (2008) *A ductile fracture analysis using a local damage model* Int J Press Vessel Pip 85(4): 219–227.

Solutions for Sustainable Development – Szita, Jármai & Voith (eds.)
© 2020 Taylor & Francis Group, London, ISBN 978-0-367-42425-1

Investigating three learning algorithms of a neural networks during inverse kinematics of robots

Hazim Nasir Ghafil
University of Miskolc, Miskolc, Hungary
University of Kufa, Najaf, Iraq

Kovács László & Károly Jármai
University of Miskolc, Miskolc, Hungary

ABSTRACT: The algorithms used in this paper was investigated while learning inverse Kinematic solution for a manipulator. In this paper, the performance of the three learning algorithms; Levenberg-Marquardt algorithm, Bayesian regularization algorithm, and scaled conjugate gradient algorithm is studied while learning single hidden layer network how to solve the inverse Kinematic problem of a three degree of freedom robot manipulator. During tests, Bayesian Regularization algorithm was found the best method among other methods with more accurate results but consuming larger time. Also, it is found that Scaled Conjugate Gradient method is inappropriate for the inverse Kinematic problem.

Keywords: Robotics, Artificial neural network, Inverse Kinematics

1 INTRODUCTION

The usage of the robots grows increasingly during the last decade for many applications in industry (Ghafil and Jármai 2018b). One of the most important functions during robot operation is to find the inverse Kinematic (IK) (Ghafil 2015) solution for a specific configuration of the end-effector of a robot. Also, one of the best methods to find this solution is the optimisation algorithms like particle swarm optimisation and artificial bee colony (Ghafil and Jármai 2018a) which can guarantee an optimal solution for the problem because of the nature of the objective function which is differentiable one. However, for real-time applications, we cannot use optimisation algorithms because they consume a moderated time. As an alternative, artificial neural networks ANN (Mosavi and Varkonyi 2017; Vargas et al. 2017) are perfect for real-time application even if their solutions are approximated and not as accurate as optimisation algorithms. We shall manage between accuracy and immediate approximated solutions especially for robots which are equipped with sensors and need fast solutions to take responsibility for sudden action. This is a reasonable justification to consider learning neural networks how to solve inverse Kinematic problem of a robot. Multilayer feedforward networks were learned to find the set of joint variables of a three degree of freedom manipulator given end-effector position (Choi and Lawrence 1992), the learning process is done using given end-effector positions and their corresponding joint angles. Multiple feedforward networks were found to be more efficient than the single layer to generalise a functional relationship for the inverse Kinematic problem (Perrusquía et al. 2019). After training, network was able to return the joint variables for arbitrary Cartesian positions. Figure 1 reveals the architecture of the networks used to solve the IK problem by assigning the Cartesian position as inputs and joint angles as output. ANN has been used with back propagation algorithm (Adigun and Kosko 2019) to solve inverse Kinematic for 6R robot manipulator with offset

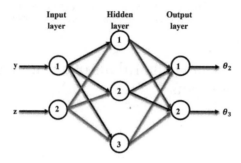

Figure 1.   ANN architecture of ANN used by (Choi and Lawrence 1992).

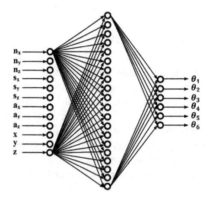

Figure 2.   ANN topology used to learn IK in.

using MATLAB Neural Networks Toolbox (Bingul et al. 2005). The network had 12 inputs representing elements of the homogenous transformation matrix and six outputs representing six joint angles. According to this study, the disadvantages of using ANN for inverse Kinematic is the high number of data set required for the training process.

In other words, the high amount of data set for training leads to a more accurate neural network; Figure 2. represent the topology the ANN used in this study. 4000 data items had been used to learn the network which is consists of 20 neurons in a single hidden layer. ANN builds models of systems by learning from examples of preset data; this makes it flexible and widely used in Robotics and automation. ANN is a linear or nonlinear function that can learn to solve and classify linear and nonlinear problems. In recent years, different researches have focused on new reliable algorithms to analyse finite data sets without increasing the computational intensity of the algorithms. For inverse Kinematic problem of the robot manipulator, many constraints should be taken into account during the learning process of a candidate network. Inverse Kinematic is a complex problem where the complexity is coming from the geometry of the robot and the trigonometric equations which relate the Cartesian space and joint space of the robot. Also, Kinematic equations are coupled, and there is the singularity problem, all these constraints increase the difficulty of the learning process. Three links planar manipulator was used as a test bed for the inverse Kinematic problem (Duka 2014) by learning a neural network how to generate a function for this issue. Non-uniform and customised topologies were used to learn the IK problem using the concept of network inversion (Tejomurtula and Kak 1999).

Kohonen and error backpropagation networks are the most common types of neural networks applied to solve inverse Kinematic of robot manipulators. Back-propagation algorithm, which is one of learning mechanism, takes a long time to find a feasible mapping from Cartesian space to the joint space of the robot. The response of all feed patterns is transported to the output layer. The difference between desired output and calculated output represent the error which is back propagated to adjust weights iteratively. If the training data set is too large, the taken amount of time for the training process will be tremendous, and this is unpractical for real-time training. Also, subnetworks are proposed (Lu and Ito 1995) to find more than one solution for a specific Cartesian position of the end-effector. In this work, the effect of using Levenberg-Marquardt learning algorithm, Bayesian regularization learning algorithm, and scaled conjugate gradient learning algorithm is studied while learning one input layer, one hidden layer, and one output layer how to find the inverse position of a RRR robot arm. The topology of the neural network has been discussed in details as well as the advantage and disadvantage of the abovementioned learning algorithms.

2 EXPERIMENTAL WORK

In this work, RRR robot manipulator with the ability to move in 3D space is proposed to conduct the effect of the number of the layers for a fixed amount of neurons. Figure 3 shows the frame assignment and robot specifications while Table 1 shows the parameters of the link.

It is known that homogenous transformation matrices of the three links in the manipulator can be estimated by substitute the four parameters of each link to Denavit-Hartenberg matrix (Spong et al. 2006). Set of forward Kinematic equations will be:

$$H_2^1 = \begin{bmatrix} \cos(\theta_1) & 0 & \sin(\theta_1) & 0 \\ \sin(\theta_1) & 0 & -\cos(\theta_1) & 0 \\ 0 & 1 & 0 & d_1 \\ 0 & 0 & 0 & 1 \end{bmatrix} \tag{1}$$

$$H_3^2 = \begin{bmatrix} \cos(\theta_2) & -\sin(\theta_2) & 0 & a_2.\cos(\theta_2) \\ \sin(\theta_2) & \cos(\theta_2) & 0 & a_2.\sin(\theta_2) \\ 0 & 0 & 1 & 0 \\ 0 & 0 & 0 & 1 \end{bmatrix} \tag{2}$$

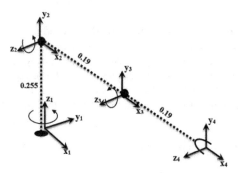

Figure 3.   RRR robot manipulator.

Table 1.   Spatial parameters of the Lab-Volt 5150 manipulator.

Link ID	Frame	φ	α	a	d	θ	limit
1	$o_1x_1y_1z_1\text{-}o_2x_2y_2z_2$	0	90	0	d_1	θ_1	-185,153
2	$o_2x_2y_2z_2\text{-}o_3x_3y_3z_3$	0	0	a_2	0	θ_2	-32,149
3	$o_3x_3y_3z_3\text{-}o_4x_4y_4z_4$	0	0	a_3	0	θ_3	-147,51

35

$$H_4^3 = \begin{bmatrix} \cos(\theta_3) & -\sin(\theta_3) & 0 & a_3 \cdot \cos(\theta_3) \\ \sin(\theta_3) & \cos(\theta_3) & 0 & a_3 \cdot \sin(\theta_3) \\ 0 & 0 & 1 & 0 \\ 0 & 0 & 0 & 1 \end{bmatrix} \tag{3}$$

$$H_4^1 = H_2^1 \cdot H_3^2 \cdot H_4^3 = \begin{bmatrix} n_x & s_x & a_x & x \\ n_y & s_y & a_y & y \\ n_z & s_z & a_z & z \\ 0 & 0 & 0 & 1 \end{bmatrix} \tag{4}$$

where

$$n_x = \cos(\theta_1)\cos(\theta_{2+3}) \tag{5}$$

$$n_y = \sin(\theta_1)\cos(\theta_{2+3}) \tag{6}$$

$$n_z = \sin(\theta_{2+3}) \tag{7}$$

$$s_x = -\cos(\theta_1)\sin(\theta_{2+3}) \tag{8}$$

$$s_y = -\sin(\theta_1)\sin(\theta_{2+3}) \tag{9}$$

$$s_z = \cos(\theta_{2+3}) \tag{10}$$

$$a_x = \sin(\theta_1) \tag{11}$$

$$a_y = -\cos(\theta_1) \tag{12}$$

$$a_z = 0 \tag{13}$$

$$x = \cos(\theta_1) * (a_2\cos(\theta_2) + a_3\cos(\theta_{2+3})) \tag{14}$$

$$y = \sin(\theta_1) * (a_2\cos(\theta_2) + a_3\cos(\theta_{2+3})) \tag{15}$$

$$z = d_1 + a_2\sin(\theta_2) + a_3\sin(\theta_{2+3}) \tag{16}$$

Equations (5) to (13) represent the orientation of the end-effector relative to the base frame while equations (14), (15), and (16) represent the Cartesian position of the end-effector. They are the mapping from the joint space to the Cartesian space; they represent the position of the end-effector corresponding to a set of joint angles. From these equations, a data set of 2000 input/output training data is generated to learn the proposed topologies of neural networks how to solve the inverse position using information out of the training data

3   NEURAL NETWORK ARCHITECTURE

Figure 4 reveals the topology of the artificial neural network used in this study where it has 12 inputs, one hidden layer, and three outputs. The input layer is the elements of the homogenous transformation matrix which relate the end-effector to the base frame and they are explained in equations (5-16). In this study, we have used one hidden layer with number of neurons n = 10 and n = 100. The output layer is consists of three neurons represent the three joint angles of the robot arm. Three learning algorithms are available for MATLAB neural network toolbox and we have employed them and compared the obtained results in case of n = 10 and n = 100. The proposed neural network has been learned using three learning algorithms; Levenberg-Marquardt algorithm LM, Bayesian Regularization algorithm BR, and Scaled Conjugate

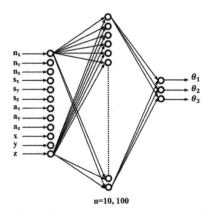

n=10, 100

Figure 4.   ANN architecture for RRR robot manipulator.

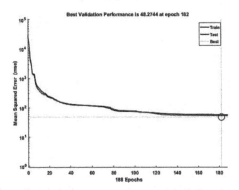

Figure 5.   Mean square error for Levenberg-Marquardt algorithm with n = 10.

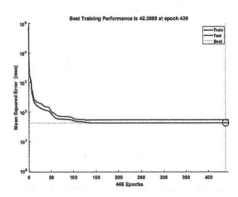

Figure 6.   Mean square error for Bayesian Regularization algorithm with n = 10.

Gradient algorithm SCG where each of these algorithms has its own advantage and disadvantage that should be explained in this study.

Figure 5 to Figure 7 shows the performance of Levenberg-Marquardt, Bayesian Regularization, and Scaled Conjugate Gradient respectively on the learning process in case of number of neurons equal to 10. Levenberg-Marquardt algorithm needs less time than Bayesian Regularization and more time than Scaled Conjugate Gradient but it needs more memory. Also, LM is more accurate than SCG and less accurate than BR. By repeating the same comparison in case of n = 100, see Figure 8 to Figure 10, it easy to configure the huge reduction in mean

37

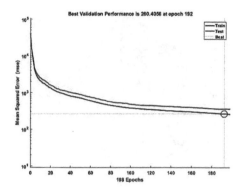

Figure 7. Mean square error for Scaled Conjugate Gradient algorithm with n = 10.

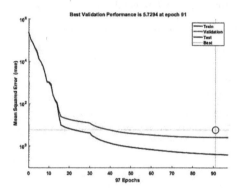

Figure 8. Mean square error for Levenberg-Marquardt algorithm with n = 100.

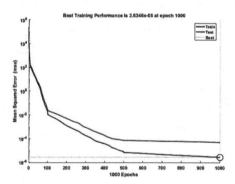

Figure 9. Mean square error for Bayesian Regularization algorithm with n = 100.

square error during test compare with the case when n = 10. A data set of 2000 items is used for learning process where 70% of the items used for training while 30% have used for testing.

Axiomatically, increasing number of neurons in the hidden layer can lead to best results with less error, but here we have to evaluate the performance of the three learning algorithms. Table 2 reveals the elapsed time to learn the network how to solve inverse Kinematic of the robot where BR needs more time than LM while SCG has taken a very short time. Back to Figure 5 to Figure 10, increasing number of neurons to 100 has led to a dramatic drop in mean square error MSE in case of LM and BR learning algorithms but SCG algorithm still cannot learn the network solving inverse Kinematic. Table 3 demonstrates MSE induced by using the three mentioned learning algorithms on the proposed problem.

38

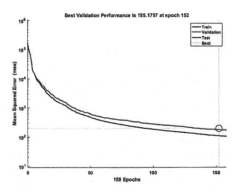

Figure 10.   Mean square error for Scaled Conjugate Gradient algorithm with n = 100.

Table 2.   Time consuming of Levenberg-Marquardt, Bayesian Regularization, and Scaled Conjugate Gradient learning algorithms.

	time (sec)	
learning algorithm	n = 10	n = 100
Levenberg-Marquardt	4	171
Bayesian Regularization	8	1840
Scaled Conjugate Gradient	0.01	1

Table 3.   Mean square error of using Levenberg-Marquardt, Bayesian Regularization, and Scaled Conjugate Gradient learning algorithms.

	Mean square error MSE	
learning algorithm	n = 10	n = 100
Levenberg-Marquardt	48.2744	5.7294
Bayesian Regularization	**42.3099**	**2.6346E-06**
Scaled Conjugate Gradient	260.4056	195.1757

4   DISCUSSION

For a simple architecture of manipulator which is shown in Figure 3, we have needed a set of 2000 individual solutions to get a reasonable comparison in this work. In other words, we need a huge amount of training data to learn the inverse problem for complex architectures like 7 DOF. Also, the error percentage get bigger while increasing the degree of the freedom of the robot with extra time to reach acceptable results. Thus, we can say that using neural network for the inverse Kinematics is restricted to the DOF of the robot. For the other side, Bayesian Regularization algorithm gives accurate results but using this method is critical for real-time applications because it takes a relatively long time to converge to a feasible solution. For realtime application, Levenberg-Marquardt method can be good choice especially we approximated solutions are adequate (less accurate and faster than Bayesian Regularization). While Scaled Conjugate Gradient method has failed to converge to an acceptable solution for this simple robot, it cannot be used for any types of robots.

5 CONCLUSION

A neural network is a powerful tool for nonlinear problems like inverse Kinematic problem. The neural network represents a function between inputs and outputs represented by distributed weights on connections among neurons. In this paper, we have studied the effect of using three different learning algorithms namely, Levenberg-Marquardt algorithm, Bayesian Regularization algorithm, and Scaled Conjugate Gradient algorithm to learn a neural network how to solve inverse Kinematic of a three revolute joints robot manipulator. The topology of the network is consisting of one input layer with twelve neurons, one hidden layer with number of neurons n = 10, 100, and one output layer with three neurons. The twelve neurons in input layer represent the given element of the homogenous transformation matrix while the three neurons in the output layer represent the desired joint angles of the robot. During this study, we have found that Scaled Conjugate Gradient learning algorithm unable to learn the network the proposed problem even with high number of neurons in the hidden layer. Bayesian Regularization learning algorithm returned the best results but with elapsed time greater than what is required for Levenberg-Marquardt algorithm.

ACKNOWLEDGEMENT

The described article was carried out as part of the EFOP-3.6.1-16-2016-00011 "Younger and Renewing University – Innovative Knowledge City – institutional development of the University of Miskolc aiming at intelligent specialization" project implemented in the framework of the Széchenyi 2020 program. The realization of this project is supported by the European Union, co-financed by the European Social Fund.

REFERENCES

Adigun O., Kosko B. (2019) Bidirectional backpropagation. IEEE Transactions on Systems, Man, and Cybernetics: Systems.

Bingul Z., Ertunc H., Oysu C. (2005) Applying neural network to inverse kinematic problem for 6R robot manipulator with offset wrist. In: Adaptive and Natural Computing Algorithms. Springer, (pp 112–115).

Choi BB., Lawrence C. (1992) Inverse kinematics problem in robotics using neural networks.

Duka A-V. (2014) Neural network based inverse kinematics solution for trajectory tracking of a robotic arm. Procedia Technology 12: 20–27.

Ghafil H. (2015) A virtual reality environment for 5-DOF robot manipulator based on XNA framework. International Journal of Computer Applications 113 (3): 33–37.

Ghafil H., Jármai K. (2018a) Comparative study of particle swarm optimization and artificial bee colony algorithms.

Ghafil HN., Jármai K Research and Application of Industrial Robot Manipulators in Vehicle and Automotive Engineering, a Survey. In: Vehicle and Automotive Engineering, 2018b. Springer, (pp 611–623).

Lu B-L., Ito K Regularization of inverse kinematics for redundant manipulators using neural network inversions. In: Proceedings of ICNN'95-International Conference on Neural Networks, 1995. IEEE, (pp 2726–2731).

Mosavi A., Varkonyi A. (2017) Learning in robotics. International Journal of Computer Applications 157(1): 8–11.

Perrusquía A., Yu W., Soria A. (2019) Position/force control of robot manipulators using reinforcement learning. Industrial Robot: the international journal of robotics research and application 46 (2): 267–280.

Spong MW., Hutchinson S., Vidyasagar M. (2006) Robot modeling and control.

Tejomurtula S., Kak S. (1999) Inverse kinematics in robotics using neural networks. Information Sciences 116 (2–4): 147–164.

Vargas R., Mosavi A., Ruiz R. (2017) Deep learning: a review. Advances in Intelligent Systems and Computing.

Solutions for Sustainable Development – Szita, Jármai & Voith (eds.)
© 2020 Taylor & Francis Group, London, ISBN 978-0-367-42425-1

Prediction of distillation plates' efficiency

V. Kállai
Department of Chemical Machinery, University of Miskolc, Hungary

P. Mizsey
Institute of Chemistry, University of Miskolc, Hungary

L.G. Szepesi
Department of Chemical Machinery, University of Miskolc, Hungary

ABSTRACT: This study deals with the prediction of distillation trays' efficiency in rectification columns, in case of low carbon content hydrocarbons (for example ethane-ethylene and propane-propylene) separation. During these technologies typically trayed columns are used, therefore the bubble cap and sieve trays were investigated in this paper.

In the previous studies, we have assumed that the leaving vapour and liquid from each tray is in equilibrium. It assumed that the trays have 100% efficiency, however, in practice, this supposition is not true. It means that it is necessary to determine the actual efficiency of the trays because this value gives the actual trays' number too. The studied methods are the O'Connell method and the AIChE method (Perry 1950).

This study's goal is to calculate and compare different trays' efficiency because it has an effect of the tower height.

In petrochemical industry the previously mentioned hydrocarbons have a high importance, these are the raw material of the plastic manufacturing. Generally, hydrocarbons are produced by distillation from higher carbon content materials, and this procedure has extremely high energy consumption. In case of higher plate efficiencies, the costs can be reduced.

Keywords: efficiency, distillation trays, rectification column

1 INTRODUCTION

Distillation is the most important separation technology in the chemical industry. (Klemola 1998) For separate from each others the propylene and propane the distillation technology is used. The benefit of this procedure is the large experience; however, it is a really cost- and energy-intensive procedure, especially for propane-propylene separation, because of their low relative volatility (approximately $\alpha = 1.188$). Seader and Henley (Seader and Henley 1998) showed that if the overall efficiency of a propylene-propane column with 150 theoretical plates decreases from 70 to 60%, it means that the number of the actual trays will increase with 34. This will increase the height and the capital cost of the column, therefore it is also economically necessary to determine the overall efficiency as accurately as possible (Salunke 2011).

During the design of a rectification column the plate efficiency has an enormous influence, it is a key parameter. One type of the efficiency is the Murphree's plate efficiency, which calculates the rate of the difference of the leaving vapour composition and the actual tray composition and the difference of the equilibrium composition and the leaving vapour composition. This is only reasonable when the liquid has not composition gradient on the plate and the liquid is well-mixed. In case of larger columns and relative volatilities the Murphree efficiency could be more, than 100% due to the variable liquid composition on a tray (Onda et al. 1971).

2 DEFINITIONS OF TRAY EFFICIENCIES

The overall tray efficiency could be defined like the ratio of the theoretical number of trays and the real number of trays (Øi 2003).

$$E_o = \frac{N_{theoretical}}{N_{real}} \tag{1}$$

The Murphree tray efficiency could be defined (Øi 2003):

$$E_M = \frac{y - y_{n+1}}{y* - y_{n+1}} \tag{2}$$

where y: mole fraction in vapour phase on the tray n,
y_{n+1}: the vapour phase's mole fraction on the tray (n+1),
y*: the vapour mole fraction, which is in equilibrium with the liquid mole fraction on the tray n (x_n).

The point efficiency could be defined (Salunke 2011):

$$E_P = \frac{y_{n,\ local} - y_{n-1,\ local}}{y*_{n,\ local} - y_{n-1,local}} \tag{3}$$

where E_P: point efficiency,
$y_{n,local}$: composition of vapour leaving the point,
$y_{n-1,local}$: the composition of vapour approaching the point,
$y*_{n,local}$: the vapour composition in equilibrium with the liquid mole fraction at the point $x_{n,local}$.

There is a connection between the overall and the Murphree tray efficiencies, it is determined by the equation below (Øi 2003):

$$E_o = \frac{1 + E_M(mV/L - 1)}{log_{10}(mV/L)} \tag{4}$$

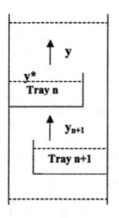

Figure 1. Definition of Murphree efficiency (Øi 2003).

where V: molar vapour flow rates (kmol/h),
    L: molar liquid flow rates (kmol/h),
    m: slope of equilibrium curve (dimensionless).

It was pointed out that in case of higher reflux ratio value the Murphree efficiency (based on the vapour in equilibrium with the leaving liquid) is decreased, while the plate efficiency (based on the vapour composition in equilibrium with the liquid entering the plate) is increased. The Murphree efficiencies are less sensitive to modifying the column parameters than the overall efficiencies (Onda et al. 1971).

2.1   *Empirical correlation of plate efficiencies*

The variables that affected the plate efficiencies can be divided into three groups:

- system property variables (type and composition of components, relative volatility, density, viscosity, surface tension),
- operating variables (rate of the phases, temperature, pressure),
- tray-design variables (tray spacing, weir height) (Klemola 1998).

Many theoretical and empirical models are available in the literature for predicting efficiency. Theoretical models were developed by Bakowski (Bakowski 1952), AIChE (AIChE 1958), Kastanek (F. Kastanek 1970), Todd and Van Winkle (Todd and Van Winkle 1972), Bolles (Bolles 1976), Garrett, et al. (Garrett, et al. 1977), Zuiderweg (Zuiderweg 1982), Lockett and Ahmed (Lockett and Ahmed 1983), Chan and Fair (Chan and Fair 1984), Scheffe and Weiland (Scheffe and Weiland 1987), Garcia and Fair (Garcia and Fair 2000), Syeda, et al. (Syeda, et al. 2007).

Empirical models to estimate plate efficiencies were developed by Drickamer and Bradford (Drickamer and Bradford 1943), O'Connell (O'Connell 1946), Chaiyavech and Van Winkle (Chaiyavech and Van Winkle 1961), English and Van Winkle (English and Van Winkle 1963), Onda, et al. (Onda et al. 1971), Tarat, et al. (Tarat et al. 1974), and MacFarland, et al. (Mac-Farland, et al. 1972) (Salunke 2011).

The most common method used for estimating the overall plate efficiency is the Drickamer and Bradford (Drickamer and Bradford 1943) method, which is based on an empirical relationship between Murphree's plate efficiency and the molar average liquid viscosity of the feed and relative volatility of the key components. This method was developed for determining the overall column efficiency of sieve and bubble-cap trayed towers by O'Connell (Burns 1966; Onda et al. 1971). Chaiyavech and Van Winkle (Chaiyavech and Van Winkle 1961) estimated some several system properties (like surface tension, viscosity, relative volatility, density and diffusivity). English and Van Winkle (English and Van Winkle 1963) based on this developed a correlation of tray efficiency (Burns 1966).

AIChE Research Committee's aim was to find a method of predicting efficiencies for bubble cap trays. The AIChE method is based on the two-film theory of Lewis and Whitman (Burns 1966).

According to the O'Connell correlation, the physical properties are more significant than the tray geometry and flow rates from the viewpoint of the efficiency. Chen and Chuang (1993) (Chen and Chuang 1993) prepared a semi-empirical model based on the physical properties and the tray geometry too. This model showed that the tray internal parameters (like weir height, pitch of holes, fractional perforated tray area) are less essential than the system's physical properties (Salunke 2011). Thereafter this study deals with the following correlation's results in a propane-propylene system.

2.2   *Correlation for the overall tray efficiency*

In 1943 Drickamer and Bradford (Drickamer and Bradford 1943) developed a correlation to predict the overall column efficiency (Salunke 2011):

$$E_o = 0.17 - 0.616 \, log_{10} \sum_{i=1}^{c} Z_i \mu_{L,i} \qquad (5)$$

where E_o: the overall column efficiency,

Z_i: the feed liquid mole fraction,

c: the number of components in the feed,

$\mu_{L,i}$: the liquid viscosity of component i on the bubble point temperature (cP).

There is another form of the Drickamer and Bradford empirical correlation:

$$E_o = 0.133 - 0.668 \, log_{10}(\mu) \qquad (6)$$

This correlation is valid for hydrocarbon mixtures in the range of 1 atm < P < 25 atm, 342 K < T < 488.5 K and 0.066 cP < μ < 0. 355 cP (Majunder and Das 2012).

In 1946 O'Connell expounded that the Drickamer and Bradford's correlation is appropriate only for hydrocarbon systems with low relative volatility. O'Connell to overcome this limitation correlated the E_o with the product's liquid viscosity and relative volatility at average column pressure and temperature (Duss and Taylor 2018).

$$E_o = 0.503(\mu_L \alpha)^{-0.226} \qquad (7)$$

where E_o: the overall column efficiency,

μ_L: the feed liquid viscosity on bubble point temperature (cP),

α: relative volatility.

In the following, this study deals with some various equation forms of the O'Connell correlation. All of them depend on the relative volatility and the liquid viscosity (Salunke 2011).

Economopoulos's equation (Economopoulos 1978) (Salunke 2011):

$$E_o = 0.485 - 0.129 \, ln(\alpha\mu_{Feed}) + 0.018(ln\alpha\mu_{Feed})^2 + 0.001(\alpha\mu_{Feed})^3 \qquad (8)$$

where E_o: the overall column efficiency,

μ_L: the feed liquid viscosity (cP),

α: relative volatility.

Lockett's equation (Lockett 1986) (it is reported only for bubble-cap trays) (Salunke 2011):

$$E_o = 0.492(\alpha\mu_L)^{-0.245} \qquad (9)$$

where E_o: the overall column efficiency,

μ_L: the feed liquid viscosity on bubble point temperature (cP),

α: relative volatility.

Kessler and Wankat's equation (Wankat 1988) (1988) (Salunke 2011):

$$E_o = 0.54159 - 0.28531 \, log_{10}(\alpha\mu_L) \qquad (10)$$

where E_o: the overall column efficiency,

μ_L: the feed liquid viscosity on bubble point temperature (cP),

α: relative volatility.

OSU (Oklahoma State University)'s equation (2011) (Salunke 2011):

$$E_o = 0.514(\alpha\mu_L)^{-0.23} \qquad (11)$$

where E_o: the overall column efficiency,
$\quad\quad\mu_L$: the feed liquid viscosity on bubble point temperature (cP),
$\quad\quad\alpha$: relative volatility.

3 THE STUDIED SYSTEM

During this study, a propane-propylene rectification column was investigated. To make the column's model the ChemCAD process simulator software was used with SRK equation of state (Némethné Sóvágó Judit 2013). It is really useful because it includes several options of continuous distillation and batch distillation too. The software is included in many thermo-dynamic models (Chemstations User Guide, 2007; Valderrama, et al. 2012).

The tower's parameters were summarized in Table 1. It was mentioned that the separation of these two compounds is very difficult due to the low relative volatility (it is shown in Figure 2., the equilibrium curve's slope is nearly equal to the 45° line's slope); therefore, the column has 140 theoretical plates and high reflux ratio. Total condenser (Nyul Gyula 1955) was used during the simulations, therefore the degrees of freedom was two, in this case, these were the reflux ratio ($R = 16$) and the bottom product's mole fraction in propylene ($x = 0.00016$), the pressure drop was neglected.

In the ChemCAD calculations, we used the equilibrium model which assumes theoretical plates by definition. Such plates' efficiency is 100%. However, if the tray design is considered, it is possible to have tray efficiencies calculated when the tray geometry and the internal flows should be known or the models described here.

ChemCAD will calculate the tray efficiency by two different methods. These methods require that the user identify the light key component and the heavy key component for fractionators. Components are identified by position in the stream list.

Table 1. The column's parameters.

Number of theoretical trays	140
Flow rate of the feed [kg/h]	1000
Heat condition of the feed (q)	1
Place of the feed*	90th
Reflux ratio	16
Pressure of the tower [bar]	8.5
Composition	
Propylene [%]	50
Propane[%]	50

* Trays are numbered from top to down.

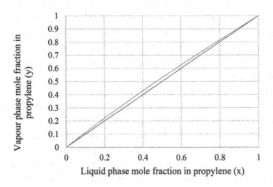

Figure 2. Propylene-propane equilibrium curve.

45

Table 2. The results of efficiency correlations.

CC	Drickamer-Bradford	O'Connell	Economopoulos	Lockett	Kessler and Wankat	OSU
0.891	0.809	0.8251	0.8452	0.8434	0.8142	0.8525

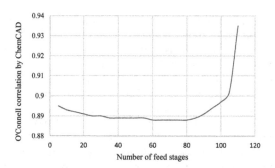

Figure 3. Connection between O'Connell efficiency and the feed stages location.

ChemCAD with Tray Sizing function calculates the O'Connell efficiency. Table 2 contains the ChemCAD (CC) efficiency value and from the previously mentioned literature correlations' equations results. In the studied system, the viscosity (μ_L) was 0.093289 cP and the relative volatility (α) was 1.188. For the Drickamer-Bradford correlation, it is necessary to know the composition of the feed, which was 50-50 mole% of propylene and propane too. According to the results, it could be pointed out that the newest empirical correlation, the OSU gives the nearest results to the simulation's result.

Simulations were made to examine what happens if the feed stages change with the same tray design parameters (Figure 3). With empirical methods, this could not be measured, because the relative volatility and the viscosity will not modify, so this analyzing only made by ChemCAD software. Between the 58th and 88th the O'Connell efficiency value is the lowest, while the feed location is getting lower the efficiency is getting higher. However, in a previous study of this column parameter examination, it was pointed out that the viewpoint of the lower energy consumption the best interval for feed stages is between the 63rd and 82nd trays (Kállai, et al. 2019).

The vertical axis in Figure 3 is naturally not a correlation but the efficiency calculated by ChemCAD.

4 CONCLUSION

The previously demonstrated methods concentrate on stream properties and these do not take into account the tray type and design. The simulation also does not calculate with the influence of the tray design to the efficiency. However, the feed location has an effect to the efficiency, lower location of the feed causes higher efficiency, but with this modification, the purity of the distillate product will not be appropriate, and the energy consumption of the heat exchangers will be higher.

Taking into consideration the cost-reducing and the efficiency-increasing, according to the correlations the solution may be the relative volatility's or the viscosity's increasing. Lower pressure causes higher relative volatility (Fonyó and Fábry 1998), but because of the cooling of the condenser, it is necessary that the technology operates on high pressure. On the other hand, liquid viscosity is reduced exponentially with temperature increasing.

Our further goal is to make simulations and calculations to another system (like ethane-ethylene separation) with these methods. From the two models' data, it could be possible to specify the optimal correlation to low carbon content separation systems.

ACKNOWLEDGEMENT

"The described article/presentation/study was carried out as part of the EFOP-3.6.1-16-2016-00011 "Younger and Renewing University – Innovative Knowledge City – institutional development of the University of Miskolc aiming at intelligent specialisation" project implemented in the framework of the Szechenyi 2020 program. The realization of this project is supported by the European Union, co-financed by the European Social Fund."

REFERENCES

AIChE. 1958. *Bubble Tray Design Manual*. New York.

Bakowski, S. 1952. "A New Method for Predicting the Plate Efficiency of Bubble-Crap Columns." *Chemical Engineering Science:* 266–282.

Bolles, W.L. 1976. "Multipass Flow Distribution and Mass-Transfer Efficiency for Distillation Plates." *AIChE Journal* 22(1): 153–158.

Burns, Michael Dean. 1966. "A Distillation Column for Tray Efficiency Studies." University of Arkansas.

Chaiyavech, P., and M. Van Winkle. 1961. "Small Distillation Column Efficiency." *Industrial and Engineering Chemistry* 53(3): 187–190.

Chan, H., and J.R Fair. 1984. "Prediction of Point Efficiencies on Sieve Trays. 1. Binary-Systems." *Industrial & Engineering Chemistry Process Design and Development* 23(4): 814–819.

Chemstations. 2007. "CHEMCAD Version 6 User Guide."

Chen, G.X., and K.T. Chuang. 1993. "Prediction of Point Efficiency for Sieve Trays in Distillation." *Industrial & Engineering Chemistry Research* 32(4): 701–708.

Drickamer, H.G, and J.R Bradford. 1943. "Overall Plate Efficiency of Commercial Hydrocarbon Fractionating Columns as a Function of Viscosity." *Transaction of American Institute of Chemical Engineers* 39: 319–360.

Duss, Markus, and Ross Taylor. 2018. "Predict Distillation Tray Efficiency." *AIChE CEP Magazine*: 24–30.

Economopoulos, A.P. 1978. "Computer Design of Sieve Trays and Tray Columns." *Chemical Engineering* 85(27): 109–120.

English, G.E, and M. Van Winkle. 1963. "Efficiency of Fractionating Columns." *Chemical Engineering*: 241–246.

F. Kastanek. 1970. "Efficiencies of Different Types of Distillation Plates." *Collection of Czechoslovak Chemical Communications* 35: 1170–1187.

Fonyó, Zsolt, and György Fábry. 1998. *Vegyipari Műveletttani Alapismeretek*. Nemzeti Tankönyvkiadó, Budapest.

Garcia, J.A, and J.R Fair. 2000. "A Fundamental Model for the Prediction of Distillation Sieve Tray Efficiency. 1. Database Development." *Industrial & Engineering Chemistry Research* 39(6): 1809–1817.

Garrett, G.R., R.H. Anderson, and M. Van Winkle. 1977. "Calculation of Sieve and Valve Tray Efficiencies in Column Scale-Up." *Industrial & Engineering Chemistry Process Design and Development* 16(1): 79–82.

Kállai, Viktória, Gábor L. Szepesi, and Péter Mizsey. 2019. "Propylene-Propane Rectification Column's Examination." In *MultiScience - XXXIII. MicroCAD International Multidisciplinary Scientific Conference,*.

Klemola, Kimmo T. 1998. "Efficiencies in Distillation and Reactive Distillation." Helsinki University of Technology.

Lockett, M.J. 1986. "Distillation Trays Fundamental." Cambridge University.

Lockett, M.J, and I.S Ahmed. 1983. "Tray and Point Efficiencies from a 0.6 m Diameter Distillation Column." *Chemical Engineering Research & Design* 61: 110–118.

MacFarland, S.A., P.M. Sigmund, and M. Van Winkle. 1972. "Predict Distillation Efficiency." *Hydrocarbon Processing* 51(7): 111–114.

Majunder, S K, and Chandan Das. 2012. "NPTEL- Chemical - Mass Transfer Operation 1 MODULE 5: DISTILLATION." In, 1–11.

Némethné Sóvágó Judit. 2013. "A Vegyipari Szimulációs Programok Működéséhez Alkalmazható Termodinamikai Modellek." *Anyagmérnöki Tudományok* 38(1): 231–243.

Nyul Gyula. 1955. *Lepárlás*. Műszaki Könyvkiadó, Budapest.

O'Connell, H.E. 1946. "Plate Efficiency of Fractionating Columns and Absorbers." *Transaction of American Institute of Chemical Engineers* 42(4): 741–755.

Øi, Lars Erik. 2003. "Estimation of Tray Efficiency in Dehydration Absorbers." *Chemical Engineering and Processing* 42: 867–878.

Onda, K, E Sada, K Takahashi, and S A Mukhtar. 1971. "Plate and Column Efficiencies of Continuous Rectifying Columns for Binary Mixtures." *AIChE Journal* 17(5): 1141–1152.

Perry, John H. 1950. 27 Journal of Chemical Education *Chemical Engineers' Handbook*. The McGraw-Hill Companies, Inc. http://pubs.acs.org/doi/abs/10.1021/ed027p533.1.

Salunke, Dadasaheb Baburao. 2011. "An O'Connell Type Correlation for Prediction of Overall Efficiency of Valve Tray Columns." Pune University.

Scheffe, R.D., and R.H. Weiland. 1987. "Mass-Transfer Characteristics of Valve Trays." *American Chemical Society* 26: 228–236.

Seader, J.D., and E.J. Henley. 1998. *Separation Process Principles*. John Wiley & Sons Inc.

Syeda, S.R, A. Afacan, and K.T Chuang. 2007. "A Fundamental Model for Prediction of Sieve Tray Efficiency." *Chemical Engineering Research & Design* 85(A2): 269–277.

Tarat, E.Y et al. 1974. "Efficiency of New Designs of Valve Trays in Rectification of Binary-Mixtures." *International Chemical Engineering* 14(4): 638–640.

Todd, W.G., and M. Van Winkle. 1972. "Correlation of Valve Tray Efficiency Data." *Industrial & Engineering Chemistry Process Design and Development* 11(4).

Valderrama, José O, Luis A Toselli, and Claudio A Faúndez. 2012. "Advances on Modeling and Simulation of Alcoholic Distillation. Part 2 : Process Simulation." *Food and Bioproducts Processing* 90(4): 832–840. http://dx.doi.org/10.1016/j.fbp.2012.04.003.

Wankat, P.C. 1988. *Equilibrium Staged Separations*. New York: Elsevier.

Zuiderweg, F.J. 1982. "Sieve Trays - A View on the State of the Art." *Chemical Engineering Science* 37 (10): 1441–1464.

CAD tools for knowledge based part design and assembly versioning

K. Nehéz
Department of Information Engineering, Institute of Information Science, University of Miskolc, Hungary

P. Mileff
Department of Information Technology, Institute of Information Science, University of Miskolc, Hungary

O. Hornyák
Department of Information Engineering, Institute of Information Science, University of Miskolc, Hungary

ABSTRACT: Increasing the efficiency of production processes is crucial in the modern manufacturing industry. This paper presents two software tools to enhance the productivity of common part/assembly design process. Both are developed for the widely used Siemens NX CAD system. The first module is an interactive multi-component software tool, which was developed for supporting similarity-based product design. This application can use product repository, makes possible to search by any specific parameters and export a new parameter file for creating new parts. The second module offers a UI driven semi-automatic way to enhance component replacement process of a complex assembly. The presented modules make the whole development process more effective because they are suitable to semi-automate the 3D model creation process. Applying these, human errors can be reduced, and the modelling process can be significantly accelerated.

1 INTRODUCTION

The growing need for highly productive industrial processes has forced industrial bodies to reduce the cost of each part including design and manufacturing processes. Product development lifecycle requires lower manpower cost and less time. To achieve this goal, efficient product modeling and design is necessary. The existing and feasible design may form a solution space. The design process can be shortened if the design can be parametrized. A generative design approach (GDA) can be utilized which provides a design framework to generate different alternatives during the product development process in order to find the best solution between the problem space and solution space. There must be however some limitation to GDA. The solution space can be an enormous result in too many instances, which is difficult to work with.

2 MOTIVATION

Our task was to investigate existing CAD designs for certain product families and create a library of their parametric, manufacturing and design properties. These parameters were subject to review. The review could spot some unwanted parameters or parameter combinations, for example, parameter d1 should allow 2 mm increments but there was an instance of 3.5 mm.

The second goal was to set up a parameter database of the allowed parametric values. A rule definition was also required to set up constraints.

The next requirement for this project was to generate a new product model variant that uses a generative design approach and also satisfies the aforementioned rules. Figure 1. depicts the motivation and the main steps of the project.

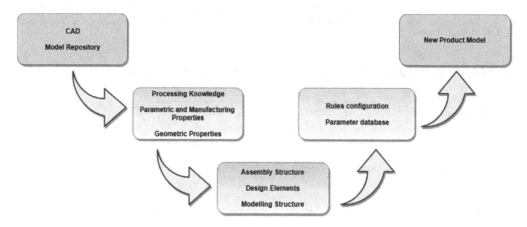

Figure 1.   Automatic parameter acquisition and parametric rule based GDA.

3   LITERATURE OVERVIEW

Grabowik et al, 2015 present methods of the computer aided design with self-generative models in NX Siemens CAD/CAE/CAM software. They identified the five methods of self-generative models preparation in NX with: parametric relations model, part families, GRIP language application, knowledge fusion and OPEN API mechanism. Their method requires a good knowledge of the relationship that appears between particular technical means parameters and particular dimension values. It also requires knowledge of the mutual dimension relations.

Zbiciak et. al, 2015 presents their software works that exclusively in NX Siemens CAD/CAM/CAE environment. It was developed in Microsoft Visual Studio with application of the .NET technology and NX SNAP library. Their software functionality allows designing and modelling of spur and helicoidal involute gears.

Their valuable problem formulation reviews the programing tools offered by Siemens NX.

- NX Open it is a collection of APIs (application programming interfaces) that allows creating software applications for NX through an open architecture using high-level programming languages such as C/C++, Visual Basic, C#, and Java. It could be used for automation complex and routine-repetitive tasks;
- knowledge fusion which is an interpreter, object-oriented language that allows adding engineering knowledge to a part element by both design and geometric creating rules. The knowledge fusion language is declarative, rather than procedural one, which means that, in general, the rules are only evaluated when referenced or demanded. In KF it is possible to access external knowledge bases, such as databases or spreadsheets;
- SNAP (Simple NX Application Programming) is an easy-to-learn programming tool intended for mechanical designers and other typical NX users – not necessarily programmers. It can speed up work by automating simple routine tasks. SNAP is based on the Visual Basic (VB.Net) language and can be used with the NX Journal Editor or with IDEs, such as Visual Studio.

Their implementation is based on SNAP programming tool as a compromise between language abilities and development cost.

Marefatt & Pitta 2007 introduce a concept of signature, which consists of features and the spatial relationships between the features. Using this signature they developed an indexing scheme to store and speedily retrieve digital part/component searching. In this way, their proposed techniques and algorithms enable automated retrieval of similar designs from a database of old designs.

Zhao et al 2017 proposed a structure-based 3D CAD model similarity assessment approach in their paper. They extracted the attribute information of the structural CAD models for the similarity comparison. Microsoft Visual Studio 2008 was used as the integrated development environment, and Open CASCADE was adopted as the geometry modeling platform.

Gao et al 2009 discuss the strategy of semi-similarity design for motorcycle brakes. Based on their work an appropriate design scheme was chosen.

Pascarelli et al 2018 discuss the evaluation of assembly plans by applying virtual reality.

4 PROPOSED SOFTWARE MODULES

Based on an in-depth study of the manufacturing processes at our industrial partner, two areas could be highlighted that contained potential for process enhancement and both areas are linked to the product design phase. The company uses the Siemens NX CAD system for product design, so the purpose of research and development is to design and develop two software components integrated into Siemens NX. With their help, the design and implementation processes of some products and complete assemblies can be made more efficient by introducing an automated solution. In order to develop software modules for the Siemens NX system, it is essential to have a detailed understanding of how the system works. Prior to the development process, the appropriate technology and the programming interface that is advantageous for the modules that can be integrated into the system should be selected. Visual Basic is the programming language most supported by Siemens NX, which contains many useful programming library packages (APIs).

In Siemens NX module developments, the most commonly used function libraries are: SNAP (Simple NX Application Programming) and NX Open. The former provides assistance in the development of simpler applications, with the latter being able to implement more complex modules. Due to the complexity of the task, we have chosen the NX Open integration solution for both software, and the completed modules are designed for the Siemens NX 11 system. The modules will be presented below.

4.1 *Automated modelling of base bodies*

If some parameters are not correctly specified or applied during product development, it may affect the production processes and the quality of the products. In the case of automotive products, the exact determination of the various parameters is very important because the components are complex and consist of many different parts. For example, a pulley, which is designed to be practical as well as more than 600 variations. Designing a new type of pulley is often started from a so-called master plan, where size and dimension data are given in a parametric way. During the planning process, design engineers change these parameters when developing a new pulley. It is a natural feature of the process that there may be pulley designs that are similar to a previously designed pulley when changing parameters. In practice, however, for engineers, it is difficult to find an existing pulley design which is very similar in size (parametric) to previous designs. If it is possible to search in a "database" which is constantly expanding in time, this part of the design process could be greatly simplified, because the design of a new pulley model could start from a plan that is very similar to it. During the research, our goal was to develop a computer software, which supports this process on several levels. It makes possible to maintain and verify existing product designs. In case of new designs, it is capable to check whether if there is an existing design that is very similar to the newly produced parts. Finally, depending on the user's decision, it creates a parametric descriptor file that can be used to generate the new pulley model. Figure 2. illustrates the connection of the developed software component to the Siemens NX system and the structure of the software:

Because of the license and Siemens NX API constraints of the industrial partner, the software had to be developed from several components. The first part is a Visual Basic based Siemens NX transfer module (Parameter Collector Application). As our industrial partner uses the Siemens NX Teamcenter solution to manage designs and models, there was a need for a solution that can

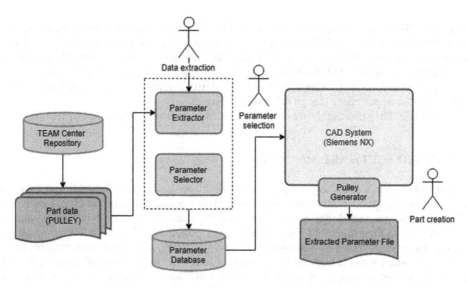

Figure 2. Siemens NX and the application integration.

Part Number	Left outer diameter (D0)	Number of Grooves (N0)	Two side groove angle (W2)	Corner angle (W3)	Middle groove angle (W4)	1st. groove distance (a1)	First groove to corner (a2)	Groove distance (a3)	Left outer diameter (d0)	Pitch-circle diameter (d1)	Internal diameter (d2)	Right outer diameter (d7)
F00M591109_005.prt	NA	NA	NA	NA	NA	21.5	2.4	3.56	72.0	62.47	48.0	33.5
F00M591109_005_x_F00M591109_D.prt	NA	NA	NA	NA	NA	NA	NA	NA	NA	NA	NA	NA
F00M691128_004.prt	NA	NA	NA	NA	NA	NA	NA	NA	NA	NA	NA	NA
F00M691128_004_x_F00M691128_D.prt	NA	NA	NA	NA	NA	NA	NA	NA	NA	NA	NA	NA
F00M691135_003.prt	NA	NA	NA	NA	NA	10.3	2.4	3.56	72.0	63.0	48.0	33.5
F00M691135_003_x_F00M691135_D.prt	NA	NA	NA	NA	NA	NA	NA	NA	NA	NA	NA	NA
F00M691136_002.prt	NA	NA	NA	NA	NA	17.3	2.4	3.56	72.0	63.0	48.0	33.5
F00M691136_002_x_F00M691136_D.prt	NA	NA	NA	NA	NA	NA	NA	NA	NA	NA	NA	NA
F00M992785.prt	86.5	8.0	40.0	40.0	90.0	NA	NA	NA	NA	NA	NA	NA
F00M992785_x_F00M992785_D.prt	NA	NA	NA	NA	NA	NA	NA	NA	NA	NA	NA	NA
F00M992788.prt	73.0	8.0	40.0	40.0	90.0	NA	NA	NA	NA	NA	NA	NA
F00M992788_x_F00M992788_D.prt	NA	NA	NA	NA	NA	NA	NA	NA	NA	NA	NA	NA
F00M992789.prt	91.5	8.0	40.0	40.0	90.0	NA	NA	NA	NA	NA	NA	NA
F00M992789_x_F00M992789_D.prt	NA	NA	NA	NA	NA	NA	NA	NA	NA	NA	NA	NA

Add/Remove search fields:

Name	Value	Tolerance	Connection
a1	15.2	7	AND
d1	63	1.2	OR
d0	72	1.0	AND
			AND

Figure 3. Search pulley designs based on parameters.

automatically download the pulleys from the TeamCenter database. The program exports the important parameters of the pulleys into a text file (*Extracted Parameters File*), which prepares them for the second module. The second software component is a JavaFX technology-based application with a graphical user interface (*Parameter Selector Application*). This program creates a database (*Parameter Database*) from the extracted data collected by the above-mentioned Visual Basic program. The user interface (Figure 3) allows the user to search by a given condition in the database and list the matching pulleys.

The program offers the opportunity to find a similar pulley for the newly developed model based on the parameters. During the export process, the editing of certain important parameters became a necessity. The user can modify these parameters within certain constraints, then the data can be exported. The third software component is also a Visual Basic based NX module (*Pulley Generator Application*), which is able to create a new pulley model in Siemens NX based on the data file created by the JavaFX application by using the master plan selected in the second phase. The module downloads the master plan from the TeamCenter, and then replaces the changed parameters with the desired ones (defined in the JavaFX program). Although we achieved our goal by developing multiple software modules, the solution still provides great help to the engineers in practice by reducing the time spent on the design phase.

4.2 *Automated modelling of assemblies*

An automotive product usually may consist of many parts (often hundreds). In the product repository, subcomponents may have several revisions. Each version can vary in size and material quality. Different versions may appear in different products. When designing a new product, the designers start with a similar earlier product and manually replace and check the component versions one by one. This is an application of group technology (Tóth et al, 2014, Hornyák & Sáfrány 2010)

At present, in the absence of an automated solution, in most cases, this is the form used in practice. This manual operation is slow, cumbersome and error-prone. During subsequent validation, a great experience is needed to eliminate errors.

The solution to the problem can be the development of a software module that can semi-automatically replace a given component. Assembly is controlled by manipulating the values describing its attributes. (Pokhilko, 2018,) Automation should be implemented in such a way that the replacement of the replacement part within the given model takes place from the TeamCenter repository, preferably in an interactive form.

We have successfully developed a Visual Basic based Siemens NX module (Assembly Component Replacer) that can achieve these goals. Figure 4 illustrates the interaction between the developed software module and the Siemens NX CAD system.

The software provides a user-friendly interface with a built-in graphical user interface. The prerequisite for its use is that the product assembly must be opened in the NX system.

The use of the software is very simple: you have to select the item you want to replace in the product assembly and then specify the part number for the replacement part on the graphical interface. The specified replacement components are retrieved through TeamCenter (product life cycle management system), with their revision and status codes.

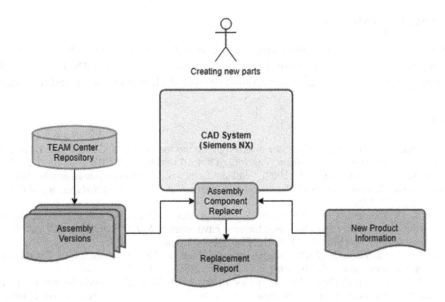

Figure 4.   Concept of our Assembly Component Replacer (ACR) module.

Figure 5. Starting screen of our ACR module in Siemens NX.

After selecting a specific revision and status of the model listed by the software, it is possible to replace the given parts. Bíró & Kusper, 2018 presents a graph formularization of the problem. Replacement of the given parts within the product assembly is achieved by maintaining the positions and constraints of the original parts. The software will produce a text report about the replacement of parts made. This report can be shown on the screen and can be saved for later auditing.

The great advantage of the solution is that the GUI interface (Figure 5) can handle the model loaded into the NX system and directly interact with them. The compactness of the window makes the replacement of parts faster as if we used the NX system built-in multi-window solution with time-consuming user clicks.

5   CONCLUSIONS

The research and development work conducted improved the capabilities of the CAD system. It forms a bridge between CAD and CAPP. As a result, the pulley modelling time was reduced significantly. Moreover, the software user has a greater influence on final product model parameters, shape. This allows manufacturing product of better quality and high degree of integration to the existing processes.

ACKNOWLEDGEMENTS

This research was supported by the European Union and the Hungarian State, co-financed by the European Regional Development Fund in the framework of the GINOP-2.3.4-15-2016-00004 project, aimed to promote the cooperation between the higher education and the industry.

REFERENCES

Bíró, Cs. & Kusper, G. (2018). Equivalence of Strongly Connected Graphs and Black-and-White 2-SAT Problems, *Miskolc Mathematical Notes*, Vol. 19(2), DOI: 10.18514/MMN.2018.2140, pp. 755–768.

Gao, F., & Xiao, G., & Yuanming, Z. (2009). Semi-similarity design of motorcycle-hydraulic-disk brake: strategy and application. *11th IEEE International Conference on Computer-Aided Design and Computer Graphics*, pp. 576–579.

Grabowik, C., & Kalinowski, K., & Kempa, W., & Paprocka, I. (2015). A method of computer aided design with self-generative models in NX Siemens environment. *In IOP Conference Series: Materials Science and Engineering* (Vol. 95, (1), p. 012123), IOP Publishing.

Hornyák, O., & Sáfrány, G. (2010). Models and Methods to Detect Similarity of Manufacturing Machines. *Hungarian Journal of Industry and Chemistry*, 38(2), 149–153.

Marefat, M. M., & Pitta, C. (2007). Similarity-based retrieval of CAD solid models for automated reuse of machining process plans. *IEEE International Conference on Automation Science and Engineering*, pp. 312–317.

Pascarelli, C., Lazoi, M., Papadia, G., Galli, V., & Piarulli, L. (2018). CAD-VR Integration as a Tool for Industrial Assembly Processes Validation: A Practical Application. In International Conference on Augmented Reality, *Virtual Reality and Computer Graphics* (pp. 435–450). Springer, Cham.

Pokhilko, A, (2018). The Design Process Data Representation Based on Semantic Features Generalization, *3rd Russian-Pacific Conference on Computer Technology and Applications (RPC)*, doi: 10.1109/RPC.2018.8482132, pp. 1–4.

Toth, T., Radeleczki, S., Veres, L., & Korei, A. (2014). A new mathematical approach to supporting group technology. *European Journal of Industrial Engineering*, 8(5), 716–737.

Zbiciak, M., & Grabowik, C., & Janik, W. (2015). An automation of design and modelling tasks in NX Siemens environment with original software-generator module. *In IOP Conference Series: Materials Science and Engineering* (Vol. 95, (1), p. 012117), IOP Publishing.

Zhao, X., & Liu, X., & Zhang, K. (2017). A structure-based 3d CAD model similarity assessment approach. *In Proceedings of the 6th International Conference on Software and Computer Applications*, pp. 281–284.

Solutions for Sustainable Development – Szita, Jármai & Voith (eds.)
© 2020 Taylor & Francis Group, London, ISBN 978-0-367-42425-1

Heat transfer analysis for finned tube heat exchangers

M. Petrik, L.G. Szepesi & K. Jármai
University of Miskolc, Miskolc, Hungary

ABSTRACT: This present paper fulfils a parametric study about heat exchangers with extended surface. These finned tube heat exchangers commonly used in industry, household and automotive industry. With the use of different type of fins a much higher heat transfer area can be achieve. However, while this extra surface advantageous for the heat transfer area, disadvantageous for the heat transfer coefficient. This current research focuses on the effect of the size of the transversal fins to the heat performance of the finned tube. Numerical CFD-simulations carried out to investigate the air side flow and heat transfer characteristics. Using the result data and previously published experimental data, new correlation for estimation of heat transfer coefficient have been established.

Keywords: finned tube, heat transfer, CFD analysis

1 INTRODUCTION

Finned tube heat exchangers play a very important role in many industrial processes, especially where gas is one of the heat exchange medium, such as process gas heaters and coolers, refrigeration systems, air conditioning systems, automotives or even electronic cooling systems. These equipment are the key components in the gas-to-gas and the gas-to-liquid heat transfer, so their use is indispensable for waste heat recovery technologies and cooling/heating systems [Chong & Tan, 2012; Verde et al., 2017; Hussein et al., 2014].

Several experimental and numerical investigations of finned tube heat exchange available in the literature. These type of devices have the greatest variety of the heat exchangers, and changing any of these factors will affect to the performance. O'Brien & Sohal (2000) investigated finned oval tube heat exchangers. They built a test equipment and the measured values were compared with CFD results. They investigated only one finned tube with different air-side Re numbers, and demonstrated that an oval tube increases the value of the heat transfer factor compared to the circular tube. Chu et al. (2013) numerically studied the heat transfer of oval finned tube, but their investigations dealt with the angle between the air flow and the fins.

The other important geometric parameter is the type of the fins and the tube layout. Singh et al. (2015) carried out numerical simulation for in-line tube arrangement, Khudheyer & Mahmoud (2011) for staggered tube arrangement. The common feature of these simulation is that the examined geometry is very small part of a real fin pattern, only one fin with two or three tubes. This is because the numerical simulating of a whole finned tube heat exchanger would be accompanied by a very high computational requirement, as the number of mesh elements would large due to the prism layer insertion of the extended surface.

This division method is also used for other types of fins. There are double-pipe heat exchangers for exhaust gas systems which contain fins between the inner and outer pipe. Hasan (2015) investigated the heat transfer properties of these fans and also divided the geometry to simulate only one fin. Perrotin & Clodic (2003) also examined angular fin of rectangular and hexagonal fins at in-line, staggered and louvered cases.

Furthermore there are very special finned tube heat exchangers too, for example the Jou-leThomson (JT) cryocoolers, which is a spiral-shaped tube with spiral-shaped fins. Tzabar (2014) have made calculations about these devices. For the analysis the author used the theory of thermal resistance and the NTU-ε methods. There are articles that publish the experimental and analytical results, as Taler & Korzen (2016). The authors showed the comparison of the heat performance from empirical correlation and measurements.

2 NUMERICAL SIMULATION

2.1 Computational geometry

For the analytical simulations three types of fins with different geometric dimensions were used. The common features were the pipe length, the inner and outer diameter of the tubes, the thickness of the fins and the fin spacing. Table 1 summarizes the values of the common features, and Figure 1 shown the 3D model.

The purpose of this article was examining the effect of the size of the fins for the heat transfer coefficient and heat performance. Three different geometries have been studied, which are shown in Table 2.

Analysing the heat transfer and the flow between the surface of the fin and the air CFD software SC-Tetra was used. However, due to the small wall thickness and fin spacing, simulation of a whole finned tube heat exchanger is not suitable for the used computers, only one tube was

Table 1. The common features of the simulated finned tube.

Geometry	Parameter	Value	Unit
thickness of fins	δ_f	0,5	mm
fin spacing	a_f	1	mm
tube inner diameter	d_i	4	mm
tube outer diameter	d_o	6	mm
heat conductivity of fins and tube	λ_f	210	W/(mK)
numbers of fins	n_f	30	pieces
length of tube	lf	60	mm

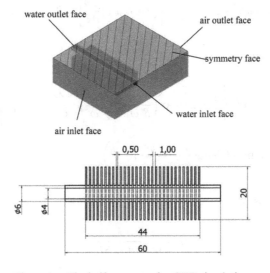

Figure 1. The half geometry for CFD simulation.

57

Table 2. The differences between the geometries.

Fin ID	Length of fins (l_f)	Width of fins (b_f)
1	15 mm	15 mm
2	20 mm	20 mm
3	25 mm	25 mm

investigated with 30 pieces fins. The other consideration is the symmetry of the geometry and the air flow, so it was enough to examine the half geometry. Using these considerations, the used mesh can be adjusted to the sufficient level. The half geometry shown in Figure 1.

2.2 Turbulence model

The simulations have been done with use of the SST k-ω turbulence model. This model is one of the Reynolds Averaged Navier Stokes (RANS) models. This turbulence model is a two-equation eddy-viscosity model which is suitable for investigate the extended surfaces because this is a hybrid model of the Wilcox k-ω and the k-ε models. The main problem with the Wilcox k-ω model is the strong sensitivity to the free stream conditions. To avoid this sensitivity, the Baseline (BSL) model was developed, which is the Wilcox model multiplied by the bending function F_1 (3), while the k-ε model multiplied by $1-F_1$. However, the BSL model is not able to fully handle the run-up and detachment on smooth surfaces, therefore overestimates the vortex viscosity. Therefore, the SST model uses F_2 coefficient (6) that limits the use of the limiter to the boundary layer. The previous model suitable for simulating flow in the viscous sub-layers, while the latter model is ideal for anticipating flow behaviour in regions away from the wall (Nemati & Moghimi, 2014). The governing equation of the SST k-ω model:

- The turbulance kinetic energy:

$$\frac{\partial k}{\partial t} + U_j \cdot \frac{\partial k}{\partial x_j} = P_k - \beta^* \cdot k \cdot \omega + \frac{\partial}{\partial x_j}\left[(\nu + \sigma_k \cdot \nu_T) \cdot \frac{\partial k}{\partial x_j}\right] \quad (1)$$

- Specific dissipation rate:

$$\frac{\partial \omega}{\partial t} + U_j \frac{\partial \omega}{\partial x_j} = \alpha S^2 - \beta \omega^2 + \frac{\partial}{\partial x_j}\left[(\nu + \sigma_\omega \nu_T)\frac{\partial \omega}{\partial x_j}\right] + 2(1 - F_1)\sigma_{\omega 2}\frac{1}{\omega}\frac{\partial k}{\partial x_i}\frac{\partial \omega}{\partial x_i} \quad (2)$$

- The blending function:

$$F_1 = \tanh\left\{\left(\min\left[\max\left(\frac{\sqrt{k}}{\beta^* \omega y}, \frac{500\nu}{y^2\omega}\right), \frac{4\sigma_{\omega 2}k}{CD_{k\omega}y^2}\right]\right)^4\right\} \quad (3)$$

$$CD_{k\omega} = \max\left(2\rho\sigma_{\omega 2}\frac{1}{\omega}\frac{\partial k}{\partial x_i}\frac{\partial \omega}{\partial x_i}, 10^{-10}\right) \quad (4)$$

- Kinematic eddy viscosity:

$$\nu_T = \frac{a_1 k}{\max(a_{1\omega}, SF_2)} \quad (5)$$

- The second blending function:

$$F_2 = \tanh\left\{\left[\max\left(\frac{2\sqrt{k}}{\beta^* \omega y}, \frac{500\nu}{y^2\omega}\right)\right]^2\right\}$$ (6)

- P_k (production limiter)

$$P_k = \min\left(\tau_{ij}\frac{\partial U_i}{\partial x_j}, 10\beta^* k\omega\right)$$ (7)

where k is the turbulence kinetic energy (m^2/s^2), ω is the turbulence frequency, S is the invariant measure of the strain rate, ρ is the density (kg/m^3), U is velocity vector, ν is kinematic viscosity (m^2/s), y is the distance of the wall (m), β, β^*, α, $\alpha_{1,\omega}$, σ_ω, $\sigma_{\omega 2}$ and $CD_{k\omega}$ are coefficients. The other important property of this model is the dimensionless wall distance y^+. At this case, the first point of the grid must be maximum 1. If this condition does not satisfied, a mesh refinement must be done.

A comparative study of turbulence models was performed for RNG k-ε and SST k-ω models. The results showed that this heat transfer case the SST k-ω model was more accurate, as Khudheyer & Mahmoud (2011) also proved.

2.3 Boundary conditions

In order to address a well-defined problem, the following boundary conditions were considered to numerically solved the governing equations. The model which used for the simulation contains 3 domains: the air, the aluminium and the water domains. To simplify the solution and later the comparison with the analytical model, the temperature dependence of the material properties was neglected and their change is small with these small temperature changes. In the future inspection will be carried out for investigate this simplification. Therefore, all of the material properties were set up at the base temperature, which was 20°C.

2.3.1 Inlet boundary conditions
In the models there are two inlet surfaces in all cases: one for the air and another for the water. To investigate the heat transfer (for example the operation of an automotive radiator), the temperature of the inlet air was set to 20°C with turbulence kinetic energy of 0.0001 m^2/s^2, while the same properties of the incoming water was set up to 60°C temperature with 0.0001 m^2/s^2 turbulence kinetic energy and 0.1 m/s water velocity. The effect of the air velocity also investigated in this study, so the simulations were performed for three different velocities, with values of 2 m/s, 5 m/s and 10 m/s.

2.3.2 Outlet boundary condition
The size of the outlet surfaces was the same as the inlet surfaces. For these, static pressure boundary conditions were set up with 0 Pa outlet overpressure and 0.0001 m^2/s^2 turbulence kinetic energy. These values came from the CFD software User's Guide.

2.3.3 Wall boundary conditions
For the simulations presented in the study, a total of 4 wall boundary condition were used. First of all, the air wall, which covers the whole model, was set up as stationary and adiabatic wall. The second was the wall on the plane of the symmetry, along the model was split. This was a free slip wall. Due to the right set up of this wall can be simplified the geometric model. The next wall was which surround the water domain. For this one, a constant heat transfer coefficient was set up. The reason is that in this study the heat transfer of the extended surface was investigated, not the heat transfer of the tube side. With this consideration, the numerical and analytical results can be comparable with each other. The constant value for the heat

transfer coefficient was selected to 2500 W/(m²K). The wall to be presented last is the wall between the aluminium and air. This was also a stationary wall, and from a thermal point of view it has no thermal resistance.

2.4 *Used mesh*

For the procedure of numerical simulation the used commercial fluid dynamics software was the SCTetra v14 based on CFD to solve the governing equations described in section 2.2, with the geometry shown in Figure 1. Since the heat transfer occurs on the surface, prism layer insertion have been set up between the water domain and the inner surface of the tube, and between the outer surface of the tube/fins and the air domain, with $5 \cdot 10^{-5}$ m thickness of first layer, 1.1 variation rate of thickness and 8 layers. Mesh independence study also performed and a mesh of total 6,433,022 elements is selected. Figure 2 shows the total cross section of the investigated model, while Figure 3 shows the prism layer insertion of the fins.

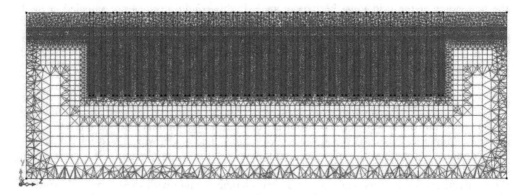

Figure 2.   The used mesh of the geometric model ID1.

Figure 3.   The prism layer insertion of the fins (all cases).

3 ANALYTIC CALCULATION METHOD

The usage of the fins as extended surface will cause a higher heat transfer area and also a lower heat transfer coefficient. These fins represent a resistance in the flow, which will cause a higher velocity and degree of turbulence, which would mean a higher value of heat transfer coefficient. However the heat is not only transferred from the water to the air by convection, but also with conduction along the fins. As a result, the surface temperature of the fins will be a lower value, so the higher theoretical heat transfer coefficient do not affect a higher transferred heat, because of the lower temperature difference. To take this into account, an efficiency value and a modified heat transfer coefficient must be calculated in the function of the geometry of the fins. The following relationships are available in the VDI-Heat atlas (2010). This modified heat transfer coefficient:

$$\alpha_v = \alpha_m \left[1 - (1 - \eta_f) \frac{A_f}{A} \right], \tag{8}$$

where α_m is the theoretical heat transfer coefficient without the fins (as a smooth tube), η_f is the efficiency, A_f is the heat transfer area of a single fin and A is the total heat transfer area. The calculation of the theoretical heat transfer coefficient is analogous to the calculation of any other types of convections: with experimental Nu-number correlation. In case of finned tube heat exchanger two different cases can be extinguished: inline (9) and staggered (10) layout:

$$Nu = 0.22 \cdot Re^{0.6} \cdot \left(\frac{A}{A_{t0}} \right)^{-0.15} \cdot Pr^{0.33} \tag{9}$$

and

$$Nu = 0.38 \cdot Re^{0.6} \cdot \left(\frac{A}{A_{t0}} \right)^{-0.15} \cdot Pr^{0.33}, \tag{10}$$

where A_{t0} is the outside surface of the tube.

The value of the efficiency highly depends on the shape of the fins. This study investigated the rectangular shape, so the showed correlation related to this type. First of all, a geometric parameter must be calculated, which depends on the diameter of the tube (d_o), the length (l_f) and width (b_f) of the fins:

$$\varphi' = 1.28 \cdot \frac{b_f}{d_0} \sqrt{ \left(\frac{l_f}{b_f} - 0.2 \right) } \tag{11}$$

and

$$\varphi = (\varphi' - 1)(1 + 0.35 \cdot \ln \varphi'). \tag{12}$$

The calculation of this φ value is different in case of other types of fins (circular or adjacent fins). The next parameter is:

$$X = \varphi \frac{d_0}{2} \sqrt{\frac{2\alpha_m}{\lambda_f \delta_f}}. \tag{13}$$

From there, the efficiency can be calculated:

$$\eta_f = \frac{\tanh X}{X} \tag{14}$$

The next figure shows the efficiency factor of the finned surfaces dependence:

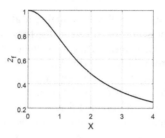

Figure 4.   The efficiency of the extended surface.

4   RESULTS, COMPARISON

4.1   *CFD results*

The temperature distributions of the ID2 geometry with 5 m/s air velocity is shown in Figure 5 and Figure 6.

4.2   *Fin efficiency and heat transfer coefficients*

Table 3 summarizes the efficiency of the finned tubes for the three suspected air velocities:

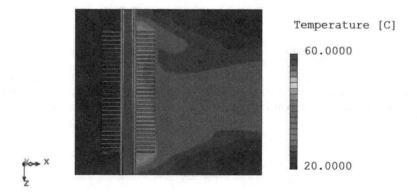

Figure 5.   Temperature distribution in the plane of the symmetry.

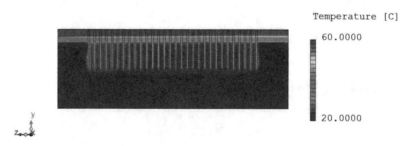

Figure 6.   Temperature distribution in the plane of the symmetry axis of the tube.

Table 3. The calculated efficiency and heat transfer coefficients of the investigated fins.

Fin ID	air velocity	$X[-]$	$\eta_f[-]$	$\alpha_m\left[\frac{W}{m^2K}\right]$	$\alpha_u\left[\frac{W}{m^2K}\right]$
	2 m/s	0.235	0.981	49.72	47.99
ID 1	5 m/s	0.309	0.969	86.17	81.05
	10 m/s	0.381	0.954	130.61	119.07
	2 m/s	0.377	0.891	48.48	44.21
ID 2	5 m/s	0.496	0.925	84.00	71.67
	10 m/s	0.611	0.955	127.32	100.25
	2 m/s	0.533	0.914	48.77	40.56
ID 3	5 m/s	0.702	0.862	84.52	61.58
	10 m/s	0.864	0.808	128.10	79.53

Table 3 shows, that the higher air velocity will cause a higher individual heat transfer coefficient, as expected, and the differences between the geometries does not have high effect to this parameter (for 2 m/s velocity the values are 49.72, 48.48 and 48.77 W/(m²K)). In contrast, the geometric dimensions have a higher effect to the fin-efficiency, its value cannot be predicted. As the results show, for the middle fins (ID2) the higher air velocity will cause the increment of the fin efficiency, while in case of the largest (ID3) and smallest (ID1) fins the efficiency will be decreasing.

4.3 Heat transfer areas and heat performances

In order to compare the numerical and the empirical results, it is necessary to determine the heat performances. These performances of the fluid dynamics analysis can be calculated from the cooling of the cooling fluid or from the warming of the air. In every simulations, this heat balance is satisfied, and the results shown in Table 4. The heat transfer area can calculated easily with simple geometric formulas, and can also be determined by the used CAD-software. It is important to note, that simulations and the calculations have done with the half geometry. The performance of the equipment highly depends on the temperature difference of the media. In basics, this difference is the logarithmic mean temperature difference (ΔT_{LOG}). But in these cases, the air flows perpendicular to water, so this is a cross-flow heat exchanger. Furthermore, the decrement in the temperature of water and increment in the temperature of the air are so low, that the average temperature difference is the difference between the average temperature of the water and average temperature of the air. The two temperature differences were important because the study examined which one fits better to the numerical results.

$$\Delta T_{LOG} = \frac{(T_{w,in} - T_{a,out}) - (T_{w,out} - T_{a,in})}{\ln\left(\frac{T_{w,in} - T_{a,out}}{T_{w,out} - T_{a,in}}\right)} \tag{15}$$

$$\Delta T_{ave} = \frac{T_{w,in} + T_{w,out}}{2} - \frac{T_{a,in} + T_{a,out}}{2} \tag{16}$$

With these considerations, the results are the next:

The results show that the temperature difference calculated from the average temperatures provides better results than the calculation by logarithmic mean difference. The Q_{LOG} performance calculated with ΔT_{LOG}, while Q_{ave} with ΔT_{ave} temperature difference.

Also interesting, that while the analytical model underestimates smaller fins at low air velocity, and produces relatively good results at higher velocities, for bigger fins it gives good results at lower velocities and overestimates at higher velocities. If the results plotted on a surface graph, the there obtained by numerical simulation were convex to the origin, while the result given by analytical calculation were concave.

Table 4.   The calculated efficiency and heat transfer coefficients of the investigated fins.

Fin ID	A [m^2]	air velocity	Q_{sim}[W]	Q_{LOG}[W]	ΔQ [%]	Q_{ave} [W]	ΔQ [%]
		2 m/s	7.04	4.73	32.81%	4.75	32.57%
ID 1	0.003057	5 m/s	8.96	7.36	17.86%	7.38	17.69%
		10 m/s	10.28	9.85	4.18%	9.87	4.03%
		2 m/s	9.00	7.25	19.44%	7.27	19.23%
ID 2	0.005681	5 m/s	10.84	10.52	2.95%	10.54	2.76%
		10 m/s	11.99	13.16	−9.76%	13.19	−10.02%
		2 m/s	9.94	9.56	3.82%	9.59	3.51%
ID 3	0.009057	5 m/s	11.86	15.94	−34.40%	12.89	−8.64%
		10 m/s	12.86	18.65	−45.02%	15.08	−17.25%

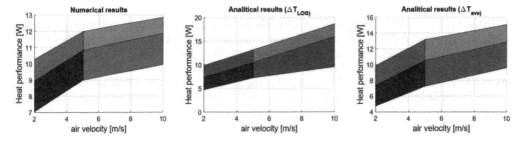

Figure 7.   Comparison of results of CFD, logarithmic mean and average temperature theorems.

5   OPTIMIZATION

After demonstrating that the heat exchangers' performance can be determined with relatively high degree of accuracy as described in Section 3, optimisation of the heat exchanger dimension can be performed. As with any optimisation process, the objective function, the condition functions and the constraints must be define.

5.1   Objective function

By default, two different objective function can be considered: minimization of the total mass and maximization of the heat performance. When it comes to optimizing an automotive radiator, as in this study, the total mass does not play a big role, however the performance is more important. Nevertheless, to calculate the heat performance, knowing the inlet and outlet temperatures are not known. So the objective function will be the product of the total heat transfer area and overall heat transfer coefficient.

5.2   Condition functions, constraints

The constraints and condition functions for the optimization are highly depends on the type of the vehicle. First, the dimensions of the device must be fixed, which determine the width and length of the fins, the length of the tubes, the number of the tubes and the fins per tube. Two of the inner and outer diameters of the tubes and the wall thickness of the tube can be freely selected, the third will be expansible size. These have to be set at lower and upper limit. The numbers of tubes must be integer, and its value depends on the total length and the diameter of the tubes. In addition, the thickness of the fins and the spacing between the fins also must be limited with an upper and a lower boundary.

5.3 *Result of the optimization*

Doing the optimization with the Microsoft-Excel Solver plug-in shows that optimization result are highly depends on the values of the constraints. Performing numerous calculations, it can be concluded the the thickness of the fins, space between the fins always reach the lower limits, the diameter of the tubes and the number of tubes reach the upper limits. From these, it can be concluded that the real limitation of optimization will be caused by the manufacturing technology. It is important to note that each geometry will only achieve its maximum performance at a certain air velocity. For fixed installations, such as air conditioners, this velocity can be easily adjusted by a fan, while in case of automotive radiators it can vary widely, and it has affects to the heat performance.

6 CONCLUSION

The aim of this study was to compare the analytical data to simulations using computational fluid dynamics software. To analyse the flow and heat transfer characteristics of the heat exchanger, three models were created using SCTetra software. Three different inlet velocities ranging from 2 m/s to 10 m/s were simulated in the three different geometric models. The SST k-ω turbulence model was used for the numerical simulations. Comparing the result with the analytical calculations, it has been found that the difference in temperatures calculated from the average temperatures gives a much better results, than the results from the logarithmic mean temperatures.

The following conclusions can be drawn from the results of the CFD analysis:

Geometry of fins						Boundary conditions			
length	l_f	13,33333333	mm	0,013333	m	Height	H	0,3	m
width	b_f	50	mm	0,05	m	Length	L	0,4	m
thickness	δ_f	0,4	mm	0,0004	m	Width	W	0,05	m
spacing	a_f	1	mm	0,001	m				
inner tube diameter	d_i	6	mm	0,006	m	Max fin length	$b_{f,max}$	0,05	mm
outer tube diameter	d_o	8	mm	0,008	m	Max tube length	$l_{f,max}$	270	mm
tube wall thickness	t_t	1	mm	0,001	m	min fin distance	$a_{f,min}$	1	mm
heat conductivity	λ_f	210	W/mK			max fin distance	$a_{f,max}$	3	mm
number of fins	n_f	192,8571429	pcs	192	pcs	min tube diameter	$d_{o,min}$	3	mm
tube length	l_f	270	mm	0,27	m	max tube diameter	$d_{o,max}$	8	mm
number of tubes	n_t	30	pieces			min tube wall thickness	$t_{t,min}$	1	mm
						max tube wall thickness	$t_{t,max}$	3	mm
Objective function						min fin thickness		0,4	mm
	kA	91,82168439	W/K			max fin thickness		2	mm

Figure 8. Screenshot from the spreadsheet with optimization results.

1. Square shaped fins with larger geometric dimensions have a higher heat performance at thesame air velocity.
2. The correlation between the air velocity and the heat performance is not directly proportional.
3. The larger the size of the fins, the less the effect of the air velocity. At 15×15 mm fins, 10 m/sair velocity instead of 2 m/s causes about 46% increment in the heat performance, at 20×20 mm fins this value is 33%, while at 25×25 mm fins is just 29%.

ACKNOWLEDGEMENT

The described article/presentation/study was carried out as part of the EFOP-3.6.1-16-201600011 Younger and Renewing University Innovative Knowledge City institutional development of the University of Miskolc aiming at intelligent specialisation project implemented in the framework of the Szechenyi 2020 program. The realization of this project is supported by the European Union, co-financed by the European Social Fund.

REFERENCES

Chong, K.K., & Tan, W.C. 2012. Study of automotive radiator cooling system for dense-array concentration photovoltaic system. *Solar Energy*, 86(9), 26322643. https://doi.org/10.1016/j.solener.2012.05.033.

Chu, W., Yu, P., Ma, T., Zeng, M., & Wang, Q. 2013. Numerical analysis of plain fin-and-oval-tube heat exchanger with different inlet angles. *Chemical Engineering Transactions*, 35(2005), 481486. https://doi.org/10.3303/CET1335080.

Hasan, A. 2015. Evaluating Mathematical Heat Transfer Effectiveness Equations Using Cfd Techniques for a Finned Double Pipe Heat Exchanger. *Advanced Energy: An International Journal (AEIJ)*, 2(1), 117.

Hussein, A.M., Bakar, R.A., Kadirgama, K., & Sharma, K.V. 2014. Heat transfer enhancement using nanofluids in an automotive cooling system. *International Communications in Heat and Mass Transfer*, 53, 195202. DOI: 10.1016/j.icheatmasstransfer.2014.01.003.

Khudheyer, A.F., & Mahmoud, M.S. 2011. Numerical analysis of fin-tube plate heat exchanger by using CFD technique. *Journal of Engineering and Applied Sciences*, 6(7), 112.

Nemati, H. & Moghimi, M. 2014. Numerical study of flow over annular-finned tube heat exchangers by different turbulent models. *CFD Letters*, 6(3), 101-112.

O'Brien, J.E., & Sohal, M.S. 2000. Local Heat Transfer For Finned-Tube Heat Exchangers Using Oval Tubes. *Proceedings of the 34th National Heat Transfer Conference (Nhtc2000)*.

Perrotin, T., & Clodic, D. 2003. Fin efficiency calculation in enhanced fin-and-tube heat exchangers in dry conditions. *International Congress of Refrigeration*, 18. Retrieved from http://www.ces.minesparistech.fr/francais/visite/chapitre5/pdf/ICR0026Hex.pdf.

Singh, S., Sørensen, K., & Condra, T. 2015. Multiphysics Numerical Modeling of a Fin and Tube Heat Exchanger. *Proceedings of the 56th Conference on Simulation and Modelling (SIMS 56), October, 7-9, 2015, Linkping University, Sweden*, 119, 383390. DOI: 10.3384/ecp15119383.

Taler, D., & Korze, A. 2016. Numerical Modeling of Transient Operation of a Plate Fin and Tube Heat Exchanger at Transition Fluid Flow in Tubes. *Procedia Engineering*, 157, 163-170. DOI: 10.1016/j.proeng.2016.08.352.

Tzabar, N. 2014. A numerical study on recuperative finned-tube heat exchangers. *Cryocoolers*, 18, 407–415.

Verde, M., Harby, K., & Corbern, J.M. 2017. Optimization of thermal design and geometrical parameters of a flat tube-fin adsorbent bed for automobile air-conditioning. *Applied Thermal Engineering*, 111, 489502. DOI: 10.1016/j.applthermaleng.2016.09.099.

Verein Deutscher Ingerieure 2010. VDI- Heat Atlas, VDI- Gesellschaft Verfahrenstechnik und Chemieingenieurwesen (GVC), Springer-Verlag Berlin Heidelberg 2010.

Solutions for Sustainable Development – Szita, Jármai & Voith (eds.)
© 2020 Taylor & Francis Group, London, ISBN 978-0-367-42425-1

Effect of soil reinforcement with stone columns on the behavior of a monopile foundation subjected to lateral cyclic loads

S.A. Rafa, I. Rouaz & A. Bouaicha
National Center for Studies and Integrated Researches of Building "CNERIB", Algiers, Algeria

ABSTRACT: Monopile foundations are usually used to support structures, such as electrical tower, foundations for low-rise building on weak soil and wind tower foundations. These structures are usually subjected to lateral cyclic loads generated by earthquakes, winds and waves. In order to support these loads, the design engineer usually chooses either to increase the pile diameter, pile length or both. However, these solutions increase the price of the project. Another solution, which is more economical, can be adopted. This second alternative consists of reinforcing the adjacent soil of the pile with stone columns. In order to evaluate the effect of stone columns on the behavior of a monopile under lateral cyclic loads, we performed a numerical Plane Strain model using the finite element software PLAXIS 2D in conjunction with the constitutive model "Hardening Soil model with small strain stiffness (HSS)". The constitutive model was first calibrated and validated using measured data from centrifuge tests on a monopile in Fontainebleau sand. In the second part, a series of analyses were then carried out where cyclic lateral loads were applied to the pile for investigating the length and the diameter of stone columns.

1 INTRODUCTION

The behavior of piles under cyclic lateral loading has aroused great enthusiasm among engineers around the world. Several studies and research programs have been developed in order to understand their responses to cyclic loads. The first design rules of monopiles under cyclic lateral loads were derived from Reese et al. (1974) and Cox et al. (1974) who performed cyclic loading tests on offshore monopiles anchored in sands. Given the local nature of these tests, the application of those laws to deep foundation projects in other parts of the world did not experience much enthusiasm. Therefore, other research programs have been developed, such as the program conducted by Cheang and Matlock (1983) on two monopiles in the area of Seal Beach (California, USA), or the SOLCYPE project which took place between 2008 and 2015 in the Laboratoire des Ponts et Chaussés (France) and had studied the behavior of monopiles under cyclic loading in different soil types.

In the case of liquefiable soil, monopiles are used in conjunction with stone columns that aims to reduce liquefaction hazard. However, those research projects focused only on the development of design rules of monopiles under cyclic loads in different geological setting and did not study the interaction between the monopile and stone columns.

To analyze the effect of soil improvement by stone columns on the behavior of a monopile under lateral cyclic loads, a series of numerical analyses using PLAXIS 2D was conducted where several soil improvement configurations were considered.

2 NUMERICAL SIMULATION OF CENTRIFUGE TESTS IN FONTAINEBLEAU SAND

2.1 *Description of the monopile load test*

Three centrifuge tests on a single pile subjected to cyclic horizontal loading were performed by Rosquoët et al (2004) at Laboratoire Central des Ponts et Chaussées (LCPC). The centrifuge

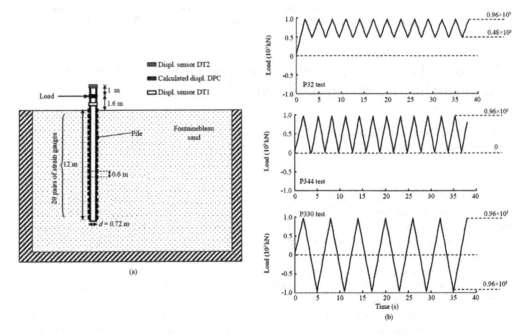

Figure 1. a) Experimental setup of the centrifuge tests conducted in LCPC. (b) Load time histories of the three tests (P32, P344 and P330). All dimensions refer to the modeled prototype.

models were 1/40 in scale and involved pile head loading with three different load-time histories. The loading time histories were: i) 12 cycles from 960 kN to 480 kN (test P32) ii) 12 cycles from 960 kN to 0 kN (test P344) iii) 6 cycles from 960 kN to -960 kN (test P330). The experimental setup and the loading time histories (in prototype scale) are portrayed in Figure 1.

The cyclic lateral load tests were conducted on vertical friction pile placed in a sand mass of uniform density. The Fontainebleau sand centrifuge "specimens" were prepared by the air sand-raining process into a rectangular container (80 cm wide × 120 cm long × 36 cm deep), with the use of a special automatic hopper developed at LCPC. The desired density of the dry sand was obtained by varying three parameters: a) the flow of sand (opening of the hopper), b) the automatically maintained drop height, and c) the scanning rate. The unit weight and the relative density of the specimen were measured to be $\gamma_d = 16.5 \pm 0.04$ kN/m^3 and $D_r = 86\%$, respectively. Laboratory results from drained and undrained torsional and direct shear tests on Fontainebleau sand reconstituted specimens indicated mean values of peak and critical-state angles of $\varphi_p = 41.8°$ and $\varphi_{cv} = 33°$, respectively.

Evidently, in this dense sand the pile used may be considered as flexible. The model pile at a scale of 1/40 is a hollow aluminum cylinder of 18 mm external diameter, 3 mm wall thickness, and 365 mm length. The flexural stiffness of the pile is 0.197 kN.m^2 and the elastic limit stress of the aluminum is 245 MPa. The centrifuge tests were carried out at 40 g.

2.2 Finite element model with PLAXIS 2D

The above mentioned centrifuge tests were modeled numerically in 2D as Plane Strain model using the finite element code PLAXIS 2D. The pile is assumed to be linear elastic while the soil behavior is described via the Hardening Soil model with small-strain stiffness (HS small). The HS small model implemented in PLAXIS is an advancement on the HS model in that, it accounts for small-strain stiffness nonlinearity and therefore this model is capable of capturing hysteresis in cyclic loading.

Figure 2 depicts the finite element discretization for the centrifuge tests. The pile was modeled by a linear element with a Young's Modulus of 16.2 GPa and an inertia of 0.0311 m^4; this was

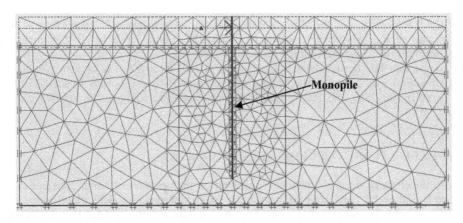

Figure 2. Numerical model on PLAXIS 2D.

chosen to give an equivalent flexural rigidity, EI, of 505 MN.m² to coincide with the properties of the pile used in those load tests. However, the soil was modeled by a triangular element with 15 nodes. The boundary conditions were rigid in the bottom, i.e., both horizontal (u) and vertical displacement (v) are zero. Standard fixities are used at the left and right boundaries of the model. These side boundaries act like rollers such that u = 0, but v ≠ 0. An Interface element was added to model the interaction between the soil and the pile. It is described with an elastic-plastic model, where the Coulomb criterion is used. The properties of the element are based on the corresponding soil and the user can reduce (or increase) the strength of the interface with the strength reduction factor R_{inter}, according to $C_{inter} = R_{inter} C_{soil}$; $\varphi_{inter} = R_{inter} \varphi_{soil}$.

The soil parameters (Table 1) for the calculation were taken from Sheil's PLAXIS 3D model (2014).

In 2D plane-strain FE analysis, it is not possible to model the 3D nature of a pile. As such, the actual properties of a 3D pile are "smeared" in the plane-strain direction to obtain the "equivalent" pile properties per meter width. To model effectively a 3D pile in 2D plane strain, D. Ong (2008) suggested to divide the axial (EA) and flexural (EI) stiffness of the pile by three times the diameter of the pile (d). Axial stiffness: $(E_p A_p)/(3d) \ldots$ [1];

Flexural stiffness: $(E_p I_p)/(3d) \ldots$ [2].

Table 1. HS-small parameters for Fontainebleau sand.

Symbol	Soil parameters	Value
γ_{unsat}	Unsaturated weight density	16.50 kN/m³
γ_{sat}	Sat weight density	18.50 kN/m³
E_{50}^{ref}	Secant stiffness in drained triaxial test,	18.00 MPa
E_{oed}^{ref}	Tangent oedometric stiffness	18.00 MPa
E_{ur}^{ref}	Unloading/reloading stiffness	45.00 MPa
G_0	Initial (small-strain) shear modulus	85.00 MPa
$\square_{0.7}$	Shear strain corresponding to 0.7G0	4×10^{-3}
v_{ur}	Unloading/reloading Poisson's ratio	0.20
C	Cohesion	1.00
φ	Friction angle	33°
ψ	Dilatancy angle	8°
P_{ref}	Reference pressure for stiffness,	100 kPa
m	Power for stress-level dependency of stiffness	0.50
R_{inter}	Interface strength reduction factor	0.90

For the interpretation of the numerical analysis results, the resulted deflections and rotations remain similar. However, in order to obtain the "actual" pile bending moment and forces, multiplication of smeared dimensions is necessary.

This approach has been validated by Rafa et al (2016) in the case of a monopile subjected to lateral cyclic loads. The results of this analysis are reported as load displacement curve (Figure 3.a) and relative displacement curve according to the number of cycle (Figure 3.b).

3   EFFECT OF STONE COLUMNS REINFORCEMENT

To study the effect of stone columns reinforcement on the behavior of a monopile under subjected to lateral cyclic loads, we took over the previous numerical model and we added stone columns around the monopile (Figure 4). The stone columns were modeled by a fifteen-node triangular element. Their behavior was modeled by the Hardening Soil Model (HSM). The parameters of the model are given in Table 2.

To determine the optimal dimensions of the stone columns, a parametric study was carried out where two different diameters of the stone columns were considered ($B_1 = 0.5$ m and $B_2 = 0.8$ m). The length of the stone column "L" also varied from 1 m to 12 m which represents a variation of the ratio L/D from 0.083 to 1. Figure 5 depicts the relative head displacement curve according to

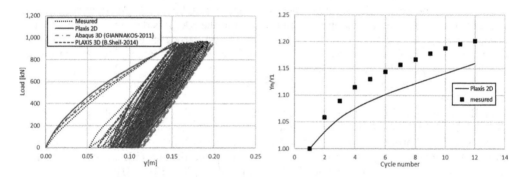

Figure 3.   a) Load displacement response curve, b) Relative head displacement curve according to the number of cycles.

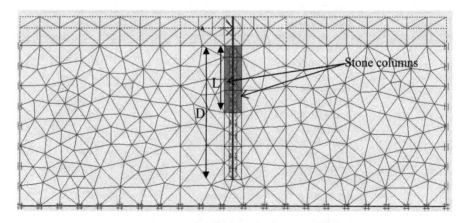

Figure 4.   Numerical model on PLAXIS 2D with stone columns.

70

Table 2. HSM parameters for stone columns.

Symbol	Soil parameters	Value
γ_{unsat}	Unsaturated weight density	19.00 kN/m^3
γ_{sat}	Sat weight density	19.50 kN/m^3
E_{50}^{ref}	Secant stiffness in drained triaxial test,	70.00 MPa
E_{oed}^{ref}	Tangent oedometric stiffness	70.00 MPa
E_{ur}^{ref}	Unloading/reloading stiffness	210.00 MPa
ν_{ur}	Unloading/reloading Poisson's ratio	0.20
C	Cohesion	1.00
φ	Friction angle	45°
ψ	Dilatancy angle	15°
P_{ref}	Reference pressure for stiffness,	100 kPa
m	Power for stress-level dependency of stiffness	0.30
R_{inter}	Interface strength reduction factor	1.00

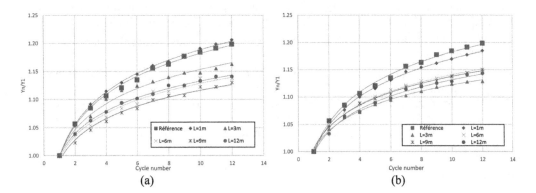

(a) (b)

Figure 5. Relative head displacement curve according to the number of cycles, (a) 0.5 m diameter stone columns, (b) 0.8 m diameter stone columns.

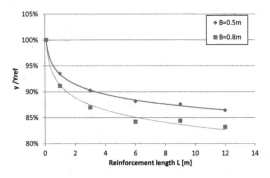

Figure 6. Ratio between the displacement of the head of unreinforced monopile (reference model) and the monopile reinforced by stone columns according to the length of stone columns at the 12th load cycle.

the number of cycles in the case of a reinforcement with stone columns with 0.5 m diameter (Figure 5.a) and 0.8 m diameter (Figure 6.b). It can be seen that in the case of reinforcement with stone columns of 1 m length and 0.5 m diameter, there is no decrease in the relative displacement compared to the reference model. Therefore, for stone columns of 0.8 m diameter, there is a decrease in relative displacement from the fourth loading cycle. It can also be seen that the relative displacement decreases according to the length of the column with a rate of 5% for the two

71

diameters up to a reinforcement length of 3 m (L = 0.25D). Beyond this length, there is a slight change in the relative head displacement curve according to the number of cycles.

Figure 6 represents the ratio between the displacement of the head of unreinforced monopile (reference model) and the monopile reinforced by stone columns according to the length of stone columns at the 12th load cycle. It can be seen that from a reinforcement length of 3 m, the contribution of stone column is no longer significant. Indeed, for a reinforcement length of 3 m, there is a 10% decrease in the relative displacement of the unreinforced monopile to the reinforced pile in the case of a 0.5 m diameter stone columns and a decrease of 13% in the case of a 0.8m diameter stone columns. On the other hand, for a reinforcement length of 12 m, there is a 14% decrease in the relative displacement of the unreinforced monopile to the reinforced pile in the case of a 0.5m diameter stone columns and a decrease of 17% in the case of a 0.8m diameter stone columns.

4 CONCLUSION

In this article, we presented the effect of stone columns soil reinforcement on the behavior of a monopile embedded in a dense send subjected to a lateral cyclic load with a particular interest on the head displacement of the monopile. The main results of the numerical analysis lead us to conclude that:

➢ The relative head displacement curve according to the number of cycles evolve in a logarithmic curve;
➢ The contribution of stone columns in the reduction of monopile head displacements reaches saturation for a length L = 0.25D;
➢ Stone columns reduce the monopile head displacement by a ratio of 15%.

REFERENCES

Cheang L. & Matlock H. 1983. Static and cyclic lateral load tests on instrumented piles in sand. *Report of the earth technology corporation.*

Cox, W.R., Reese L.C. & Grubbs B.R. 1974. Field testing of laterally loaded piles in sand. *Proceedings, 6th annual offshore technology conference, OTC 2079.* Houston: Texas.

Ong D.E.L. 2008.Benchmarking of FEM technique involving deep excavation, pile soil interaction and embankment construction. *proceeding of 12th international conference of International Association for Computer and Advances in Geomechanics*: 154–162. Goa: India.

Puech A. 2017. Présentation générale du projet SOLCYP. *SOLCYP, Journée de restitution.*

Reese L.C., Cox, W.R. & Koop F.D. 1974. Analysis of laterally loaded piles in sand. Proceedings, *6th annual offshore technology conference, OTC 2080.* Houston: Texas.

Rosquoët F., Garnier J., Thorel L. & Canepa Y. 2004. Horizontal cyclic loading of piles installed in sand: Study of the pile head displacement and maximum bending moment. In T. Triantafyllidis (Ed.) *Proceedings of the International Conference on Cyclic Behaviour of Soils and Liquefaction Phenomena*: 363–368., Bochum.

Rafa S., Rouaz I. & Bouaicha A. and Kahlouche F. 2017. Comparison between 2D and 3D analysis of a mono-pile under lateral cyclic load. *Revue Scientifique et technique de la construction,* N(135–136),: 15–18.

Sheil B. 2014. On the multi-directional loading of a monopile foundation: finite element modeling. *Proceedings of the 23rd European Young Geotechnical Engineers Conference.* Barcelona.

Solutions for Sustainable Development – Szita, Jármai & Voith (eds.)
© 2020 Taylor & Francis Group, London, ISBN 978-0-367-42425-1

Performance of Cold Formed Steel Shear Wall Panel with OSB Sheathing under lateral load

I. Rouaz, S. Rafa & A. Bouaicha
Centre National d'Etudes et de Recherches Intégrées du Bâtiment (CNERIB)-Cité Nouvelle El Mokrani-Souidania, Algiers, Algeria

ABSTRACT: Cold-Formed Steel Shear Wall Panel (CFS-SWP) is one of a lateral load resisting system in seismic and/or wind area. In this paper, a detailed modeling technique of a Shear Wall Panel (SWP) with OSB sheathing is presented by using finite element, in order to evaluate these performances under lateral load. Material and assembly nonlinearities are introduced in the FE model. The orthotropic material of the OSB sheathing is also considered. The results obtained by the finite element model have been validated with corresponding experimental data. A good arrangement in the nonlinear response has been achieved. Moreover, the performance in terms of shear strength and ultimate displacement has been assessed, confirming that the proposed finite element model can be used to predict the nonlinear behavior and the shear strength of cold formed steel shear wall panels with OSB sheathing.

1 INTRODUCTION

The use of cold-formed steel (CFS) framing in low and mid-rise building construction has become very popular in recent years due to its many beneficial aspects. Some of its desirable traits include cost-effectiveness, non-combustibility, recyclability, and excellence in material consistency. Also, CFS has high durability and high strength despite the fact that it is significantly lighter than traditional framing materials such as concrete and hot rolled steel (Ding, 2015 & Noritsugu, 2013).

Along with CFS framing, wood and steel sheets are used as sheathing materials to provide lateral strength and stiffness to the framing. This type of structural component is called Shear Wall Panel (SWP) and it is widely employed in CFS construction as a lateral force resisting system against wind and earthquake forces (Branston, 2006 & Rouaz, 2015).

Several experimental tests have been carried out to assess the shear strength of these SWP (AISI, 2007, Branston, 2004, 2006 & Ding, 2015), hence, a tabulated nominal shear strength values are presented for three types of sheathing materials: 11.90 mm (15/32 in) structural plywood, 11 mm (7/16 in) OSB, and 0.46 mm (0.018 in) and 0.68 mm (0.027 in) steel sheet, which is based on full-scale shear wall tests.

However, those experimental provisions provide nominal shear strength values for a specified and limited wall configuration in terms of geometrics and mechanics characteristics. In order to provide more design options of SWP to the engineers without conducting full-scale testing, it is necessary that numerical models developed be able to predict nominal shear strength of CFS shear walls and its behavior (Ding, 2015 & Rouaz, 2015).

The purpose of this work is to present a numerical method based on Finite Element model (FE) to assess the shear strength of CFS shear walls panel with Oriented Strand Board (OSB) sheathing. Although the global behavior and shear strength of the SWP are governed by screw connection between the OSB sheathing and the steel framing (Branston, 2004), the steel framing connections non-linearity is also determined by test and introduced in the FE model.

The orthotropic material characteristics of the OSB sheathing and non-linearity of the steel material are taken into account this analysis. The numerical results are compared to those corresponding experimental done by Branston (2004).

2 GEOMETRICAL CHARACTERISTICS OF THE SWP

According to the SWP tested by Branston (2004), the stud dimensions are 88.9 mm web, 41.3 mm flange, and 12.7 mm lip. However, steel tracks dimensions are 92.1 mm web and 31.8 mm flange. No.8 gauge is 12.7 mm (d = 4.8 mm). Self-drilling wafer head Phillips drive screws were used to connect the studs to the track. OSB sheathing (2440 × 1219 mm) is fastened to the framing members on one side; a field stud is also placed at a spacing of 610 mm and back-to-back studs are fixed at each ends of the wall (Figure 1). Two fasteners spacing (101.6 mm, 76.2 mm) at the edge of the SWP are investigated. However, the screw spacing at the intermediate stud is 305 mm.

The geometrical characteristics of all members of this shear wall panel are summarized in Table 1.

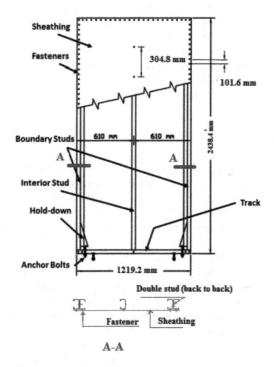

Figure 1. Shear wall panel.

Table 1. Cross section members.

Fastener	Stud	Track	OSB sheathing
Spacing (101.6 mm/ 76 mm) at the edge.	41.3 mm 12.7 mm 88.9 mm 12.7 mm	31.8 mm 92.1 mm	(h x w) 2438.4 x 1219.2 mm
Diameter (4.83mm)	Thickness (1.09 mm)	Thickness (1.09 mm)	Thickness (11 mm)

3 MECHANICAL CHARACTERISTICS

3.1 *Steel framing*

In order to take into account the material nonlinearities of the steel framing, a tensile test has been carried out at National Center of Studies and Integrated Research on Building Engineering (CNERIB) on the same steel grade used in the tested SWP. Figure 2 shows the strain-stress curve results.

According to this experimental result, the f_y = 242 *MPa* yield stress was obtained by using the 0.2% nominal proof stress and the tensile stress, f_u = 335 *MPa*. Elastic modulus, Es = 2.1*105 MPa. However, the shear modulus was taken as Gs = 8.1*104 MPa, Poisson's ratio μ_s = 0.2, mass density ρ_s = 7800 kg/m^3 (Branston, 2004).

3.2 *OSB sheathing*

Oriented Strand Board (OSB) material is an orthotropic material with many stiffness in different directions. The APA Panel Design Specification (APA 2008) considers this characteristic by specifying panel strength in the direction parallel and perpendicular to the strength axis (Ding, 2015). The mechanical characteristics of the OSB sheathing used in the experimental SWP are summarized in Table 2 (Branston, 2004).

3.3 *Mechanical screw connection*

The shear strength behavior of connections between OSB sheathing and steel farming of the shear wall panel was taken from tests conducted by APA Panel Design Specification and presented by Noritsugu Yanagi (2013). Figure 3 shows the specimen under tensile test assessing the shear strength of this connection. Figure 4 shows load-displacement relationship of the connection between OSB sheathing- stud and OSB sheeting-track.

However, due to the absence of experimental data regarding force – displacement relation between steel framing members (stud-to-track), a series of experimental tests were undertaken in laboratory of (CNERIB) according to the European convention for constructional steelwork

(ECCS TC7 TWG 7.10, 2009), with the same thickness members, tensile grade of steel sheathing and framing, and screw diameter. Figure 5 shows the maximum shear capacity around 4.1 kN.

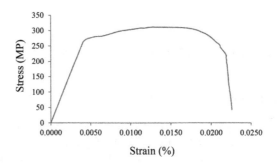

Figure 2.   Stress-strain response.

Table 2.   Mechanical properties of the OSB sheathing.

Material	Elasticity modulus E_x (MPa)	Elasticity modulus E_y (MPa)	Shear modulus G (MPa)	Sheathing's Poisson ratios v_{sh}
OSB	1983	9917	925	0.23

Figure 3. Testing setup sheathing to framing connection.

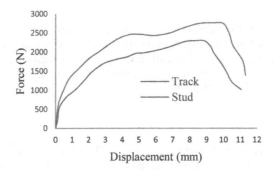

Figure 4. Load–displacement behavior connection framing to sheathing. #8 (4.12 mm).

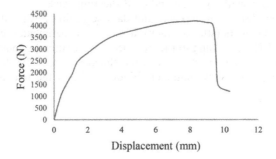

Figure 5. Load – displacement behavior of stud-to-track connection.

4 NUMERICAL MODELING

All elements of the shear wall panel; stud, track and OSB sheathing are modelled with deformable shell elements using 4-node S4R with reduced integration (Figure 6). This element has three transnationals and three rotational degrees of freedom at each node. A static nonlinear analysis has been conducted using displacement control method. Material and assembly nonlinearity are implemented in the FE model. The full Newton–Raphson method is considered for solving nonlinear equations in the analysis.

After comparing several research results based on the change in size and type of mesh, a better convergence in terms of time and accuracy in the results was adopted, having a dimension of mesh 50 x 50 mm type.

As far the boundary conditions are concerned, the displacements along the X, Y and Z-directions and rotations along Y and X directions of bottom track were restrained.

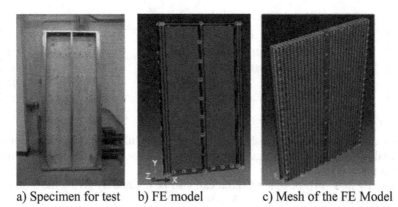

| a) Specimen for test | b) FE model | c) Mesh of the FE Model |

Figure 6.   View of the specimen.

However, the top track was assumed to have no displacement in Z, Y and rotation along the Y and X-directions. A lateral displacement was applied on the top track nodes.

4.1   *Modelling of mechanical characteristic of steel member*

The input of the stress-strain steel material curve in Abaqus software is required in terms of true stress versus true plastic strain. The true stress and true strain were converted from the engineering stresses and engineering strains using the following equations (Abaqus, 2010):

$$\sigma_{tru} = \sigma_{nom}(1 + \varepsilon_{nom}) \tag{1}$$

$$\varepsilon_{tru} = \ln(1 + \varepsilon_{nom}) \tag{2}$$

$$\varepsilon_{pl} = \varepsilon_{tru} - \frac{\sigma_{tru}}{E} \tag{3}$$

4.2   *Modelling of mechanical characteristic of OSB sheathing*

Under plane stress conditions, such as in a shell element (OSB sheathing), only the values of E_1, E_2, ν_{12}, G_{12}, G_{13} and G_{23} are required to define an orthotropic material in Abaqus. The shear moduli G_{13} and G_{23} are included because they may be required for modeling transverse shear deformation in a shell. The Poisson's ratio, ν_{21} is implicitly given as:

$$\nu_{21} = (E_1/E_2)\nu_{12} \tag{4}$$

$$\begin{bmatrix} \sigma_{11} \\ \sigma_{22} \\ \sigma_{33} \\ \sigma_{12} \\ \sigma_{13} \\ \sigma_{23} \end{bmatrix} = \begin{bmatrix} D_{1111} & D_{1122} & D_{1133} & 0 & 0 & 0 \\ & D_{2222} & D_{2233} & 0 & 0 & 0 \\ & & D_{3333} & 0 & 0 & 0 \\ & & & D_{1212} & 0 & 0 \\ & & & & D_{1313} & 0 \\ & & & & & D_{2323} \end{bmatrix} \begin{bmatrix} \varepsilon_{11} \\ \varepsilon_{22} \\ \varepsilon_{33} \\ \varepsilon_{12} \\ \varepsilon_{13} \\ \varepsilon_{23} \end{bmatrix} = \left[D^{-el} \right] \cdot \begin{bmatrix} \varepsilon_{11} \\ \varepsilon_{22} \\ \varepsilon_{33} \\ \varepsilon_{12} \\ \varepsilon_{13} \\ \varepsilon_{23} \end{bmatrix} \qquad (5)$$

Due to the fact that the out-of-plane shear deformation moduli (G_{23}, G_{13}) is not significant in shear wall analysis, the shear modulus corresponding to out-of-plane direction is taken to be the same as in-plane (G_{12}).

Linear elasticity in an orthotropic material can also be defined by giving the nine independent elastic stiffness parameters (D_{iijj}), as (Abaqus, 2010):

$$D_{1111} = E_1(1 - \nu_{23}\nu_{32})\Upsilon \qquad (6)$$

$$D_{2222} = E_2(1 - \nu_{13}\nu_{31})\Upsilon \qquad (7)$$

$$D_{3333} = E_3(1 - \nu_{12}\nu_{21})\Upsilon \qquad (8)$$

$$D_{1122} = E_1(\nu_{21} + \nu_{31}\nu_{23})\Upsilon \qquad (9)$$

$$D_{1133} = E_1(\nu_{31} + \nu_{21}\nu_{32})\Upsilon \qquad (10)$$

$$D_{1133} = E_2(\nu_{31} + \nu_{21}\nu_{32})\Upsilon \qquad (11)$$

$$G_{1212} = G_{12} \qquad (12)$$

$$G_{1313} = G_{13} \qquad (13)$$

$$G_{1212} = G_{23} \qquad (14)$$

Where:

$$\Upsilon = \frac{1}{1 - \nu_{12}\nu_{21} - \nu_{23}\nu_{32} - \nu_{31}\nu_{13} - 2\nu_{21}\nu_{32}\nu_{13}} \qquad (15)$$

$$\nu_{21} = \frac{E_2}{E_1}\nu_{12} \qquad (16)$$

4.3 *Modeling screw assembly*

Mesh-independent fasteners technical available in Abaqus software is used to set a connection between two elements, allowing each area or surface of element to be connected to another by a fastening element defined as connector. The connector is defined as Cartesian elements; taking into account, spacing, diameter, orientation and influence radius of screw (Figure 7).

The shear strength of the connector is defined in the plan directions, while the pullout force is defined in out of plan as given by the Eq. 17:

$$P_{not} = 0.85(t_c \times d) \cdot F_{u2}, \qquad (17)$$

78

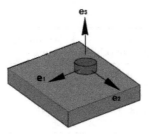

Figure 7.   Cartesian connector.

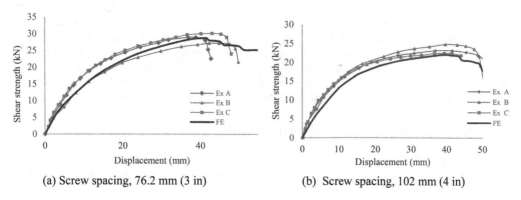

(a) Screw spacing, 76.2 mm (3 in)   (b) Screw spacing, 102 mm (4 in)

Figure 8.   Comparisons of numerical and experimental results.

where d = nominal screw diameter, t_c = lesser of the depth of penetration and thickness, P_{not} = nominal pull-out strength per screw, F_{u2} = tensile strength of member not in contact with screw head or washer.

5   RESULTS AND DISCUSSION

In order to assess the accuracy of the proposed micro-modeling method, the FE models are compared with corresponding experimental results (Branston, 2004). Two different spacing (76.2 mm and 102 mm) of screws at the edge of the SWP panel are investigated.

Figure 8 shows a good concordance in term of nonlinear response "shear strength-lateral displacement" of the two shear wall panels under monotonic load between FE model and experimental results.

Table 3 and 4 summarize the comparison in terms of shear strength and ultimate lateral displacement of each shear wall panel. For SWP having 76 mm screws spacing, the difference

Table 3.   76 mm screw spacing (3 in).

	Displacement (mm)	Shear strength (kN)	Average displacement (mm)	Average Shear strength (kN)
FE model	41.17	28.71	41.17	28.71
Ex A	38.88	28.9		
Ex B	44.92	27.17	42.30	28.72
Ex C	43.11	30.1		
			ΔD (mm) %	ΔF (kN) %
			-2.68	-0.05

Table 4. 101.6 mm screw spacing (4 in).

	Displacement (mm)	Shear strength (kN)	Average displacement (mm)	Average Shear strength (kN)
FE model	37.17	21.88	37.17	21.88
Ex A	45.55	22.73		
Ex B	45.45	24.38	43.51	23.19
Ex C	39.53	22.46		
			ΔD (mm) %	ΔF (kN) %
			-13.89	-5.65

between the shear strength and the ultimate displacement are 0.05% and 2.68% respectively; which is acceptable. For the SWP having 101.6 mm screws spacing, the difference between the FE model and the experimental results increases especially in the ultimate displacement. Furthermore, the FE model underestimates this performance response by 13.89%. This means that the FE has more stiffness than the tested SWP, this is mainly due to the absence of modeling some devices such as the hold-down with its lateral stiffness instead of a fixed bottom track.

6 CONCLUSIONS AND RECOMMENDATIONS

- According to this proposed technical modelling, the error between FE models and those of the experimental results in term of shear strength and corresponding displacement is acceptable (small);
- The nonlinear response of the SWP is correlated to the experimental results. The small discrepancies are mainly due to the difference of the real mechanical characteristic of the framing and sheathing;
- The comparison between numerical and experimental results have demonstrated the accuracy of this micro modeling technique and the influence of narrow spacing of screws in the shear panels "SWP";
- This technique, based on micro-modeling can also be used to study other parameters influencing the strength of the SWP with other geometric characteristics as an alternative to the experimental approach.

REFERENCES

ABAQUS/Standard, Version 6.11. (2010).
AISI, "North American Specification for the Design of cold formed Steel structural members", CAS-S136-07, (2007).
Branston, A.E., Boudreault, F.A., Chen, C.Y. & Rogers, C.A. "Light Gauge Steel Frame/Wood Panel Shear Wall Test Data: Summer 2003. Research Report, Department of Civil & Applied Mechanics, McGill University, Montreal, Canada, (2004).
Branston, A.E., Chen, C.Y., Boudreault, F.A., and Rogers, C.A., "Testing of Light-Gauge Steel Frame - Wood Structural Panel Shear Walls. *Canadian Journal of Civil Engineering*, Vol. 33 No. 5, 573–587. (2006).
Ding, C., "Monotonic and Cyclic Simulation of Screw-Fastened Connections for Cold- Formed Steel Framing", Master of Science In Civil Engineering. Dissertation, Virginia Polytechnic Institute, Virginia, (2015).
Noritsugu Y., "Analytical Model Of Cold-Formed Steel Framed Shear Wall With Steel Sheet And Wood-Based Sheathing", Thesis prepared for the degree of master of science, University of north Texas, May (2013).
Rouaz, I., kahlouche, F. and sakhraoui, S. and Rafa, S., "Behavior of cold formed steel shear wall panel", Research report n: 02, National Center of Studies and Integrated Research on Building Engineering, Algeria (2015).
ECCS TC7 TWG 7.10., "The testing of connections with mechanical fasteners in steel sheeting and sections", ECCS-European Convention for Constructional Steelwork, (2009).

Solutions for Sustainable Development – Szita, Jármai & Voith (eds.)
© *2020 Taylor & Francis Group, London, ISBN 978-0-367-42425-1*

The use of the linear sliding wear theory for open gear drives that works without lubrication

Ferenc Sarka
Institute of Machine and Product Design, University of Miskolc, Hungary

ABSTRACT: In this paper, the author tries to introduce calculation formulas of the linear sliding wear theory. It also deals with the pitfalls and difficulties that are present while the use of the formulas. In the last two years calculation of the wear appeared in more research and development projects of the Institute of Machine and Product Design at the University of Miskolc. This paper shows the experiences about the calculation of the wear. In the publication, the estimation of the wear will be presented by an example. The example is a gear drive of a starter motor, that is used to start a combustion engine. This drive is an open gear drive and it has no lubrication. The pinon has a great extent of wear, which is not acceptable in an average gear drive, but in these conditions is adaptable. The reason for this, that the gear drive of the starter motor is capable to provide its task: starting the combustion engine. Microscopic pictures were used to determine the size of the wear and the new form of the teeth. The hardness of the tooth surface was measured as well, to exclude that the reason of the wear was the insufficient hardness or not. The publication shows more phenomenon, that occurs during and after the contact process of the wheels. This kind of phenomena is the change of the common face width or the friction of the faces of the gear wheels.

1   INTRODUCTION

Wear is one of the typical damages and destruction form of machine components. During wear, displaced parts remove material from each other's surface. Thus, the dimensions and consequently tolerances and joints, geometric dimensions change. During wear interference fit can become loose fit. When designing machine elements, we aspire to keep the value of wear at a lower value. As the wear rate increases, after a while, the machine element becomes inoperative and needs to be replaced. The worn part will be a waste that is one of the greatest problems of our time (Takács). Therefore, it is important that we can better and simpler examine, predict life expectancy for different machine components. One way to keep abrasion on a low limit is applying lubrication (Szabó). The lubricant is in most cases a harmful substance to the environment and should be carefully designed for use. Another option is to reduce the extent of wear by using hardened surfaces. Hardening and heat-setting treatment of surfaces involve energy investment, so this technological step also has an adverse effect on the environment. In terms of wear, there are cases where lubrication cannot be applied for various reasons. Such a typical case is the non-lubricated open gear units used for starter engines with the internal combustion engines. The article is intended to illustrate the wear patterns of such engines with the help of linear wear theory.

2   THE LINEAR WEAR THEORY, MEASURES OF THE WEAR

The easiest way to determine the extent of wear is to categorize it according to the wear pattern. This is how the category of mild and severe wear was developed in technical practice (Williams). Unfortunately, this type of very simplistic classification in technical practice is not

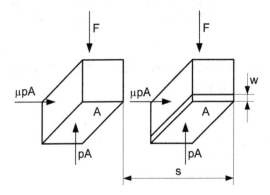

Figure 1.   Determining the magnitude of wear for surfaces moving on each other (Shigley).

enough, so it is necessary to determine the magnitude of wear with a well-defined measure. In current engineering practice, wear as a measure of length or wear rate as a ratio are the measures of the degree of wear. A detailed description of the metrics is given in the following chapters.

2.1   Wear (w) and the wear factor (K)

To determine the magnitude of wear the markings in Figure 1. are used. Take a flat surface of size A. The pressure between the surfaces is p. The friction coefficient between the surfaces is µ. The measure of the magnitude of wear is w, which is expressed in mm or inches.

   The amount of work done by the frictional force is proportional to the volume of the worn. The proportionality factor is called the wear factor K, which includes the coefficient of friction between the surfaces. Using the speed of movement and the time of movement instead of distance gives Equation (1). (Shigley).

$$w \cdot A = K \cdot p \cdot A \cdot v \cdot t \qquad (1)$$

This calculation can be easily done if the K factor is known. The magnitude of wear can be easily calculated the amount of displacement or even the time of operation can be predicted for a given wear value. Unfortunately, the literature data are very narrow for the K factor. There are only twelve materials found in literature (Shigley) with a value of K and these materials are not the ones that are commonly used in engineering practice.

2.2   The wear rate and the Archard wear coefficient

Looking for the relevant literature, we also find another amount to express the extent of wear. This amount is the wear rate. Because it is a rate, we should make some kind of ratio. The wear rate is the quotient of the volume of material worn and the distance of displacement. For this reason, its unit of measure is [m²]. The wear rate can be calculated with Equation (2). (Hamrock).

$$w_r = K_A \cdot p \cdot A \qquad (2)$$

The W_r is the wear rate in equation (2), the K_A is the Archard's wear coefficient [Pa^{-1}], p is the pressure between the contact surfaces, A is the size of the contact surfaces. For the K_A factor, literature (Hamrock) gives values for a wide range of technical materials. The calculation is easy, the selection of the Archar's or Reye – Archard – Khrushchov wear coefficient from the (Hamrock) diagram already shows uncertainty. In further research of the literature, we found another equation with the calculation of wear. According to the literature (Williams), the amount of wear is directly proportional to the load (F) and the wear factor (K) and is inversely proportional to the hardness of the test surface (H) (3).

$$w = K \frac{F}{H} \tag{3}$$

The value of factor K is given in literature (Williams) for eight different technical materials. Equation (3) does not give good approximation in all load ranges. The literature (Williams) provides data in the form of a diagram.

3 EXAMINATION OF THE WEAR AND DETERMINING IT'S AMOUNT ON A REAL GEAR DRIVE

We started our work by examining the drive that we got from the customer. There was significant wear in the pinon of the gearbox, while on the driven wheel was hardly visible any mark of the meshing.

The transmission ratio of the gearing is higher than 10, which can be said to be great for cylindrical wheels. As a result, the teeth of the pinon were connected many times more than the driven wheel. We took photos of the wear marks by using the Zeiss Discoverey v12 microscope in our institute. We took more shots of the wear marks. The applied microscope and related software (Axio Vision) gave us the opportunity to perform distance measurements on the pictures. The size of the wear was measured using the software and the microscope.

Photographs of the measurement are shown in Figures 2 and 3. The microscopic image shows that the magnitude of wear varies along the tooth profile. At the end of the tooth (tooth tip) is smaller 0.14 mm, while at the foot end of the profile is 0.215 mm (Figure 2). In the case of a drive that working under normal conditions, the gear is already classified as

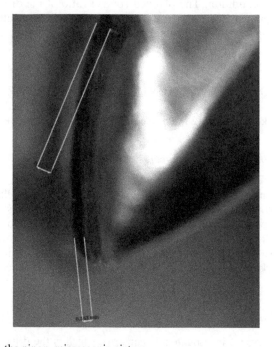

Figure 2. The wear on the pinon. microscopic picture.

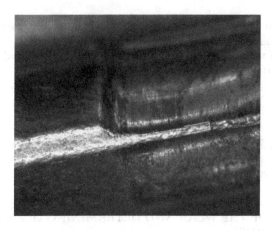

Figure 3.   The change of the top land width.

a defective machine element because of the amount of wear. In the case of starter motors not yet, because the drive still performs its function: it is able to drive the internal combustion engine for the time it takes to start. The measured values on the microscope do not go beyond the depth of the hardened layer. The thickness of the hardened layer is greater than 0.7 according to the regulation of the part drawing, and the hardness should be 600 – 800 Vickers Whether it is really so we have verified it by checking hardness (*). The wear on the top land of the pinon is also well visible (Figure 3).

Because of wear, practically the top land disappeared. The marks of the narrow thickness of the top land are already visible on the pinon before installation, which is caused by the applied profile shift. The profile shift is used to avoid undercutting. The resulting high wear further increases the decrease of the top land width.

There are other disadvantageous changes due to wear. One of them is the change of the contact ratio, which is further reduced due to the wear (Bihari). The value of the contact ratio is also low in wear-free condition. The planned contact ratio of the drive may be below 1.15 (the recommended is greater than 1.15 (Erney)), may decrease further due to wear. As a result of wear the backlash increases, which is also disadvantageous to the smooth running and noise of the drive.

4   EXPLORATION OF THE MOVEMENTS THAT CAUSE WEAR

We continued our work by exploring the movements that could cause wear on the gear teeth.

4.1   *Movements from the meshing of the gears*

It is known from the theory of tooth meshing that pure rolling between the two tooth surfaces is only in the pitch point of the meshing. In every point outside the pitch point, there is a slip beside rolling. The further we are from the pitch point, the greater the slip rate is. The reason for the slip is that the point of contact on the evolvent curve of the two tooth surfaces makes a different distance (Erney). The speed of the slip along the tooth profile can be calculated based on the literature already cited (Erney) or with the help of engineering software (e.g. KISSsoft). In our work, we calculated an average slipping speed, which was taken constant during the meshing.

4.2   *Movements from the connection process*

In the course of our work, we investigated a gear drive which is the drive of a starter motor of cars. The two gears of the drive meshing only during the start-up process and are separated in another time. The driven wheel is located on the flywheel of the internal combustion engine while the pinion is on the shaft of the starter motor. The pinion performs axial movement during the coupling process. The common tooth width between the two gears while moving is constantly changing from zero to maximum. Unfortunately, after the common tooth width reaches its maximum, the axial movement of the pinion does not disappear. This movement remains axial but moves towards the reduction of the common tooth width. The common tooth width does not disappear completely and then the direction of axial movement changes again. By the time the pinion comes to a standstill and does not perform any axial movement, there are three back and forth cycles with decreasing amplitude. This kind of movement between the gears takes place under heavy load, which is very abrasive for the teeth. A good way to determine the speed and length of this motion is a high-speed camera shot. Unfortunately, we did not have such a device. We were unable to determine exactly the impact of this kind of movement of the axial alternator on wear, but we believe that it has a significant impact on the wear.

Another important phenomenon of wear is when the connection of the pinion to the driven gear is not smooth. In this case, the front surfaces of the pinon and the driven wheel contact and act like a "friction drive" for a while. This lasts until the tooth of the pinion jumps into the groove of the driven gear. In this case, the two hardened surfaces seem to be cutting each other. The peeling particles can be placed directly between the meshing tooth surfaces, which further deteriorate the friction conditions between the two gears.

Another phenomenon, when the tooth of the pinon gets into the tooth groove but cannot get into mesh with the driven gear. The tooth of the pinion gear jumps out from the groove and gets into mesh in the next groove. This phenomenon has also a major abrasion effect for the tooth.

5 SUMMARY

Summarizing our work, we can conclude that the theories of wear in the literature are very uncertain. Mostly in the field of the wear factor and its choice. However accurate data are available the use of linear wear theory for open lubrication free gear drives can be a good help for an engineer to predict the gear life.

ACKNOWLEDGEMENT

The described article/presentation/study was carried out as part of the EFOP-3.6.1-16-2016-00011 "Younger and Renewing University – Innovative Knowledge City – institutional development of the University of Miskolc aiming at intelligent specialization" project implemented in the framework of the Szechenyi 2020 program. The realization of this project is supported by the European Union, co-financed by the European Social Fund.

REFERENCES

Bihari, J.: The Effect of The Gear Wear for The Contact Ratio, Solutions for Sustainable Development, Proceedings of the 1st International Conference on Engineering Solutions Development, ISES²D 2019, 3–4 October 2019, Miskolc, Hungary.

Erney Gy.: Fogaskerekek, Műszaki könyvkiadó, Budapest, 1983, ISBN 963 10 5099 0.

Hamrock B.J. – Jacobson B – Schmid S.R.: Fundamentals of Machine Elements, McGraw Hill, 1999, ISBN 0-256-19069-0.

Shigley J.E.– Mischke C.R. – Budynas R.G.: Mechanical Engineering Design, McGraw Hill, 2004, ISBN 007-123270-2.

Szabó, F.J.: Journal Bearing Optimization for Minimum Lubricant Viscosity, DESIGN OF MACHINES AND STRUCTURES 6:(1) pp. 56–62. (2016).

Takács, Á.: Green principles, DESIGN OF MACHINES AND STRUCTURES 4:1 pp. 99–104.,. (2014).

John Williams: Engineering Tribology, Cambridge University Press, 2005, ISBN-13 978-0-521-60988-3, ISBN-10 0-521-60988-7.

Solutions for Sustainable Development – Szita, Jármai & Voith (eds.)
© 2020 Taylor & Francis Group, London, ISBN 978-0-367-42425-1

Numerical simulation methods of stress corrosion cracking

B. Spisák & Z. Siménfalvi
University of Miskolc, Miskolc, Hungary

Sz. Szávai & Z. Bézi
Bay Zoltán Nonprofit Ltd. for Applied Research, Miskolc, Hungary

ABSTRACT: Stress corrosion cracking (SCC) can appear in numerous components used in different industries. SCC is a localized failure method which is more severe under the combined action of stress and corrosion than would be expected from the sum of the individual effects acting alone. Three indispensable factors – environment, material condition, mechanical loading – has to be present for the SCC to occur. Numerical modeling has become a widely used analysis technique in the engineering world. Therefore also in case of the stress corrosion cracking numerous models have been made, and are still being developed. These models are based on a different kind of mechanism. In the article, the different kinds of numerical methods are going to be introduced in case of intergranular stress corrosion cracking.

1   GENERAL INTRODUCTION

Stress corrosion cracking is a local failure method which can appear in several industries, it includes the airline-, oil and gas- and also the nuclear industry. There are several examples where stress corrosion cracking caused severe accidents. One of the first accidents happened in Ohio, where an eyebar-chain suspension bridge, which was named Silver Bridge collapsed because of a small stress corrosion crack. It was around 2.5 mm long when it reached the critical value of it and broke in a brittle mode. Researchers also found this type of crack in the components of the nuclear industries for example in baffle former bolts, control rod drive mechanism, steam generator (Anon, 2016), and nowadays because of the ageing of them this failure can occur more frequently. Therefore it is necessary to be able to predict the time to failure. For this numerous models have been generated, however, because of the complexity of this mechanism this theme is still a research topic even today.

Stress corrosion cracking is a local phenomenon, which occurs if three factors are present at the same time, which are the environment, the material conditions and the mechanical loading. If any of these effects are missing then stress corrosion cracking cannot occur. The environment includes the flow conditions – if the flow rate is slow then the corrosion can appear easier – the corrosion potential, the temperature, the pH-value, the impurities and also the radiolysis. The material conditions as its name says involves those parameters which are somehow related to the material of the components. It contains the chemical composition, the microstructure, and the irradiation damage. Above all of this even the manufacturing method has an influence on the material, most importantly the cold working (bending, cutting, machining, etc.). It has to be highlighted that in the case of the mechanical loadings only the tensile stresses will cause stress corrosion cracking. These loadings can originate from operational, residual tensile stresses or dynamic straining.

Stress corrosion cracking can be classified in several ways. In the case of the oil and gas industry, it appears in the pipelines where two types of SCC are differed based on the pH value of the environment on the pipe surface at the crack location. If the pH value of the

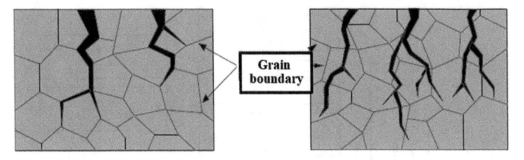

Figure 1. Schematic image of intergranular and transgranular stress corrosion cracking (Guerre, 2012).

electrolyte in contact with the pipe surface is between 8 and 11 and the concentration of carbonate is very high then it is called high pH SCC. The other type is the near-neutral pH SCC. In this case, the environment contains dissolved CO_2 (Zheng et al., 2011). Another way of classifications is based on the propagation way of the crack. If the grain is weaker than the grain boundary then the crack will propagate through the grains, this type of SCC is called transgranular stress corrosion cracking and if the grain boundary is the weaker component then the crack will propagate along it, therefore it is named intergranular stress corrosion cracking. The schematic picture of them can be seen in Figure 1. In the followings, those numerical methods and models are going to be shown which are related to the intergranular stress corrosion cracking.

2 NUMERICAL METHODS

Since the appearance of stress corrosion cracking several models were created to describe the mechanism of SCC and to predict the crack growth rate. These models include the basic mechanisms which were created in the second half of the twenty century. The mechanisms can be categorized according to the driving force, the model type and the basic phenomenons. The nature of the driving force can be electrochemical, mechanical or mechanochemical. The model type can also be separated into three groups, the empirical, where the parameters of the model are based on measured data, semi-empirical or theoretical. The main types of the basic mechanisms are the anodic dissolution-and mechanical fracture–based mechanism. In the case of the first one, the crack will advance at its tip because of preferential dissolution. The basic idea is that if precipitated phases or segregated solutes appear at the grain boundary then locally electrochemical heterogeneity will form, and with the appropriate environment and stress the preferential dissolution will occur resulting in the advancement of the crack. In case of the mechanical fracture-based SCC it is assumed, that the as in the previous case the crack propagates by dissolution, but the failure would occur as a mechanical fracture, basically at the corrosion pits or the mechanical defects the stress concentration would rise to the point of ductile deformation which leads to the mechanical fracture (Zheng et al., 2011). However if in the system hydrogen is also present, then one of the hydrogen embrittlement mechanism should be used.

Numerical modeling has become a widely used analysis technique in the engineering world; therefore in case of SCC, numerous models have been made, and are still being developed (Choi et al., 2007). However, it is very hard to study the SCC process by the method of numerical simulation, because besides the simulation of cracking propagation under mechanical load the impact of electrochemical corrosion has to be taken into account. Almost all of the numerical simulations which can be found in the literature use the anodic dissolution or the hydrogen embrittlement methods as their basic mechanisms. In Figure 2 the different numerical methods are separated based on these phenomena however there are cases where it can not be determined from the given articles, these methods are included in the group of other

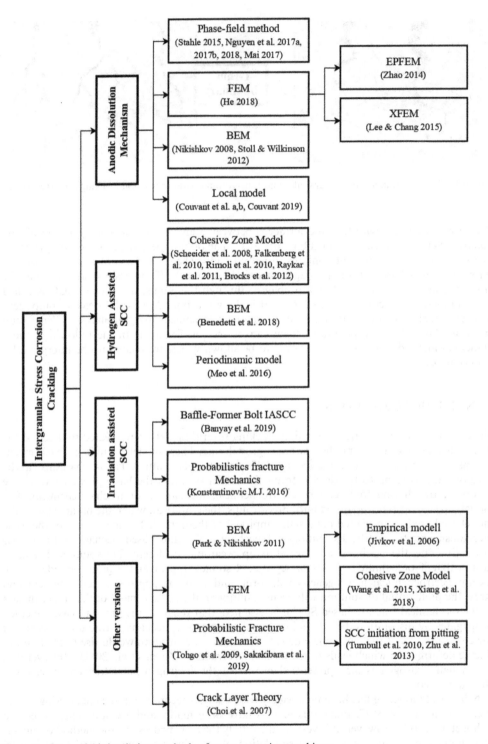

Figure 2.   Numerical simulation methods of stress corrosion cracking.

versions. The irradiation assisted SCC was also separated, as in this case the characterization of the evolution of stress in a reactor environment, as well as the redistribution of stress has to be included (Banyay et al., 2019). The same method can be found in more categories, for example, the phase-field method, the probabilistic fracture mechanics, however there are cases where only one or two articles were published with the given solution like in the case of peri-odinamic model (De Meo et al., 2016) or the local model(Couvant et al., 2015; Couvant, 2019; Couvant et al., 2019a). In the followings, the listed numerical methods are going to be introduced.

2.1 *Finite element method*

The finite element analysis is a well-known numerical method, therefore even in case of the stress corrosion cracking examples can be found in the literature (He et al., 2018; Lee & Chang, 2015; Zhao et al., 2014). The first attempt to use FEA for SCC was made by Jivkov et al., (2006) who described the 2D and 3D models of cracking in case of Type 304 stainless steel. The microstructure of the model was built up and they also made a strategy for crack advance. The results proved that the finite element method is applicable when the process besides the mechanical effect contains chemical or corrosion based factors too.

The initiation of stress corrosion cracking can occur from corrosion pittings and material defects. The simulation of it is a major challenge as the pit growth has to be estimated too. The basic concept of it is that under loading if the pits growth reaches a critical value then it will transform into cracks. Zhu et al. (2013) used an ultra-low elastic load for the simulations and also made a corresponding pitting test for it. The simulations give a more accurate predic-tion for the theory of Kondo (1989) and Macdonald et al. (2004). They proposed that the pit-to-crack transition will occur only if the pit depth is greater than a threshold depth and that the crack growth rate exceeds the pit growth rate. Another example of this phenomenon is the work of Turnbull et al., (2010). They showed that if the stress and strain is distributed around a single pit where high stress was applied, then the plastic strain will not appear at the pit base but on the pit walls below the pit mouth, however in this case the different shape of the pit was not considered, it was a simple deep bullet-shaped pit. In these cases the authors did not simulate the advance of the crack, the focus was more on the initiation of SCC, therefore with this solution, the stress and strain concentration places can be determined and the way of the crack not. Also in these simulations, the chemical parameters were not taken into consideration.

The basic linear elastic fracture mechanics does not write down well the process which occurs at the cracking tip. Because of this different kind of mechanical damage models were developed. One example is the cohesive zone model (CZM) which is based on the elastic-plastic fracture mechanics. The basic concept of it is that at the cracking tips there is a cohesive zone, and the material failure will occur when the rigidity of this zone reaches zero. With the improved version of this model even the initiation and propaga-tion of the crack can be simulated. This method was applied in numerous research works, for example by Xiang et al. (2019) combined the cohesion zone model with the rupture process of corrosion scale and made the FE analysis in ABAQUS. The results showed that in case of the models they examined the cracks nucleated from the corrosion scale surface and the crack would only propagate into the matrix when the corrosion scales break due to the continuous action of tensile load. Wang W.W. et al., (2015) also used this model. They investigated the effect of corrosion product films on the crack ini-tiation and propagation in SCC. Their results are similar to the previously mentioned one, as they also concluded that the stress corrosion crack initiates first in the corrosion product film, and they also showed that the fracture strain is lower if the corrosion prod-uct film exists, and higher if there is no corrosion product film.

The cohesive zone model is also used in case of hydrogen assisted stress corrosion cracking, examples in the literature are the work of Scheider et al. (2008), Falkenberg et al. (2010), Rimoli & Ortiz (2010), Raykar et al. (2011), Brocks et al. (2012). In these cases, the cohesive zone model was extended or was coupled with other solution methods so the effect of hydro-gen embrittlement could be included. The articles mainly focus on the relationship between the hydrogen concertation and the mechanical quantities, the results show the advantage of the numerical simulations as in this case the various influences can be separated.

From these examples it can be seen that the finite element analysis can be used well for this type of failure mechanism, however, to get an accurate prediction it is better to combine it with other methods.

2.2 Boundary element method

The boundary element method is a numerical computational method which has been used mainly in fluid mechanics, acoustics, electromagnetics, fracture mechanics and contact mechanics. In case of problems with small surface/volume ratio, the model is more efficient than other numerical methods. In the case of SCC, it is mainly used for crack propagation in microstructures. With the model of Stoll & Wilkinson (2012) the variation of stress intensity factor can be calculated during the crack growth. The crack propagates along the grain boundary path, therefore the intergranular stress corrosion cracking can be simulated with this, just like in case of the other BEM models of SCC too.

Another type of the boundary element method is the SGBEM, which stands for the symmetric Galerkin boundary element method. The modeling of SCC is separated into two parts, for the uncracked structural component finite elements were used and the SGBEM was only used for the modeling of the crack. The model alternates between these two methods. The crack growth procedure starts with the analysis of an initial crack. During the crack growth, the crack shape and the stress intensity factor distribution is obtained (Nikishkov, 2008; Park Jai Hak & and Nikishkov, 2011). Benedetti et al., (2018) also combined the boundary element method with the finite element model, however in this case the intergranular hydrogen diffusion is modeled with FEM. The authors made several numerical tests, which showed that the proposed technique provides a qualitatively good representation of the hydrogen assisted intergranular stress corrosion cracking.

2.3 Phase-field method

The phase-field models are used for interfacial problems. It has been applied for solidification dynamics, viscous fingering, fracture mechanics and also hydrogen embrittlement problems. In case of the stress corrosion cracking the method was combined with the anodic dissolution mechanism. With the models, the evolution of microstructures with complex morphologies can be simulated. One of the first publication where SCC was studied with the phase-field model is from Ståhle & Hansen (2015). In the research an order parameter was used to separate the empty space from the solid because there are cases when the distance between the structural inhomogeneities has the same order of magnitude as the thickness of the surface, therefore if the surface thickness is not taken into consideration then the prediction may lead to unreal results. With this solution the evolution of the corroding body can be obtained, however, the evolved model does not incorporate the mass transport in the formulation. Mai & Soghrati (2017) simulated the initiation of SCC from surface pits. They also examined the shape of the pits, from the results they concluded that that the pit morphology has a huge influence on the crack initiation

In another research, the phase-field model was combined with a robust algorithm and a diffusion model, which is capable to simulate the stress corrosion cracking in a nickel-based alloy (Nguyen et al., 2017a; b; Nguyen et al., 2018). The results of the proposed multiphysics model are in good agreement with the observed SCC effects, where the driving force was the anodic dissolution crack growth.

2.4 Probabilistic fracture mechanics

One of the tools used for the probabilistic fracture mechanics is the Monte Carlo simulation. Based on experimental results the distribution of an input parameter can be determined, and the Monte Carlo method is used to predict the probability distribution of a parameter from the known distributions (Besuner & Tetelman, 1977). One of the first simulations with this method for SCC was made by Wang et al. (1995). The time dependence of crack nucleation, the dormancy of cracks at various intervals, the statistical distribution of growth rates and the coalescence of appropriately spaced cracks were taken into consideration during the simulations, and the results were in reasonable agreement with the experimental results. Tohgo et al. (2009) and Sakakibara et al. (2019) also used the Monte Carlo method. The basic concept of Tohgo et al. is that the for a given space the possible number of microcracks were set and the initiation time for every crack was assigned by random numbers, in this case, the exponential distribution was used for the type 304 stainless steel. Sakakibara et al. first made a uniaxial constant loading test on alloy 600, the environment temperature was 400°C. The results showed that after every 450h the number of cracks observed on the surface of a specimen is similar to the Poisson distribution. The difference between the two cases is the material and the working environment. In both cases, the predictions showed good agreement with the experimental results, however because of the differences in the environment and the material two different distribution models were used.

In another example, **Weaknesses:**

There are a lot of written mistakes, grammar errors in the text. How, where and what kind of device was used during the measurements? If the dynamic thermal simulation results figures (like temperature distribution), then it should be included.

Comments and Recommendations:

Correct all of the grammar errors, and include the measurement device description.

the irradiation assisted stress corrosion cracking was examined with the help of probabilistic fracture mechanics (Konstantinović, 2016). It was assumed that the oxidized part of the examined specimen had an important role in the propagation of the crack. The center of this research was the determination of the time-to-failure. For this, the inert strengths of the oxide were distributed according to the Weibull distribution. The results were in excellent agreement with the measured data.

3   SUMMARY

In the article the different types of numerical simulation methods used for SCC were introduced, the finite element method, the boundary element method, the phase-field method and finally the probabilistic fracture mechanics. It was shown that the simulation of stress corrosion cracking is a very complex task, where not only the initiation of the crack causes problems but the numerous parameters too. It is hard to take into account all of the effects, and the biggest challenge is that these parameters are not independent of each other.

The introduced models mainly use empirical parameters, as the development of an overall model which can be used in every case is very difficult. The recently created models give a better view of this type of failure method, however, there are still several unanswered questions in this topic.

ACKNOWLEDGEMENT

The described article/presentation/study was carried out as part of the EFOP-3.6.1-16-2016-00011 "Younger and Renewing University – Innovative Knowledge City – institutional development of the University of Miskolc aiming at intelligent specialisation" project implemented in the framework of the Szechenyi 2020 program. The realization of this project is supported by the European Union, co-financed by the European Social Fund.

REFERENCES

Anon. 2016. Historical views on stress corrosion cracking of nickel-based alloys: The Coriou effect. *Stress Corrosion Cracking of Nickel Based Alloys in Water-cooled Nuclear Reactors*: 3–131.

Banyay, G.A., Kelley, M.H., McKinley, J.K., Palamara, M.J., Sidener, S.E. & Worrell, C.L. 2019. Predictive Modeling of Baffle-Former Bolt Failures in Pressurized Water Reactors. In *Proceedings of the 18th International Conference on Environmental Degradation of Materials in Nuclear Power* Systems – *Water Reactors*. Portland: Springer: 1573–1588.

Benedetti, I., Gulizzi, V. & Milazzo, A. 2018. Grain-boundary modelling of hydrogen assisted intergranular stress corrosion cracking. *Mechanics of Materials*, 117: 137–151.

Besuner, P.M. & Tetelman, A.S. 1977. Probabilistic fracture mechanics. *Nuclear Engineering and Design*, 43(1): 99–114.

Brocks, W., Falkenberg, R. & Scheider, I. 2012. Coupling aspects in the simulation of hydrogen-induced stress-corrosion cracking. *Procedia IUTAM*, 3: 11–24.

Choi Byoung-Ho, Chudnovsky, A. & Kalyan Sehanobish. 2007. Stress Corrosion Cracking in Plastic Pipes: Observation and Modeling. *International Journal of Fracture*, 145(1): 81–88.

Couvant, T. 2019. Prediction of IGSCC as a Finite Element Modeling Post-analysisPrediction of IGSCC as a Finite Element Modeling Post-analysis. In *Proceedings of the 18th International Conference on Environmental Degradation of Materials in Nuclear Power* Systems – *Water Reactors*. Springer: 1535–1550.

Couvant, T., Caballero, J., Duhamel, C., Crépin, J. & Maeguchi, T. 2019. Calibration of the Local IGSCC Engineering Model for Alloy 600. In *Proceedings of the 18th International Conference on Environmental Degradation of Materials in Nuclear Power* Systems – *Water Reactors*. Springer: 1511–1534.

Couvant, T., Wehbi, M., Duhamel, C., Crépin, J. & Munier, R. 2015. Development of a local model to predict SCC: preliminary calibration of parameters for nickel alloys exposed to primary water. In *17th International Conference on Environmental Degradation of Materials in Nuclear Systems-Water Reactors*. Ottawa: 1955–1973.

Engelhardt, G. & Macdonald, D.D. 2004. Unification of the deterministic and statistical approaches for predicting localized corrosion damage. I. Theoretical foundation. *Corrosion Science*, 46(11): 2755–2780.

Falkenberg, R., Brocks, W., Dietzel, W. & Schneider, I. 2010. Simulation of Stress-Corrosion Cracking by the Cohesive Model. In *Advances in Fracture and Damage Mechanics VIII*. Key Engineering Materials. Trans Tech Publications Ltd: 329–332.

Guerre, C. 2012. Stress corrosion cracking of nickel base alloys in PWR primary water. In *MINOS Workshop, Materials Innovation for Nuclear Optimized Systems*. CEA – INSTN Saclay, France.

He, X., Yinghao, C., Gangbo, L. & Shuai, W. 2018. Crack Growth Driving Force at Tip of Stress Corrosion Cracking in Nuclear Structural Materials at Initial Stage. *Rare Metal Materials and Engineering*, 47(8): 2365–2370.

Jivkov, A.P., Stevens, N.P.C. & Marrow, T.J. 2006. A three-dimensional computational model for intergranular cracking. *Computational Materials Science*, 38(2): 442–453.

Kondo, Y. 1989. Prediction of Fatigue Crack Initiation Life Based on Pit Growth. *CORROSION*, 45(1): 7–11. http://corrosionjournal.org/doi/10.5006/1.3577891.

Konstantinović, M.J. 2016. Probabilistic fracture mechanics of irradiation assisted stress corrosion cracking in stainless steels. *Procedia Structural Integrity*, 2: 3792–3798.

Lee, S.-J. & Chang, Y.-S. 2015. Evaluation of primary water stress corrosion cracking growth rates by using the extended finite element method. *Nuclear Engineering and Technology*, 47(7): 895–906.

Mai, W. & Soghrati, S. 2017. A phase field model for simulating the stress corrosion cracking initiated from pits. *Corrosion Science*, 125: 87–98.

De Meo, D., Diyaroglu, C., Zhu, N., Oterkus, E. & Siddiq, M.A. 2016. Modelling of stress-corrosion cracking by using peridynamics. *International Journal of Hydrogen Energy*, 41(15): 6593–6609.

Nguyen, T.-T., Bolivar, J., Réthoré, J., Baietto, M.-C. & Fregonese, M. 2017. A phase field method for modeling stress corrosion crack propagation in a nickel base alloy. *International Journal of Solids and Structures*, 112: 65–82.

Nguyen, T.-T., Bolivar, J., Shi, Y., Réthoré, J., King, A., Fregonese, M., Adrien, J., Buffiere, J.-Y. & Baietto, M.-C. 2018. A phase field method for modeling anodic dissolution induced stress corrosion crack propagation. *Corrosion Science*, 132: 146–160.

Nguyen, T.T., Réthoré, J., Baietto, M.-C., Bolivar, J. & Fregonse, M. 2017. A phase field method for modeling stress corrosion crack propagation induced by anodic dissolution. In *14th International Conference on Fracture*. Rhodes.

Nikishkov, G. 2008. SGBEM-FEM Modeling of Stress Corrosion Cracking. In *16th Pacific Basin Nuclear Conference (16PBNC)*.

Park Jai Hak & Nikishkov, G.P. 2011. Growth simulation for 3D surface and through-thickness cracks using SGBEM-FEM alternating method. *Journal of Mechanical Science and Technology*, 25(9): 2335.

Raykar, N.R., Maiti, S.K. & Singh Raman, R.K. 2011. Modelling of mode-I stable crack growth under hydrogen assisted stress corrosion cracking. *Engineering Fracture Mechanics*, 78(18): 3153–3165.

Rimoli, J.J. & Ortiz, M. 2010. A three-dimensional multiscale model of intergranular hydrogen-assisted cracking. *Philosophical Magazine*, 90(21): 2939–2963.

Sakakibara, Y., Shinozaki, I., Nakayama, G., Nan-Nichi, T., Fujii, T., Shimamura, Y. & Tohgo, K. 2019. Monte Carlo Simulation Based on SCC Test Results in Hydrogenated Steam Environment for Alloy 600. In *18th International Conference on Environmental Degradation of Materials in Nuclear Systems-Water Reactors*. Springer: 1551–1562.

Scheider, I., Pfuff, M. & Dietzel, W. 2008. Simulation of hydrogen assisted stress corrosion cracking using the cohesive model. *Engineering Fracture Mechanics*, 75(15): 4283–4291.

Ståhle, P. & Hansen, E. 2015. Phase field modelling of stress corrosion. *Engineering Failure Analysis*, 47: 241–251.

Stoll, A. & Wilkinson, A.J. 2012. Use of a dislocation-based boundary element model to extract crack growth rates from depth distributions of intergranular stress corrosion cracks. *Acta Materialia*, 60 (13–14): 5101–5108.

Tohgo, K., Suzuki, H., Shimamura, Y., Nakayama, G. & Hirano, T. 2009. Monte Carlo simulation of stress corrosion cracking on a smooth surface of sensitized stainless steel type 304. *Corrosion Science*, 51(9): 2208–2217.

Turnbull, A., Wright, L. & Crocker, L. 2010. New insight into the pit-to-crack transition from finite element analysis of the stress and strain distribution around a corrosion pit. *Corrosion Science*, 52(4): 1492–1498.

Wang, Wen-Wen, Luo, J., Lei-Chen Guo, Zhi-Meng Guo & Yan-Jing Su. 2015. Finite element analysis of stress corrosion cracking for copper in an ammoniacal solution. *Rare Metals*, 34(6): 426–430.

Wang, Y.-Z., Hardie, D. & Parkins, R.N. 1995. The behaviour of multiple stress corrosion cracks in a Mn-Cr and a Ni-Cr-Mo-V steel: III—Monte Carlo simulation. *Corrosion Science*, 37(11): 1705–1720.

Xiang, L., Wei, X. & Chen, S. 2019. Numerical investigation on the stress corrosion cracking of FV520B based on the cohesive zone model. *Results in Physics*, 12: 118–123.

Zhao, L., He, X., Yang, F. & Suo, Y. 2014. Numerical investigation on stress corrosion cracking behavior of dissimilar weld joints in pressurized water reactor plants. *Frattura ed Integrità Strutturale*, 29: 410–418.

Zheng, W., Elboujdaini, M. & Revie, R.W. 2011. Stress corrosion cracking in pipelines. *Stress Corrosion Cracking*: 749–771.

Zhu, L.K., Yan, Y., Qiao, L.J. & Volinsky, A.A. 2013. Stainless steel pitting and early-stage stress corrosion cracking under ultra-low elastic load. *Corrosion Science*, 77: 360–368.

Solutions for Sustainable Development – Szita, Jármai & Voith (eds.)
© 2020 Taylor & Francis Group, London, ISBN 978-0-367-42425-1

Implementation of a customized CAD extension to improve the calculation of center of gravity

M. Szabó, P. Mileff & K. Nehéz
Department of Information Engineering, Institute of Information Science, Faculty of Mechanical Engineering and Informatics, University of Miskolc, Hungary

ABSTRACT: This paper presents an automated system that can effectively support human work, thereby speeding up the Computer Aided Designing process. The motivation of the research and development task was to implement a module in an existing CAD system which can calculate the center of gravity, which requires investigating the distribution of the mass. There were complex CAD models designed in the application called Siemens NX. It has some built-in functions for designers for supporting the calculations. The first chapter of the paper overviews the mathematical background of the calculation. For software development point of view, the authors investigated the programming facilities of Siemens NX. Investigations showed that the first task is to assign material properties for the components that are in the design. Our goal was to automate this task as the manual material attribute assignment is time-consuming and potentially leads to errors. To perform this task, the relevant documentation and literature have been reviewed. A recently published model carries out material assignments by identifying the relationship between shape, functionality, and material in parts. There is a study that analyses the basic concepts and relationships affecting all aspects of material selection. In this paper, a new software component is presented, providing an approach for material assignment. The center of gravity calculation based on the valid material data then provides a reliable result. The Siemens NX software plug-in developed by the authors has been installed and actively used by our industrial partner.

1 INTRODUCTION

For the modelling of assemblies, hierarchical or tree structures have been used in CAD systems for a long time, which contain geometric and other additional information such as material properties, common/shared center of gravity, etc. (L. Kunwoo, Gossard et al. 1985) Siemens NX (Siemens NX 11 Documentation 2016) is an extremely high-level software that allows us to design quickly and efficiently. We need to know the exact material properties of the parts to calculate the center of gravity. These properties can also be defined in the NX system, but in the absence of a CASE tool, one-to-one parameterization is time-consuming, and there is a possibility of human error. The more the number of parts that build up the assemblies, the less effective the solution. So it is necessary to develop a solution that can help the process in an automated form. This does save not only time but also reduces the possibility of human errors.

We measured the industrial needs for designing and implementing a prototype model. The model realizes an automated system that can effectively support human work, thereby making the planning process faster. The task is not trivial, it is necessary to use several technologies together, which is covered in several points of the publication. Siemens NX offers several options for creating custom components. The question may arise as to why these solutions are needed. NX is a professional design software with complex functionality, but in practice, industrial processes require additional specific functions and complements.

Before the implementation process, we learned the basic use of the Siemens NX CAD system (Siemens NX 11 Documentation 2016). Our most important task was to find the right programming interface and explore the possibilities to accomplish the intended tasks. We performed the development on two levels. First, we created a plug-in based on Java technology and can be embedded in the NX environment, allowing us to explore the possibilities and problems of the CAD system. Based on this experience, we developed a more robust software module in the Visual Basic programming language. In this case, efficiency means that Visual Basic is the main program language of NX, so more library packages are available, and the implementation process can be simplified. NX systems can be accessed through multiple interfaces. The most common of these are SNAP and *NXOpen* packages (Siemens NX 11 Programming Tool 2016). The former is a simplified programming function package for non-IT professionals, but with the latter, more complex applications can be designed. *NXOpen* is an API collection that allows us to create custom applications that can be integrated into NX. The package is open source and offers programming interfaces for multiple languages, including C/C++, Visual Basic, C#, Java, and Python. Considering industrial needs and opportunities, we decided to create our software component in Visual Basic.

2 IMPLEMENTATION OF THE MATERIAL ASSIGNMENT COMPONENT

The application targets an automated system that can effectively support human work, thereby making the planning process faster. The task is not simple, it requires the use of several technologies. Our goal was to create an embedded module for Siemens NX that supports to calculate the center of gravity of the assemblies (Figure 1) by simplifying the assignment of the material. Due to the number of parts, the manual material assignment is time-consuming and potentially leads to errors.

There are some solutions in the literature about this topic. For example, a recently published model performs material assignment by identifying the relationship between shape, functionality, and material of the parts (Zhang, B. et al. 2014). Also, a study (Zhang, Y. et al. 2015) analyzes the basic concepts and relationships of all aspects of material selection. A new ontology-based framework has been introduced too. Ontology-based semantic web technology is introduced into the semantic representation of material selection knowledge. The implicit material selection knowledge serves as a group of labelled samples and RDF samples for the conceptual model, providing a formal approach for organizing the fixed material selection knowledge.

If we have not programmed in a Siemens NX environment before, we should start the work with the *journal* (Siemens NX 11 Journaling Tool 2016) module that built into NX for this purpose. The journal recorder is used to automate simple tasks, such as creating, modifying, deleting, colouring, etc. primitive shapes. By manually changing the generated code, we can create more sophisticated tools, depending on what we exactly need to complete our current task. It is necessary to modify the code because the structure of the code and some methods generated by the journal recorder are difficult to understand and also contain unnecessary lines. Also, in

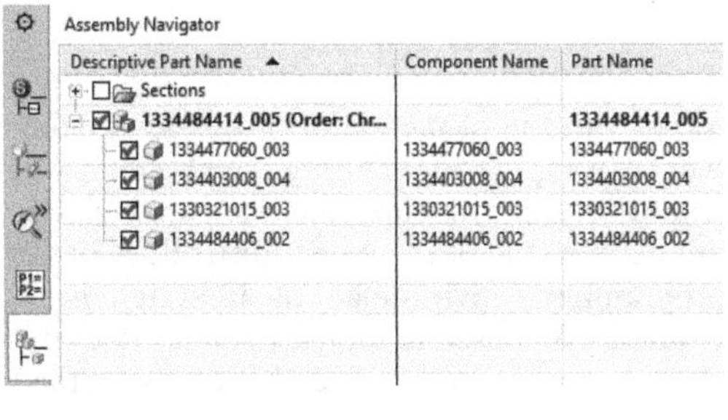

Figure 1. A common assembly tree structure (simplified).

95

some cases, for example, when we have to perform a specific task or subtask several times, we have to create the cycles manually, as journaling does not support this feature.

The first step in the assignment process is to specify an input Excel file based on a predefined structure. The file has to contain all the information needed during the assignment process (i.e. part number, the id of the material property, etc.). These data differ from manufacturer to manufacturer, so the file must always be provided by the actual automotive company. The next task is to compare the current material properties with the properties stored in the Excel file. The process examines each element one by one. We have defined four different outputs:

1. If the data matches, then no further action will be performed on the part.
2. If the material properties are not the same, it indicates what material is required based on the Excel file.
3. If the part does not have any material property, it also indicates what material is required to be assigned based on the Excel file.
4. If the current part number is not found in the file, it indicates that correction is required based on the drawing.

With our software component, based on the output of the above processes, the automatic material assignment can be easily achieved. Siemens NX contains only the predefined, default material types, so in most cases, it does not contain the specific types used by a particular automotive company. However, this problem can be solved with a specially built XML file, usually provided by the company. The application does not make any changes on the parts where the previously described process's output is (1) or (4) (i.e. if the type of material is appropriate, no further action occurs). This can effectively speed up the assignment process.

Table 1. Report structure.

Part number	Assigned material	Needed material	Result
xxxx.xxx.xxx	0f_bras	0f_steel	✓
xxxx.xxx.xxx	0f_magnet	0f_aluminium	×
xxxx.xxx.xxx	-	0f_aluminium	×
xxxx.xxx.xxx	0f_steel	N/A (Excel)	No Data
xxxx.xxx.xxx	-	N/A (Excel)	No Data

Based on the result, the application generates a report that can be displayed in Excel or on a graph. The report can be generated before, and after the automated material assignment process, so the changes can be easily checked. It is important to note that the implemented module can handle assemblies with any complexity.

Finally, the developed component was integrated with the NX CAD system. Figure 2 shows the structure of the software and connection between the component and the Siemens NX system.

3 SUMMARY

The paper presents the design and implementation processes of the component created for support automatic mass center computation. During the research, we have had the opportunity to look into real industrial design software, the Siemens NX system. Research has been prevented by many factors, such as the development difficulties of the system. The most critical problem is that the programming environment of the system is not documented correctly, which is because the software is often used in industrial areas due to its high price. Therefore, it is hard to find any suitable development aid or sample codes on the Internet.

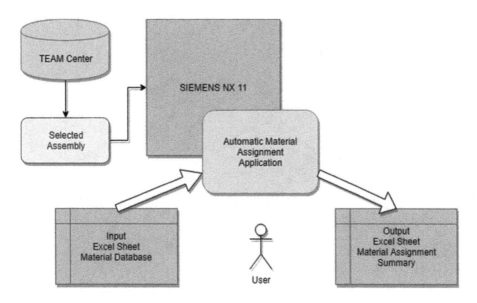

Figure 2. Siemens NX Integration model.

The application accomplishes the goal by being able to handle assemblies of any complexity, regardless of the number of the part elements. Finally, it generates a report, defined by our industrial partner, that contains all the relevant information for the material assignment process. The component can be used to automate the assignment of materials so that the task can be done quickly, efficiently, without any human intervention.

ACKNOWLEDGEMENT

This research was supported by the European Union and the Hungarian State, co-financed by the European Regional Development Fund in the framework of the GINOP-2.3.4-15-2016-00004 project, aimed to promote the cooperation between the higher education and the industry.

REFERENCES

L. Kunwoo & Gossard, C., David, 1985. A hierarchical data structure for representing assemblies: part 1. In Computer-Aided Design, volume 17, issue 1.

Siemens NX 11 Documentation 2016. *docs.plm.automation.siemens.com/tdoc/nx/11/nx_help*.

Siemens NX 11 Journaling tool, 2016. http://www.nxjournaling.com.

Siemens NX 11 Programming tool, 2016. *docs.plm.automation.siemens.com/tdoc/nx/11/nx_api*.

Zhang, B. & Rai, R., 2014. Materials Follow Form and Function: Probabilistic Factor Graph Approach for Automatic Material Assignments to 3D Objects. In Volume 7: *2nd Biennial International Conference on Dynamics for Design; 26th International Conference on Design Theory and Methodology* Buffalo. New York, USA.

Zhang, Y., Luo, X., & Zhao, Y., 2015. An ontology-based knowledge framework for engineering material selection. Advanced *Engineering Informatics*, volume 29, issue 4, PP. 985–1000.

Part B: Sustainable and Renewable Energy and Energy Engineering

Solutions for Sustainable Development – Szita, Jármai & Voith (eds.)
© 2020 Taylor & Francis Group, London, ISBN 978-0-367-42425-1

Towards CHP system: Preliminary investigation and integration of an ORC cycle on a simple gas boiler

M. Amara
Centre National d'Études et de Recherches Intégrées du Bâtiment, Cité Nouvelle El-Mokrani, Souidania, Algiers, Algeria

H. Semmari & A. Filali
Ecole Nationale Polytechnique de Constantine, Constantine, Algeria

ABSTRACT: In order to meet both the increasing demand of electricity and the drastic reduction of fossil oil, the Algerian government has implemented the national energy efficiency program. This program focuses mainly on improving the use of renewable energies, for the purpose of producing 22 GW of electricity by 2030. The program promotes also the energy efficiency in both residential and industrial sectors. In this context, the Organic Rankine Cycle (ORC) seems to be a suitable solution either for exploring renewable resources or to be integrated in industrial process to produce electricity from waste heat recovery. Thus, a possible integration of an ORC system within a simple boiler of the national polytechnic school is investigated. Such solutions allow converting the gas boiler into a Combined Heating and Power (CHP) system. Thus, the new system with a higher energy performance will produce simultaneously both heating and electricity. A thermodynamic analysis of a sub critical ORC cycle using the Cool Prop free tool is presented. The presented work is a case study applied to the gas boiler installed within the national polytechnic school of Constantine. The proposed ORC cycle is activated by the boiler's exhaust gas of about 400°C to produce mechanical work while condensing the working fluid is performed by air condenser.

Keywords: Combined Heating and Power, Organic Rankine Cycle, Waste heat, Electricity, Energy efficiency

NOMENCLATURE

CHP	Combined Heating and Power
GHG	Green House Gas
h_i	Specific enthalpies at the different points of the thermodynamic cycle
IEA	International Energy Agency
ORC	Organic Rankine Cycle
q_i	Specific thermal energy exchanged by the different components
WHR	Waste Heat Recovery
w_i	Specific work exchanged by the spinning components
η_i	Efficiency

Subscripts

p,T	Pump and Turbine
Evap, cd	Evaporator and condenser
l,sat;v,sat;is	Saturated liquid; saturated vapour; isentropic

1 INTRODUCTION

The energy and environmental issues have become more important and the subject of many research projects all over the world. Most of the consumed energy throughout the world is generated mainly from fossil sources which are considered non-renewable resources. The energy consumption is growing every year and if no serious consideration and prompt actions of reducing this consumption are undertaken, the world will soon face a severe energy shortage. Moreover, this energy consumption is leading to growing greenhouse gas (GHG) emissions which are considered one of the most significant drivers of the climate change problem. Interesting efforts so far to mitigate climate change have been adopted in international conferences such as Kyoto Protocol which is an international treaty, adopted in Kyoto, Japan, on 11 December 1997 in which 192 countries committed to reducing their greenhouse gases emissions (Fact sheet the Kyoto protocol, 2011).

In Algeria, the building sector, including residential and commercial buildings, consumes about 41% of the energy and is responsible for about 21% of CO_2 emissions (Aprue, 2009). In this context, an important effort must be carried out to improve the energy efficiency of the residential and commercial buildings.

The first economic solution proposed is to design less-energy consuming buildings. Committed to solve these problems, the Algerian government promulgated the N°99-09 law of July 28, 1999 on the energy efficiency whose main goal was to implement an optimized energy consumption process (Loi, 1999), and later the N° 04-09 of August 14, 2004 law on promoting the renewable energies in the frame of the sustained development (Loi, 2004). The Algerian government also promulgated the N°2000-90 of April 24th, 2000 decree on thermal regulation in new buildings (Décret exécutif, 2000). For the thermal regulation, the National Center of Integrated Studies and Researches of Buildings (CNERIB) has developed the Regulatory Technical Document "DTR C.3.2/4: Thermal regulation of buildings" in 2016 (DTR C.3.2/4, 2016).

In 2007, at the CNERIB, the Med-Enec project (MED ENEC Energy Efficiency in the construction sector in the Mediterranean) was finished. This project has been accomplished with the help of the European Union (EU). A rural house was built inside the centre in the frame of this project, the additional investment was 12% and the savings in energy consumption were 54%. The house was built with local materials (Compressed and Stabilised Earth Bricks), the roof was insulated with 16 cm of expanded polystyrene (EPS), the external walls with 9 cm of EPS and the ground with 6 cm of extruded polystyrene (XPS). The house was south oriented. This project demonstrated that external thermal insulation combined by some passive solutions can reduce significantly the energy consumption of a building (Derradji et al. 2013).

The second economic solution proposed, is the use of high efficiency energy systems for both heating and cooling when the exploitation of renewable energy is not possible. Among these systems, the Combined Heating and Power (CHP) system should take all interest for both residential and tertiary buildings owing to its ability to produce both heat and power from the same primary energy. In the present study, the integration of an Organic Rankine Cycle (ORC) on a simpler gas boiler is investigated. This application concerns energy efficiency measures that can be advantageously adapted for the tertiary buildings. Generally, this ORC integration deals with small-scale waste heat recovery application to produce electrical power at a range of 10 kWe (Tartière & Astolfi, 2017).

2 AN INTRODUCTION TO THE ORGANIC RANKINE CYCLE

The Organic Rankine Cycle is a Rankine cycle which uses an organic working fluid instead of water. It allows the use of heat for power generation from sources as low as 65°C. The technologies to build ORC systems were available as early as 1976. The economics of ORCs weren't favourable until the use of modified off the shelf refrigeration components lead to packaged systems. The term "waste heat recovery" means the use of any heat which is rejected to the environment. WHR cycles are interesting for the fact that they require no additional fuel for power generation (Cirincione, 2011).

The ORC is still the matter of numerous studies all over the world. The authors Carlo Carcascia and Lorenzo Winchlera (Carcascia & Winchlera, 2016) presented the study of an ORC combined with an intercooled gas turbine in order to convert the gas turbine waste heat into electrical power. ORC is generally a good choice for waste heat recovery at low/medium temperatures and an intercooled gas turbine is characterized by low exhaust temperature. In their study, the air temperature exiting from the first compressor was about 160–220°C, and the Organic Rankine Cycle was an interesting solution to improve the efficiency of the power plant. This waste heat can be recovered by an Organic Rankine Cycle to convert the low-temperature heat source into mechanical energy and increase the global power plant efficiency (Carcascia & Winchlera. 2016).

In such applications where the temperature of waste heat is in the 300–450°C range, they stand as a very interesting option to enhance the system performance. ORC cycles can have high thermodynamic efficiency with low mechanical stresses and longer component life compared to other cycles. The low-temperature heat can be converted into mechanical energy from the Organic Rankine Cycle and so the electric efficiency can increase. They found even that an ORC power plant can substitute completely or partially the cooling system, which shows that an ORC, not only the expenses of a cooling system can be avoided but still waste heat is recovered and transformed into useful electricity. They found that the power of combined power plant increased by about 20.4 MW and the electrical efficiency increase by 10.4% in respect to simple cycle gas turbine (Carcascia & Winchlera, 2016).

3 DESCRIPTION OF THE GAS BOILER

Currently, the gas boiler (Figure 1) is converting the chemical energy of the natural gas into useful heat to meet the heating demand during the winter period. At the same time, an important amount of heat is wasted to the atmosphere at higher temperature of about 400°C.

This exhaust gas was investigated to produce electricity by the means of ORC technology. The study is performed on the base of specific energy due to the lack of information about the flow rate of the exhaust gas. The exhaust gas temperature is the only data that was experimentally measured.

a) Front view   b) Side view   c) Back view

Figure 1.   CHP Gas boiler installed at the National Polytechnic School of Constantine.

4 DESCRIPTION OF THE THERMODYNAMIC MODEL

For a normal gas boiler, the exhaust gas is wasted in the atmosphere causing thermal pollution at high temperature (400°C) and GHG emission. In our experiment, the exhaust gas from boiler is directed towards the ORC system where the heat contained in the exhaust gas will feed successively the economiser (A→$A_{l,sat}$), the evaporator ($A_{l,sat} \rightarrow A_{v,sat}$) and then the superheater ($A_{v,sat} \rightarrow B$). Subsequently, the organic working fluid R134a of the ORC system

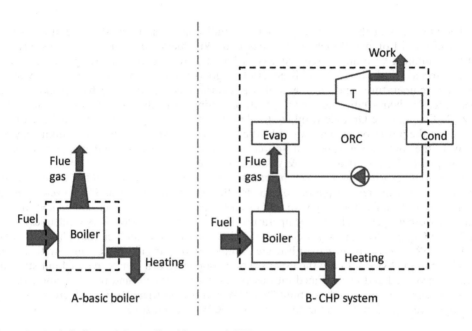

Figure 2. Basic boiler and the predicted improved CHP system.

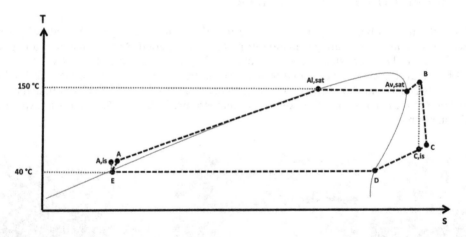

Figure 3. T-S Diagram of the Organic Rankine cycle using R245ca as working fluid.

will be respectively pre heated, evaporated and then superheated. After that, the generated vapour at high temperature and pressure is expanded (B→C) within the scroll turbine (a converted scroll compressor) producing a useful shaft work. The working fluid leaving the expander device is condensed within an air condenser until saturation point (C→E). After that, the working fluid is pressurized to higher pressure by the feed pump (E→A). Subsequently, the new configuration mentioned in Figure 2 allows to produce electricity from the same primary energy leading to improving the efficiency of the whole energy system.

Moreover, the released thermal energy within the condenser is at low temperature compared to the initial exhaust temperature estimated at 400°C leading directly to reduce thermal pollution and indirectly the emission of greenhouse gases.

During the different steps of the thermodynamic cycle described above (Figure 3), the energy balance is carried out by the calculation of the specific enthalpies of each key point of the cycles. Based on enthalpies calculations performed with the free tool CoolProp (Bell et al.

104

2014), and by assuming steady state conditions and neglecting the pressure drops, the governing equation of energy balance can be written down as follow:

Heat absorption: by considering an evaporation temperature of 150°C and neglecting pressure drops for all exchangers, the absorbed heat by the ORC cycle (economiser, evaporator and superheater) is expressed as:

$$Q_{abs} = h_B - h_A \tag{1}$$

Expansion: considering the irreversibility during the expansion, the isentropic efficiency of converted scroll turbine is about 0.8. Thus, the specific shaft work recovered by the turbine is defined by:

$$w_T = h_B - h_C \tag{2}$$

Where: $h_C = h_B - \eta_T \cdot (h_B - h_{C,is})$

Condensation: in order to close the cycle and to protect the feed pump, the working fluid coming from the turbine is condensed and the released heat rejected by the air condenser. Assuming that the isothermal condensation take place at 40°C, the exchanged thermal energy is then quantified as:

$$q_{cd} = h_C - h_E \tag{3}$$

Pressurization of the working fluid: the saturated working fluid is pressurized at high pressure by consuming an amount of specific energy calculated as:

$$w_p = h_A - h_E \tag{4}$$

Where: $h_A = h_E + \frac{h_{A,is} - h_{eE}}{\eta_p}$
With
h_E: the specific enthalpies of the saturated point of condenser a.e 40°C
$h_{A,is}$: the specific enthalpies of the point at higher pressure calculated from the isentropy of the point $h_E h_{A,is} = f(P_h, s_E)$

Finally, the efficiency the ORC sub-system is formulated by the next expression:

$$\eta_{ORC} = \frac{W_T - W_P}{q_{abs}} = \frac{(h_B - h_C) - (h_A - h_E)}{h_B - h_A} \tag{5}$$

Knowing that the average efficiency defined as the ratio of useful heat and the potential energy (equation 6) of the gas boiler is about 70%, the new configuration the CHP system requires a new definition of the performance expressed by the equation 7 (Mascuch et al. 2018):

$$\eta_{boil} = \frac{q_{heat}}{q_{fuel}} \tag{6}$$

$$\eta_{chp} = \frac{w_T + q_{heat}}{q_{fuel}} \tag{7}$$

We note that the performance of the CHP system will be greater than the basic gas boiler as expressed by equation (7). The enthalpies of different points of the ORC cycle and the whole performance of the CHP system are resumed inTable 1.

Table 1. ORC cycle's enthalpies and CHP system efficiency.

	R245ca (T_{evap} = 150°C, T_{cd} = 40°C), DT super heating = 30°C
h_A	255.12 kJ/kg
$h_{A,lsat}$	425.14 kJ/kg
$h_{A,vsat}$	518.66 kJ/kg
h_B	567.97 kJ/kg
h_C	519.6 kJ/kg
h_D	445.6 kJ/kg
h_E	252.91 kJ/kg
$h_{A,is}$	254.68 kJ/kg
$h_{C,is}$	507.51 kJ/kg
η_{ORC}	15.46%
η_{comb}	70%
η_{CHP}	85.46%

5 RESULTS AND DISCUSSION

The integration of an ORC subsystem transforms the simple gas boiler into a CHP energy system. The direct consequence of this new configuration is the improvement of the whole efficiency passing from 70% to 85%. Obviously, the amelioration of the performance of the basic system will allow reducing the energy bill. However, the transformation into CHP power system requires some investment cost which must be estimated in order to calculate the gain of the proposed solution. Moreover, it can be noted that the important energy gap between the exhaust gas and the evaporation occurring at only 150°C is leading to lost exergy which can be improved either by considering a super critical cycle or an advanced cascade configuration using several working fluids.

6 CONCLUSION

A preliminary study has been performed to assess the integration potential of an ORC system targeting to improve the efficiency of the basic gas boiler installed at the National Polytechnic School of Constantine, Algeria. The obtained results showed that the transformed CHP system allows improving the boiler efficiency by 15%. This improvement will lead to reducing the energy bill at the cost incurred by the addition of an ORC to the boiler. This cost must be estimated in order to evaluate the gain of the proposed solution in the Algerian context. This step will be performed next winter season in order to measure the flow rate of the exhaust gas to size correctly all ORC cycle components (exchangers, pump, scroll compressor...). Once the ORC components dimensions are defined, the investment cost can be assessed with the profitability of this integrated ORC solution.

Different approaches have been highlighted by the present study to improve and optimise the new CHP system. The first approach is to consider a supercritical configuration of the ORC cycle using the same working fluid R245ca to recover more energy from the exhaust gas. This will allow reducing the temperature of effluents and enhancing the whole performance of the CHP plant.

The second approach is to implement a thermal cascade by using a couple of working fluid instead of one organic fluid as investigated in the present paper.

More research work is being programmed at the National Polytechnic School of Constantine and the CNERIB, for a purpose of fully implementing these proposed different approaches and contribute to the local and worldwide energy consumption reduction. This research work aims to widespread the use of ORC system as a solution for the Algerian market especially for waste heat recovery at lower, medium and large scales.

REFERENCES

Aprue, 2009. Consommation Energétique Finale de l'Algérie, Chiffres clés Année 2007 (eds).

Carcascia, C. & Winchlera, L. 2016. Thermodynamic analysis of an Organic Rankine Cycle for waste heat recovery from an aeroderivative intercooled gas turbine, *Energy Procedia* 101: 862–8869.

CNERIB, 2016. Document Technique Règlementaire DTR C.3.2/4 "Réglementation thermique du bâtiment", Ministère de l'Habitat, de l'Urbanisme et de la Ville, Alger, Algérie.

Fact sheet the Kyoto protocol, February 2011.

Ian H.B., Jorrit W., Sylvain Q., & Vincent L., 2014. Pure and Pseudo-pure Fluid Thermophysical Property Evaluation and the Open-Source Thermophysical Property Library CoolProp. *Industrial & Engineering Chemistry Research* 53(6): 2498–2508.

Décret exécutif N°2000-90 du 24 Avril 2000 Portant Réglementation Thermique dans les Bâtiments Neufs, J.O.R.A., N°25, 30 Avril 2000, Alger, Algérie.

Derradji, L. Errebai, F. B Amara, M. Maoudj, Y. Imessade, K. & Mokhtari, F. 2013. Etude expérimentale du comportement thermique d'une maison prototype en période d'été, *Revue des Energies Renouvelables* 16(4): 709–7719.

Loi 1999, N° 99-09 du 28 juillet 1999, relative à la maîtrise de l'énergie, p.3. (N° JORA: 051 du 02-08-1999).

Loi 2004, N°04-09 du 14 août 2004, relative à la promotion des Energies Renouvelables dans le cadre du développement durable.

Mascuch, J. Novotny, V. Vodicka, V. Spale, J. & Zeleny, Z. 2018. Experimental development of a kilowatt-scale biomass fired micro – CHP unit based on ORC with rotary van expander. *Renewable Energy*, Article in press (2018) 1–14.

Cirincione, N. 2011. Design, construction and commissioning of an Organic Rankine Cycle waste heat recovery system with a Tesla-hybrid turbine expander. Thesis, In partial fulfilment of the requirements For the Degree of Master of Science Colorado State University Fort Collins, Colorado Fall.

Tartière, T. & Astolfi, M. 2017. A world overview of the organic Rankine cycle market. *Energy Procedia*, 129: 2–9.

Solutions for Sustainable Development – Szita, Jármai & Voith (eds.)
© 2020 Taylor & Francis Group, London, ISBN 978-0-367-42425-1

Examination the effect of environmental factors on a photovoltaic solar panel

I. Bodnár, L.T. Tóth, J. Somogyiné Molnár, N. Szabó, D. Erdősy & R.R. Boros
Institute of Electrical and Electronic Engineering, University of Miskolc, Miskolc-Egyetemváros, Hungary

ABSTRACT: In this study, the energy generation of a solar panel in case of different surface contaminants is experimentally examined. The energy production of a solar cell is significantly influenced by the contamination of the panel's surface. Not only the objects in the environment, but pollutants on the panel's surface can also create a shadowing effect. The energy production and the efficiency of a solar panel are experimentally investigated in case of different types of contamination, such as sand, soil, cement, road salt and fly ash. Measurements are done with different polluted areas and the amount of pollution. A correlation between the power generation and the contamination is observed. Following the observation, it is possible to determine the loss of energy production, caused by the surface contaminants, while the optimal cleaning period can be determined. Power production and type of contamination show a certain amount of correlation. Since environmental factors directly affect the output voltage, it is possible to determine the voltage and power production drop depending on the surface contaminants.

1 INTRODUCTION

Solar electricity production has been on the rise since the 1970s. Hungary has made progress over the last 10 years. In 2010, the capacity of solar power plants in Hungary was 2 MW, in 2018 it has reached 640 MW. Additional 35 MW expansions are expected in 2019. However, it should not be overlooked that solar cells operate under extreme weather conditions, which greatly affects their efficiency and lifetime. Temperature is one of the most important life affecting factors. The higher the temperature of the solar cell is, the lower the output voltage and thus the power and efficiency. The crystalline structure of overheated solar cells is changing, which also has a lifetime-reducing effect. One reason for the temperature rise is the heating effect of sunlight; another reason is that the current flowing through the solar cells results in Joule heat. In addition, the thermal insulation effect of impurities on the surface of the solar cell has to be taken into account. The deposits form a heat insulating layer on the surface of the solar cell, thereby reducing the heat emission of the solar cell and they are further heating (Abderrezek et al. 2017, Bhattacharya et al. 2015).

2 ENVIRONMENTAL EFFECTS AFFECTING SOLAR PANEL OPERATION

The environmental effects that have an influence on the operation of the solar cell can be divided into two main groups. One group is the territorial features; the other group is the pollutant sources. Both main groups cause shadowing effect (Adinoyi et al. 2013, Malik et al. 2003). The most common regional factors are:

- buildings,
- vegetation – trees,
- self-shielding of solar cells and their supporting structures.

The most common sources and types of pollutants are:

- transport - tire debris and soot particles,
- heating - fuel ash,
- vegetation - leaf and pollen,
- agriculture - soil/sand,
- agriculture and industry - dust
- fauna - bird droppings.

Factors in the first group only cause shading problem and can be eliminated partly or completely during the design process. Factors belonging to the second group are difficult to quantify because they come into direct contact with the solar cell. In addition to the shadowing effect, they also raise thermal insulation problems, which are not known precisely and dynamically change. The real effects can be recognised foremost experimentally.

The most common pollutant in Hungary is dust. Approaching the Equator the desert sand-, in big cities, the air pollutants are the biggest problem. Since there are significant differences in the particle size, density and opacity of each pollutant source, their effects on the output power of the solar cell are also dissimilar. In Europe, pollution can cause up to 5–20% annual loss of production. In desert environments, the solar cell production capacity can be reduced by 75% after a sandstorm (Abderrezek et al. 2017, Malik et al. 2003, Bhattacharya et al. 2015, Gürtürk et al. 2018).

Some contaminants, such as sand, have a scratching, abrasive effect on the surface of the solar cell depending on their shape and hardness. This results in change of surface structure and residual degradation of efficiency. Removal by dry cleaning methods is not advised. Water or special cleaning fluid is recommended for such pollutants.

A significant part of the dust is removed from the surface of the solar cell after a rainfall. However, it may clog up on the corners of the solar cell and form a muddy layer. This can lead to additional problems, which is usually overheating. Bigger problem is the leaf and bird droppings. If only a small part of the solar cell is contaminated, By-Pass diodes will not be activated and the Hot-Spot phenomenon will occur. This will cause irreversible damage to the solar panel in long term. Since the shaded cell behaves as a resistor, the current flowing through it generates additional (Joule) heat. The pollutant inhibits the emission of this heat, so the cell overheats. Local overheating can cause the cells to burn out, affecting the operation of the entire solar panel. The output current, voltage, and power of the panel are reduced, so does its efficiency. If there are more burned out cells inside a panel, it will cause the panel to become inoperative and it is inevitable to replace it. Therefore it is advisable to remove these impurities as soon as possible. Removing bird dust is the most difficult task. It can only be removed efficiently by using solar cleaning fluid or high pressure softened water. Hard water leaves limescale on the surface of the solar cell after drying, which can also reduce its performance. Depending on the size of the power plant, different methods have been developed for cleaning solar panels. There are varieties of choices available from small robots to cleaning vehicles (Bodnár et al. 2019).

Thermal imaging tests can detect damaged cells, thereby mapping the status of solar panels. In case of solar failure caused by shadowing effects, we cannot expect a warranty replacement because the failure is not a consequence of the manufacturer but an operator error. Therefore it is advisable to check the state of the solar panels and possible shadowing phenomena by appropriately sensitive thermal imaging camera at installation and as well as on a yearly basis. If cell defects are detected during the first use after installation, it is advisable to contact the distributor/manufacturer immediately, as the warranty may still be valid.

3 EXPERIMENTAL SETUP

A solar panel was placed on a table of the same size during measurements. The temperature of the solar panel's surface was measured by a four-channel PL-125-T4 digital thermometer. Four sensors were placed on four different parts of the solar panel. Previous measurements proved that the back surface of solar panels heats up likewise the absorber surface (Ndiaye et al. 2018, Gürtürk et al. 2018). The voltage and current of the solar panel were measured at the same time by a Protek DM-301 and a METEX M-365OD digital multimeters. The temperature of the solar panel stabilized in around 20 minutes. The temperature of the clean solar panel was about 65.5°C. The ambient temperature was 20°C. The dimension of the solar panel: 505 x 353 x 25 mm. The type of solar panel is polycrystalline. Figure 1. shows the measurement system and Table 1. gives the electrical parameters of the solar panel. The weight of the contaminants was measured by an Ohaus type scale of ± 1g accuracy. At the end of each measurement the panels were cleaned properly.

Figure 1.   The experimental setup.

Table 1.   Electrical parameters of the solar panel.

Parameter	Symbol	Value	Measurements
Year of manufacture	-	2019	-
Intensity of illumination	I_{ill}	861	W/m^2
Peak Power	P_{max}	20	W
Production tolerance	Tp	3	%
Max. power current	I_M	1.14	A
Max. power voltage	U_M	17.49	V
Short circuit current	I_{sc}	1.22	A
Open circuit voltage	U_{OC}	21.67	V
Nominal fill factor	φ	0.7542	-
Serial resistance	R_s	0.0035	Ω
Parallel resistance	R_P	10,000	Ω
Number of serial connected cells	N_S	18	piece
Number of parallel connected cells	N_P	2	piece
Temperature co-efficient for P_{max}	K_{PM}	-0.1137	W/°C
Temperature co-efficient for I_{sc}	K_{ISC}	0.00057	A/°C
Temperature co-efficient for U_{oc}	K_{UOC}	-0.06963	V/°C
Percentage Temperature co-efficient for P_{max}	μ_{Pm}	-0.43	%/°C
Percentage Temperature co-efficient for I_{sc}	μ_{Isc}	0,047	%/°C
Percentage Temperature co-efficient for U_{oc}	μ_{Uoc}	-0.32	%/°C
Efficiency (maximal power)	η	15.	%
Nominal operating temperature	T_N	45±2	°C
Operating Temperature range	T_{Nrange}	-40...85	°C

4 MEASURMENTS RESULTS

In our laboratory tests, we polluted the solar cell with powder having five different properties. These were sand, fly ash, soil, road salt and cement. The dust concentration was between 0–28 g/panel (0–157 g/m^2) at 3 g/panel scale. Concentration of 9 g/panel is 50 g/m^2 and 18 g/panel is equivalent to 100 g/m^2. Figure 2 shows a solar cell contaminated with various powders at a concentration of 28 g/panel. Details of the applied pollutants are given in Table 2. Our basic assumption is that the larger Blaine specific surface contaminant will result in the higher coverage, resulting in a greater loss of solar cell production.

Figure 3 illustrates power reduction as a function of individual powders and concentrations. It can be observed that pollutants typically have a linear effect on performance reduction. The exception is ash and earth. The cause of nonlinearity in case of the ash was that the surface of the solar cell began to saturate at a concentration of 9 g/panel (50 g/m^2), and in some areas, the thickness of the powder layer was higher. In case of earth, the variation in grain size may be the cause of nonlinearity. The greatest reduction in performance was observed for ash, the smallest nonlinearity was measured with road salt. At a concentration of 18 g/panel (100 g/m^2), the difference is close to 20 times. The results can be traced back to the fact that the specific surface area of the

Figure 2. Polluted solar panel: dust, fly ash, soil, road salt and cement 28 g/panel (157 g/m^2).

Table 2. Physical properties of pollutants.

Pollutant	Density [g/cm^3]	Specific surface area [mm^2/g]	Grain size [mm]
Sand	1.45	~0.30	0.2...0.8
Fly ash	0.6	~0.66	0.1...1.4
Soil	0.8	~0.35	0.1...6
Road salt	2.16	~0.06	0.4...4
Cement	3.15	~0.32	0.1...0.8

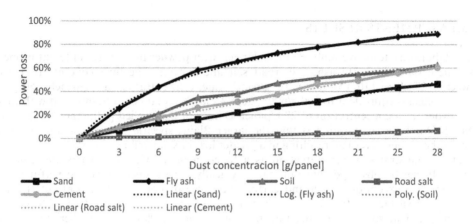

Figure 3.   Power reduction with different dust contaminated solar panel.

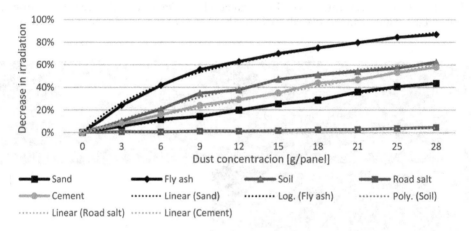

Figure 4.   Decrease of irradiation at various pollutants and concentrations.

road salt is tenth as high as the ash and has a higher opacity. Similar results can be observed in the works of Rao et al. (2013) and Ndiaye et al. (2018).

Figure 4 shows the degree of radiation reduction. The same trend can be observed as in Figure 3. The performance loss consists of two parts. The first part is the voltage drop from the decrease in irradiation and the effect of the temperature increase caused by the dust layer. As shown in Figure 4, the vital part of the weakening performance of the dirty solar cell is caused by the decrease in irradiation. The radiation reduction is proportional to the opacity and the specific surface area.

Figure 5 shows the temperature increase of the solar cell as a function of dirt. It can be observed that the reason for the increase in temperature is related to the thermal insulation and density of each material. The lower density materials at the same dust concentration form a thicker insulating layer on the surface of the solar cell, thereby better insulating it. Since the solar cell cannot radiate the Joule heat, it has a warming effect. At a concentration of 18 g/panel (100 g/m^2), the fly ash-contaminated solar cell was nearly 13°C warmer than in clean state. The temperature of the solar cell contaminated with road salt was almost unchanged during the measurement (Ndiaye et al. 2018 and Siddiqui et al. 2016). Since the voltage drop was caused by the increase in temperature, the same tendency can be seen in Figure 6.

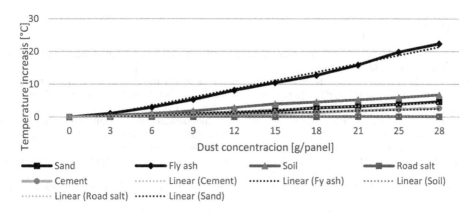

Figure 5. Rate of increase in temperature depending on the concentration of dust.

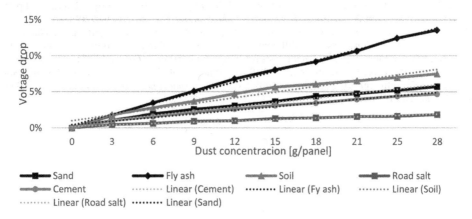

Figure 6. Voltage drop by various pollutants in different concentration.

5 CONCLUSIONS

It can be stated that measurements confirmed our basic hypothesis. Pollutants with a larger Blain specific surface area will reduce the solar cell performance proportionally to surface area at the same concentration. In case of Fly ash, the Blain specific area is two times as large as for the sand and at a concentration of 28 g/panel the same decrease in irradiation can be observed. The smallest specific surface area of the road salt does not affect the operation of the solar cell at the tested concentrations. At the concentration level of 28 g/panel the power loss was about 8%. In case of Fly ash whose Blain specific surface area is 10 times as large as for the road salt, the power loss was also 10 times as large. The highest specific surface area of ash has a significant impact on the operation and life time of the solar cell. Lower density pollutants form a thicker insulating layer on the solar panel, resulting in more heating of the solar cell. If the solar cell operates at a higher temperature, the service life is reduced due to the nature of the electronics.

ACKNOWLEDGEMENT

"This research was supported by the European Union and the Hungarian State, co-financed by the European Regional Development Fund in the framework of the GINOP-2.3.4-15-2016-00004 project, aimed to promote the cooperation between the higher education and the industry."

REFERENCES

Abderrezek, M., Fathi, M. 2017. Experimental study of the dust effect on photovoltaic panels' energy yield. *Solar Energy* 142: 308–320.

Bhattacharya, T., Chakraborty, A.K., Pal, K. 2015. Influence of Environmental Dust on the Operating Characteristics of the Solar PV Module in Tripura, India. *International Journal of Engineering Research* 4(3): 141–144.

Rao, A., Pillai, R., Mani, M., Ramamurthy, P. 2013. An experimental investigation into the interplay of wind, dust and temperature on photovoltaic performance in tropical conditions, *Proceedings of the 12th International Conference on Sustainable Energy Technologies*. 2303–2310.

Adinoyi, M.J., Said, S.A.M. 2013. Effect of dust accumulation on the power outputs of solar photovoltaic modules. *Renewable Energy* 60: 633–636.

Malik, A.Q., Damit, S.J.B.H. 2003. Outdoor testing of single crystal silicon solar cells. *Renewable Energy* 28: 1433–1445.

Ndiaye, A., Kébe, C.M.F., Bilal, B.O., Charki, A., Sambou, V., Ndiaye, P.A. 2018. Study of the Correlation Between the Dust Density Accumulated on Photovoltaic Module's Surface and Their Performance Characteristics Degradation. *Innovation and Interdisciplinary Solutions for Underserved Areas*. pp. 31–42.

Gürtürk, M., Benli, H., Ertürk, N.K. 2018. Effects of different parameters on energy – Exergy and power conversion efficiency of PV modules. *Renewable and Sustainable Energy Reviews* 92(9): 426–439.

Siddiqui, R., Kumar, R., Jha, K.G., Morampudi, M., Rajput, P., Lata, S., Agariya, S., Nanda, G., Raghava, S.S. 2016. Comparison of different technologies for solar PV (Photovoltaic) outdoor performance using indoor accelerated aging tests for long term reliability. *Energy* 107(15): 550–561.

Bodnár, I., Iski, P., Koós, D., Skribanek, Á. 2019. Examination of electricity production loss of a solar panel in case of different types and concentration of dust. *Advances and Trends in Engineering Sciences and Technologies III*. pp. 313–318.

Solutions for Sustainable Development – Szita, Jármai & Voith (eds.)
© 2020 Taylor & Francis Group, London, ISBN 978-0-367-42425-1

Thermal behaviour of PCM-gypsum panels using the experimental and the theoretical thermal properties-Numerical simulation

F. Boudali Errebai
Faculty of Mechanical and Process Engineering, LTPMP, University of Science and Technology Houari Boumediene (USTHB), Algiers, Algeria
National Center of Building Integrated Research and Studies (CNERIB), Algiers, Algeria

S. Chikh
Faculty of Mechanical and Process Engineering, LTPMP, University of Science and Technology Houari Boumediène (USTHB), Algiers, Algeria

L. Derradji
National Center of Building Integrated Research and Studies (CNERIB), Algiers, Algeria
Department of Mechanical Engineering, University of Blida, Algeria

M. Amara
National Center of Building Integrated Research and Studies (CNERIB), Algiers, Algeria

A. Terjék
Építésügyi Minőségellenőrző Innovációs (ÉMI), Nonprofit Kft. Budapest, Hungary

ABSTRACT: The determination of thermophysical properties of the PCM is a very important step in the process of thermal management by assessment of energy saving performance of PCM composites and for evaluation of the energy-saving performance. In this paper, measured and theoretical values of thermophysical properties of a composition of 50% of PCM mixed with gypsum were used to perform numerical simulations by two methods for five thicknesses of panels (10, 30, 50, 70 and 90 mm). The first method is a numerical simulation of heat transfer by conduction in the panels by using experimental values of thermophysical proprieties and the second method is a numerical simulation of heat transfer by conduction in the panels by using theoretical values of thermophysical proprieties. The results of these numerical simulations revealed that the maximum difference in the temperature swing between the two methods can reach 1.1°C.

NOMENCLATURE

Latin letters

Cp	specific heat (J/kg.K)
f	mass fractions
H	specific enthalpy (J/kg)
h	convective heat transfer coefficient (m^2.K/W)
L	length of the panel (m)
S	energy source term (W/m^3)
T	temperature (K)
t	time (s)
x_j	direction vectors (m)

Greek letters

λ	thermal conductivity (W/m.K)
ρ	density (kg/m^3)

Subscripts

e	equivalent
E	energy
ext	exterior air
gyp	gypsum
int	interior air
PCM	phase change materials

1 INTRODUCTION

In several countries all over the world, the building sector is considered as the largest consumer of energy among the economic sectors. The essential function of a building is to provide an indoor climate well suited to our needs and our comfort (Boudali Errebai et al., (2017); Derradji et al., (2011)). Using the Phase Change Materials (PCMs) for the storage of thermal energy is a preferred means for optimal management of energy. These materials can increase the thermal inertia of the building walls and help to improve thermal comfort (Boudali Errebai et al., (2018); Derradji et al., (2017); Soares et al., (2013)). Indeed, the use of PCM will allow adapting production to the needs of thermal energy by achieving a constant relationship between the energy demand and the supplied energy. Indeed, the use of PCMs can dampen and smooth the rooms heat fluctuations and maintain a comfortable average temperature. These materials have a thermal phase shift effect by delaying the heat wave to a less hot period of the day (Verbeke & Audenaert, (2018)). This has the advantage of limiting the rise in temperature for a number of hours and therefore reduce the need for air conditioning (Derradji et al., (2014)).

The determination of thermophysical properties of PCM is very important step in the process of thermal management by assessment of energy saving performance of PCM composites and for evaluation of the energy saving performance (Cheng et al., (2013)).

According to the different works found in literature, the main characteristics required for a PCM are: a melting temperature in the desired operating temperature range, a high specific heat capacity, a high thermal conductivity and high density (Oró et al., (2012)).

The best way to have a good simulation results is the determination of thermal properties by experimental measurements. In order to avoid at each time to measure these thermal properties for different compositions possible, several studies have been carried out to determine analytically the thermophysical properties of gypsum-PCM composites (Karkri et al., (2015)). Among these works, the research presented by Porfiri et al., (2009) developed an analytical model to predict the thermal conductivity of composites using homogenization techniques.

Toppi & Mazzarella, (2013) developed a correlations based on experimental measures which give the composite material thermal properties (density, thermal conductivity and specific heat capacity) for gypsum based composite materials with micro encapsulated PCM.

Borreguero et al., (2011) investigated the effect of different PCMs content on the thermal performance and used the theoretical method for the determination of the apparent specific heat capacity as a function of temperature.

Shi et al., (2014) have used a theoretical analysis of effective thermal conductivity method of a novel form-stable fiber composite concrete containing dispersed phase change materials (PCMs). The method was based on serial and parallel connection models and compared with experimental model. The results showed that the deviation between the theoretical analysis and the experimental model was 10% and the average error was 3.6% which led to the conclusion that the existing model is able to account for the effective thermal conductivity of complex structure.

Among the studies cited above, some papers indicate that there are big differences between the values obtained by these methods (Georgi Krasimirov Pavlov, (2014); Pomianowski et al.,

(2014)). However, these studies focus solely on these thermophysical properties and do not study the influence of these differences (between measured values and calculated values) on the determination of these properties on thermal behaviour. For this reason, this work contributes to studying the influence of these differences on the thermal behaviour of a PCM-gypsum board.

2 MATERIALS AND METHODS

2.1 *Phase change material*

In this paper, a commercial PCM, the Micronal DS 5001 (from the BASF company) in dry powder was used (Figure 1a). The main properties of the PCM microcapsules are presented in Table 1.

2.2 *Gypsum*

The plaster is mainly made of beta hemihydrate ($CaSO4$, $1/2H_2O$) and anhydrite ($CaSO_4$). The other present components originate from impurities initially contained in the gypsum (Derradji et al., (2014)).

The plaster used has a thermal conductivity of 0.534 W/m.K, a density of 1156 kg/m^3 and Specific heat capacity of 1133 J/kg.K.

2.3 *Elaboration of samples for the determination of thermophysical proprieties*

To prepare the panel samples (Figure 1b), gypsum-PCM mixture was manufactured using 50% of PCM amounts (by total weight of gypsum).

3 MEASUREMENT OF THERMOPHYSICAL CHARACTERISTICS OF SAMPLES

3.1 *Thermal conductivity measurements*

The thermal conductivities of samples with 50% of microencapsulated paraffin were measured with a Taurus TCA 300 DTX equipment. The device measures the thermal conductivity of a specimen (Figure 1b) with the heat flux sensor.

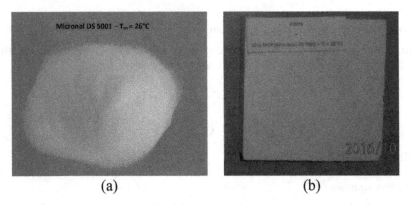

(a)   (b)

Figure 1. Material and samples used: (a) PCM Powder (Micronal DS 5001), (b) Samples for determination of thermal properties.

Table 1. Thermophysical characteristics of PCM provided by manufacturer (Source: BASF).

	Apparent density (kg/m^3)	Melting temperature (°C)	Latent heat of melting (kJ/kg)	Overall storage capacity (kJ/kg)
Micronal DS 5001	250-350	26	110	145

3.2 *Specific heat capacity measurements*

The determination of the specific heat capacity and phase change temperature of the composite of gypsum and microencapsulated PCM were measured with a Differential Scanning Calorimeter (DSC 214 Polyma). The samples of gypsum PCMs composite were heated and cooled to the temperature range between 15°C and 32°C at the rate of 2°C/min.

3.3 *Density measurements*

The density of PCMs panels was determined by mass-to-volume ratio. The "AND ER-180A" apparatus with a precision of 0.1 mg was used to weight the samples, whose size was 100 × 100 × 20 mm (Figure 1b). For the determination of volume, a Digital Caliper was used for measuring the dimensions of sample the accuracy of the measures of this caliper being 1/50 mm.

4 THEORETICAL DETERMINATION OF THE THERMOPHYSICAL PROPERTIES

4.1 *Determination of the theoretical equivalent thermal conductivity*

The simplest of the models used assumes a simple distribution of the components with respect to the direction of the heat flow. A disposition of the components in series leads to the maximum thermal resistance given by:

$$\lambda_e = \lambda_{PCM} \cdot f + \lambda_{gyp} \cdot (1 - f) \tag{1}$$

4.2 *Determination of the theoretical equivalent specific heat capacity*

The theoretical method for specific heat capacity determination of the gypsum-PCM mixture uses the weight average. This equation leads to the maximum specific heat capacity.

$$Cp_e = Cp_{PCM} \cdot f + Cp_{gyp} \cdot (1 - f) \tag{2}$$

4.3 *Determination of the theoretical equivalent density*

The determination of the density of the gypsum-PCM mixture by the theoretical method uses the weight average. This equation leads to the maximum density.

$$\rho_e = \rho_{PCM} \cdot f + \rho_{gyp} \cdot (1 - f) \tag{3}$$

5 NUMERICAL MODELING

5.1 *Governing equations*

There exist in the literature many numerical models about latent heat evolution to study the thermal behaviour of PCMs. Among these models, we can cite the enthalpy method, the heat capacity method, the temperature transforming model and the heat source method (Al-Saadi & Zhai, (2013)).

In this study, the "heat source/sink model" which considers separately the process of melting and solidification was used. This model uses an approach of specific enthalpy–temperature relationships of melting and solidification within the CFD model and allows the consideration of the hysteresis behaviour of PCM.

The temperature field through a wall is numerically solved with the use of the Computational Fluid Dynamics (CFD). To obtain a solution for the temperature field, the heat equations linearized and discretized by a finite volume method on a cartesian grid in the solver.

The energy equation:

$$\frac{\partial}{\partial t}(\rho H) = \frac{\partial}{\partial x_j}\left(\lambda \frac{\partial T}{\partial x_j}\right) + S_E \tag{4}$$

with S_E representing other sources of energy as latent heat corresponding to the case of melting and freezing.

5.2 Assumptions

In this study, the following assumptions have been made for the mathematical models:

- The PCM was considered homogeneous and isotropic.
- The radiation heat transfer was neglected.
- The density and thermal conductivity of PCM are the same for the solid and liquid phases.

5.3 Modeling phase change materials

To study the temperature behaviour of the PCM panels, a transient simulation model is used. The model considered is used to analyze conduction simulations (Figure 2). The computational domain was meshed by hexagonal elements in the wall zone.

To solve the equations presented above, the commercial software Ansys Fluent was used to calculate the energy stored/released in the PCM directly from the temperature through the use of the User Defined Function (UDF).

The details of this UDF are presented in Boudali Errebai et al., (2018).

5.4 Initial and boundary conditions

In this work, the heat transfer through the walls is supposed to be two-dimensional occurring by transient convection and conduction. These boundary conditions are summarized in Figure 2.

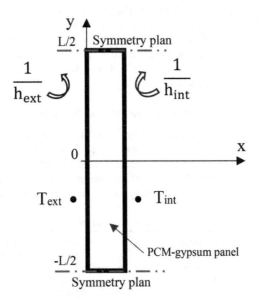

Figure 2. Schematic representation of boundary conditions.

The boundary conditions are respectively:

– The coefficient $\frac{1}{h_{int}} = 10 \ \frac{W}{m^2.K}$.
– The coefficient $\frac{1}{h_{ext}} = 16.67 \ \frac{W}{m^2.K}$.
– Interior air temperature is constant: $T_{int} = 24°C$.
– Along the symmetry lines:

$$\left.\frac{\partial T}{\partial y}\right|_{y = \frac{L}{2}} = \left.\frac{\partial T}{\partial y}\right|_{y = -\frac{L}{2}} = 0 \qquad (5)$$

– Exterior air temperature varies according to a sinusoidal function:

$$T_{ext} = 29 + 7.\sin\left(\frac{2 \times \pi \times t}{86400}\right) \qquad (6)$$

5.5   Studied models

The model studied in this paper and shown in Figure 2, corresponds to numerical simulations of heat transfer by conduction through the panel for 50% of a weight percentage of PCM, for different thicknesses (10, 30, 50, 70 and 90 mm).

The panel was meshed by hexagonal elements and generated by using the pre-processing software ANSYS-Meshing. The computational domains are divided into uniform grids with structured mesh cells.

The numerical model was validated with an experimental setup as presented in Boudali Errebai et al., (2018).

6   RESULTS AND DISCUSSION

6.1   Determination of thermophysical properties

The results obtained from the measurement and theoretical properties (thermal conductivity, density and heat capacity) of the panels of gypsum-PCM mixtures are presented in Table 2.

It was observed that the difference between the determinations of the thermal conductivity can reach 96.8%, which represents a big difference. For the determination of the specific heat capacity, a difference of 4.8% is found. This last difference is small and doesn't always reflect the reality because the use of DSC equipment is very complicated and the weight and the rate of heating strongly influence the values determined. For the density, the difference founded is 48.4%, which also represents a big difference.

6.2   Numerical modeling for comparison between the two methods

The comparison between results of simulation by using the measured properties method (dotted line) and simulation by using the theoretical properties method (solid line) is reported in Figure 3 for the cases with 50% of mass fraction of microcapsules PCM and the cases with thickness of 10, 30, 50, 70 and 90 mm respectively. From these simulation graphs, it can be seen that there is a difference in the fluctuation of the average temperature between these two methods.

Table 2.   Thermal properties determined experimentally and theoretically.

	Thermal conductivity (W/m.K)	Specific heat capacity at T<18°C (kJ/kg.K)	Density (kg/m³)
Theoretical values	0.350	1644.2	762
Measured values	0.178	1568.5	513
Percentage difference (%)	96.8	4.8	48.4

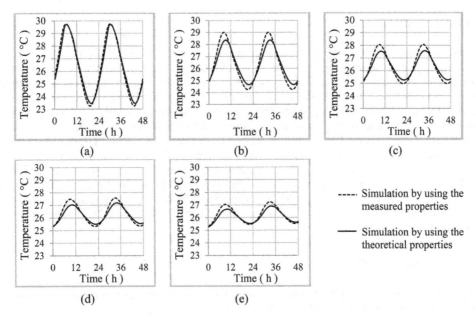

Figure 3. Thermal behavior of PCM-gypsum panels with percentage of PCM of 50% for:
a) Thk = 10 mm, b) Thk = 30 mm, c) Thk = 50 mm, d) Thk = 70 mm and e) Thk = 90 mm.

6.3 *Comparison of the temperature swing*

For more details, Figure 4 shows the temperature swing of panels of PCM-Gypsum and the difference between these values of temperature swings. As remarked in Figure 4a, the temperature swing of PCM-Gypsum panel simulated by using measured properties is higher than the temperature swing for the numerical simulation using theoretical properties. This means that the panels of PCM-Gypsum simulated with theoretical properties present important thermal inertia compared to the panels of PCM-Gypsum simulated with measured properties. The difference between the values of these two methods can be 0.3°C for a panel of 10 mm of thickness and can increase to 1.1°C for a panel of 30 mm of thickness.

To better explain the difference between the two temperature swing curves, the percentage values can tell us the magnitude of these differences (Figure 4b). It has been observed that the percentage of the differences between the two methods are 22% and 26.3% for panels between 30 mm and 90 mm thick and less than 4% for a panel 10 mm thick.

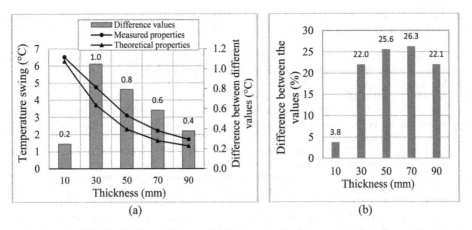

Figure 4. Panels of 10, 30, 50, 70 and 90 mm of thickness: a) temperature swing, b) difference in percentage.

7 CONCLUSION

In this paper, experimental testing of thermophysical properties (thermal conductivity, specific heat capacity, and density) and numerical simulations of thermal behaviour of panels of a mixture of PCM and gypsum were investigated. Two methods of simulations were carried out: the first method is a numerical simulation of the panels conduction heat transfer, using measured values of thermal proprieties and the second method is a numerical simulation of conduction heat transfer of the panels using theoretical values of thermal proprieties. Our conclusion is that the numerical simulation using the equivalent theoretical thermophysical properties of the PCM-gypsum panels gives a bad estimation of the thermal behaviour of these panels compared to the method using measured properties. For the temperature swing of PCM-gypsum panel of 30 mm of thickness, it has been found that the difference between the values of these two methods can reach 1.1°C for 50% of PCM in gypsum, and the percentage of the differences between the two methods can be between 22% and 26.3% for panels of 30 mm and 90 mm thick respectively.

REFERENCES

Al-Saadi, S.N., & Zhai, Z. (2013). Modeling phase change materials embedded in building enclosure: A review. *Renewable and Sustainable Energy Reviews, 21,* 659–673.

Borreguero, A.M., Luz Sánchez, M., Valverde, J.L., Carmona, M., & Rodríguez, J. F. (2011). Thermal testing and numerical simulation of gypsum wallboards incorporated with different PCMs content. *Applied Energy.*

Boudali Errebai, F., Chikh, S., & Derradji, L. (2018). Experimental and numerical investigation for improving the thermal performance of a microencapsulated phase change material plasterboard. *Energy Conversion and Management, 174,* 309–321.

Boudali Errebai, F., Derradji, L., & Amara, M. (2017). Thermal Behaviour of a Dwelling Heated by Different Heating Systems. *Energy Procedia.*

Cheng, R., Pomianowski, M., Wang, X., Heiselberg, P., & Zhang, Y. (2013). A new method to determine thermophysical properties of PCM-concrete brick. *Applied Energy.*

Derradji, L., Boudali Errebai, F., Amara, M., Maoudj, Y., Chikh, S., & Mokhtari, F. (2011). Etude expérimentale du comportement thermique d'une maison rurale à faible consommation d'énergie. *15èmes Journées Internationales de Thermique (JITH 2011).* Tlemcen, Algérie.

Derradji, L., Errebai, F.B., & Amara, M. (2017). Effect of PCM in Improving the Thermal Comfort in Buildings. *Energy Procedia, 107,* 157–161.

Derradji, L., Hamid, A., Zeghmati, B., Amara, M., Bouttout, A., & Errebai, F.B. (2014). Experimental Study on the Use of Microencapsulated Phase Change Material in Walls and Roofs for Energy Savings. *Journal of Energy Engineering.*

Georgi Krasimirov Pavlov. (2014). *Building Thermal Energy Storage. PhD. Thesis.* Technical University of Denmark (TUD).

Karkri, M., Lachheb, M., Albouchi, F., Nasrallah, S. Ben, & Krupa, I. (2015). Thermal properties of smart microencapsulated paraffin/plaster composites for the thermal regulation of buildings. *Energy and Buildings, 88,* 183–192.

Oró, E., de Gracia, A., Castell, A., Farid, M.M., & Cabeza, L.F. (2012). Review on phase change materials (PCMs) for cold thermal energy storage applications. *Applied Energy.*

Porfiri, M., Nguyen, N.Q., & Gupta, N. (2009). Thermal conductivity of multiphase particulate composite materials. *Journal of Materials Science, 44*(6), 1540–1550.

Shi, J., Chen, Z., Shao, S., & Zheng, J. (2014). Experimental and numerical study on effective thermal conductivity of novel form-stable basalt fiber composite concrete with PCMs for thermal storage. *Applied Thermal Engineering.*

Soares, N., Costa, J.J., Gaspar, A.R., & Santos, P. (2013). Review of passive PCM latent heat thermal energy storage systems towards buildings' energy efficiency. *Energy and Buildings.*

Toppi, T., & Mazzarella, L. (2013). Gypsum based composite materials with micro-encapsulated PCM: Experimental correlations for thermal properties estimation on the basis of the composition. *Energy and Buildings, 57,* 227–236.

Verbeke, S., & Audenaert, A. (2018). Thermal inertia in buildings: A review of impacts across climate and building use. *Renewable and Sustainable Energy Reviews.*

Solutions for Sustainable Development – Szita, Jármai & Voith (eds.)
© 2020 Taylor & Francis Group, London, ISBN 978-0-367-42425-1

Experimental and numerical study of the thermal behavior of a building in Algeria

Lotfi Derradji
National Center of Studies and Integrated Research on Building Engineering (CNERIB), Cité Nouvelle El Mokrani, Souidania, Algiers, Algeria
Department of Mechanical Engineering, University of Blida, Blida, Algeria

Farid Boudali Errebai, Mohamed Amara & Amel Limam
National Center of Studies and Integrated Research on Building Engineering (CNERIB), Algiers, Algeria

Anita Terjék
ÉMI Non-profit LlcforQuality Control and Innovation in Building, Szentendre, Hungary

ABSTRACT: This work presents an experimental and numerical study of the thermal behavior of the High-Energy Performance Housing (HEPH) made within the framework of the national energy management program (PNME) of the 600 dwellings of the collective rental type (LPL) compared to a control housing. To study the influence of the energy efficiency measures, which have been introduced on the HEPH project, a measuring equipment has been installed to quantify the energy saving in gas and electricity as well as to measure the thermal comfort parameters, which are mainly defined by temperature and humidity. Dynamic thermal simulation work was carried out with the thermal simulation software TRNSYS 17, to study the effect of thermal insulation on the improvement of the energy performance of the two dwellings. The results showed that the insulation had effectively contributed to improving the thermal comfort in the summer period. It has also been shown that HEPH saves 42% on cooling energy compared to a control housing.

1 INTRODUCTION

As part of the National Program for the Control of Energy (PNME), a pilot project of 600 dwellings of collective rental type (LPL) with high energy performance was launched in 2010, in which passive energy efficiency solutions have been used to reduce energy consumption.

Different passive energy efficiency solutions can be used to reduce the energy consumption of heating and cooling (Boudali Errebai et al., 2017; Derradji et al., 2017a; Derradji et al.,). Many studies have investigated the effect of building materials and insulation on energy efficiency of buildings (Derradji et al., 2017b; Dudás et al., 2015; Dudás et al., 2014). Among these studies, Balaji et al. (2019) conducted a study on the thermal performance parameters of building wall envelopes. In this study, two wall systems were considered (homogeneous and composite wall). The study is based on theoretical research by adopting admittance method and finite difference methods. The results showed the impact of various thermal properties of building materials and their configurations on the thermal performance of buildings. Junghun et al. (2018) carried out a diagnosis on approx.40 000 buildings of low-income households in South Korea, under the Energy Welfare program, using the normative method. More than 2500 cases were simulated and analyzed by combining the thermal performance of each part of the building. The results showed that in buildings with similar indoor temperature patterns, the coefficient of variation of the root mean squared error of energy demand falls within the acceptable error range.

RongDan et al. (2018) conducted a theoretical and experimental study to evaluate the thermal performance of the different building materials in China. The results of this work showed that the parameters such as the thermal conductivity of building materials, wall thickness and mortar thickness successively decreased grey correlation degrees with the wall thermal exchange coefficient.

This work is interested in the experimental and numerical study of an apartment realized as part of the national energy management program (PNME) of the 600 dwellings of the collective rental type (LPL) with high energy performance compared to a control housing.

Measuring devices were installed in the HEP housing and control housing to measure the indoor and outdoor temperature, relative humidity and energy consumption related to heating, air conditioning and lighting. An analysis of the results was performed to compare the thermal behavior of the high-energy housing to a control housing. This research work quantified energy savings achieved by using energy efficiency solutions in social housing compared to a control housing.

Moreover, dynamic thermal simulation work was carried out with the thermal simulation software TRNSYS 17, to study the effect of thermal insulation on the improvement of the energy performance of the two housings.

2 DESCRIPTION OF THE TWO HOUSINGS

The high energy performance housing which is the subject of this study has a living area of 72 m² and it is on the top floor of a building located in the region of Djelfa, Algeria (Figure 1).

The HEP housing contains two rooms, room 1 is positioned on the south-east side and room 2 is on the north-east side. The living room has a large south-facing window to capture natural light to the maximum extent, and finally, the kitchen and the bathroom are located on the north side. The walls of the HEP housing are composed of two (10 cm) hollow brick walls separated by a 5 cm expanded polystyrene insulation.

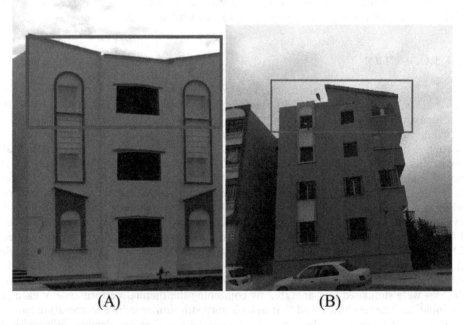

(A)   (B)

Figure 1. General view of the studied housings.
(A) HEP housing, (B) Control housing

The roof is made of a layer of 20 cm of reinforced concrete and insulated from the outside with a 5 cm insulation of expanded polystyrene. The floor is made of 20 cm reinforced concrete layer. The control housing has an area of about 78 m² and is located on the top floor of a building. The housing consists of two rooms, room 1 is positioned on the north-west side and room 2 is on the south-east side. The living room is oriented west, and finally, the kitchen and the bathroom are on the east side. The walls of the control housing consist of two hollow brick walls (10 cm) separated by a 5 cm air gap. The roof and the floor are made of a layer of 20 cm of reinforced concrete.

3 IN SITU MEASUREMENTS

Measuring devices were installed in the HEP housing and control housing to measure the indoor and outdoor temperature, relative humidity and energy consumption related to heating, air conditioning and lighting. Air temperature and humidity measurements in different zones of the two housings were done with the TESTO 175-H1thermo-hygrometers. The measurements were carried out over a two-year period to study the thermal behavior of the HEP housing and the control housing, both located in Djelfa, Algeria. In each housing, 4 thermo-hygrometers were installed to analyze the evolution of the temperature and to compare the thermal performances of the two different housing. Air temperature and humidity measurements were recorded in every hour. Gas and electricity meters were used to measure the heating, air conditioning and lighting consumption of the two housings.

4 TRNSYS SIMULATION

Dynamic thermal simulation of the two housings was realized with the TRNSYS software Version 17 (a transient thermal energy modeling software developed at the University of Wisconsin, Madison). The dwelling was modeled in TRNYSY with Type 56 using 5 zones: 3 rooms, (kitchen and bathroom) and hall. The simulation was performed using Djelfa weather data. Internal gains were around 5W/m² and an infiltration rate was set at 0.6 vol/h. The set temperature was21°C for heating and 24°C for cooling. Six persons, representing the average of the members of an Algerian family, live in the house and are divided into 2 persons per room. The model developed under TRNSYS was used to predict and compare the thermal performances of the HEP housing with the control housing.

5 RESULTS OF IN SITU THERMAL MEASUREMENTS

Figure 2 shows the effect of the insulation of the housing envelope on the interior temperature in different thermal zones (living room, room 1, room 2) during the 3rd week of August. The temperature of the outside air fluctuates due to climatic parameters when the indoor air temperature is more stable and uniform in the whole building. The indoor air temperature varies, most of the time, between 26 and 28°C. The introduction of polystyrene insulation in hollow brick walls and concrete roof has reduced the heat exchange between the exterior and the interior of the housing. It was found that insulation effectively helped to improve the temperature at the interior of the house by decreasing it from 3°C with indoor temperatures that do not exceed 28°C during the day. This meets thermal comfort in summer.

Figure 3 shows the evolution of the outdoor air temperature and that of the indoor air in different thermal zones of the control housing (living room, room 1, room 2) during the 3rd week of August. It shows that the indoor temperature of the housing varies between 27°C and 30°C, which generates a sensation of thermal discomfort for the occupants. The control housing exhibits high temperatures due to the poor orientation. The temperatures of the living room and the room1 are high compared to the temperature of the room 2 because they are oriented on the west side. The lack of insulation in the roof and walls increased the heat input, so increasing the temperature inside the housing.

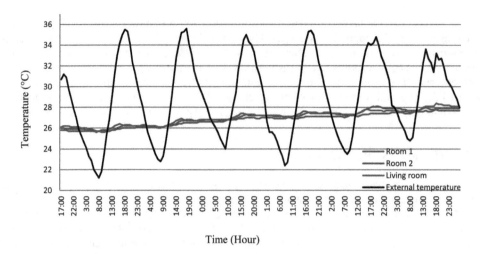

Figure 2. Evolution of ambient temperature of the HEP housing.

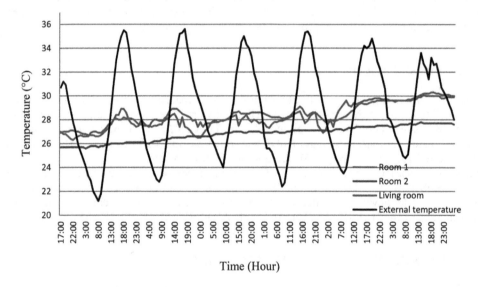

Figure 3. Evolution of ambient temperature of the control housing.

6 RESULTS OF THERMAL SIMULATION

A thermal simulation in dynamic regime was carried out with the thermal simulation software TRNSYS 17, to study the effect of the thermal insulation on the energy performances of a housing. The impact of thermal insulation of walls made with different building materials (concrete, stone, cinderblock and hollow brick) on the energy consumption of buildings was studied for the climate region of Djelfa, Algeria. Figure 4 presents the annual consumption of air conditioning for a non-insulated housing, so it is found that materials with low thermal conductivity display lower energy consumption than other materials. The stone has the highest consumption with 2571 kWh. On the other hand, the brick consumes the least energy, because it has a low thermal conductivity.

Figure 5 illustrates the annual cooling consumption for the studied housing with exterior wall insulation using expanded polystyrene insulation. The stone and concrete have the lowest

126

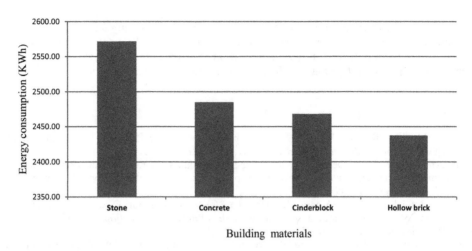

Figure 4. Cooling consumption for different building materials without insulation.

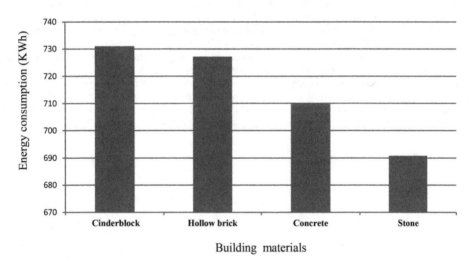

Figure 5. Cooling consumption for different building materials with insulation.

energy consumption. This is due to the high thermal capacity of these two materials that improve the thermal inertia of the house envelope by providing thermal comfort with reduced cooling consumptions. The good thermal inertia of the building materials associated with the thermal insulation avoids the indoor temperature peaks by storing the punctual overheating and spreading it in the time. Strong inertia has helped to achieve acceptable temperatures in the summer period, which reduced the cooling consumptions.

7   CONCLUSION

This work consists of an experimental and numerical study of the thermal behavior of the High-Energy Performance housing (HEP) made within the framework of the national energy management program (PNME) and to compare its performance to a control housing. The analysis of the experimental results showed that thermal insulation has effectively helped to improve the temperature at the interior of the HEP housing by decreasing with 3°C with indoor temperatures that do not exceed 28°C during the day. The numerical results showed that stone and concrete have the lowest energy consumption. The good thermal inertia of building materials combined with thermal insulation improved the thermal inertia of the housing envelope by providing thermal comfort with reduced cooling consumption.

REFERENCES

Balaji N.C., Mani M., Reddy B.V., 2019, Dynamic thermal performance of conventional and alternative building wall envelopes. *Journal of building ingenering, Vol 21, 373–395.*

Boudali Errebai F., Chikh S., Derradji L., 2018, Experimental and numerical investigation for improving the thermal performance of a microencapsulated phase change material plasterboard. *Energy Conversion and Management, 174, 309–321.*

Derradji L., Boudali Errebai F., Amara M., Maoudj Y., Chikh S., Mokhtari F., 2011, Etude expérimentale du comportement thermique d'une maison rurale à faible consommation d'énergie, 15ème journées internationales de thermique, Tlemcen, Algeria.

Derradji L., Imessad K., Amara M., Boudali Errebai F., 2017, A study on residential energy requirement and the effect of the glazing on the optimum insulation thickness. *Applied Thermal Engineering, 112, 975–985.*

Derradji L., BoudaliErrebai F., Amara M., 2017, Effect of PCM in Improving the Thermal Comfort in Buildings. *Energy Procedia (Elsevier), 107, 157–161.*

Dudás A., Terjék A., 2015, Efficiency assessment of posterior waterproofing systems of renovated porous limestone masonry work. *Technical Gazette, Vol 22, 1225–1236.*

Dudás A., Farkas L., 2014, Building physical, energetical and hygrothermal analysis of earth-sheltered building constructions. *Advanced Materials Research, Vol 899, 369–373.*

Diao R.D., Sun L., Yang F., 2018, Thermal performance of building wall materials in villages and towns in hot summer and cold winter zone in China. *Applied thermal engenering, Vol 128, 517–530.*

Lee J., Kim S., Kim J., Song D., Jeong H., 2018, Thermal performance evaluation of low-income buildings based on indoor temperature performance, *applied energy, Vol 221, 425–436.*

Solutions for Sustainable Development – Szita, Jármai & Voith (eds.)
© 2020 Taylor & Francis Group, London, ISBN 978-0-367-42425-1

Comparison of Sakiadis and Blasius flows using Computational Fluid Dynamic

M.M. Klazly & Gabriella Bognár
Department of Machine and Product Design, University of Miskolc, Hungary

ABSTRACT: CFD method has been applied using ANSYS FLUENT for the Blasius and Sakiadis flows over flat plate. The velocity profiles and the wall shear stresses for both flows have been presented and the difference in the local shear stress at the wall has been evaluated. The absolute value of the wall shear stress in Sakiadis flow is greater by 32.35% comparing to Blasius flow. The effect of Reynolds number on both flows in term of the average Nusselt number, the average heat transfer coefficient and the drag coefficient is also calculated. The results of the Nusselt number and heat transfer coefficient are higher in Sakiadis than what was obtained for Blasius flow.

1 INTROUDCTION

Blasius flow is a classical fluid mechanics problem that has been solved by Blasius in 1908 (Blasius. 1908). The fluid motion above a steady flat surface flow is produced by the uniform free stream. The solution to the Blasius problem has been given applying different numerical methods. These required excellent skills for obtaining the suitable algorithms and computing the equations of the flow. After Blasius, the flow over flat plate has been solved using Runge-Kutta method, where hand computation is performed by L. Howarth (Howarth. 1938), also finite-difference method was presented in (Zhang et al. 2009) (Asaithambi. 2005) (Bataller. 2010) to Falkner-Skan equation, and recently, many numerical methods are performed on the classical Blasius equation (Najafi et al. 2005) (Wang. 2004) (Abbasbandy. 2007). On the other hand, the motion of fluid can be produced by a flat plate continuously moving with constant velocity through quiescent fluid. This kind of flow has attracted considerable attention due to its application in several industrial processes, such as fabrication of adhesive tapes and extrusion of plastic sheets, etc. (see (Agassant et al. 1991)). The solution of this problem has been investigated first by Sakiadis (Sakiadis. 1961) using the boundary layer theory to determine the wall shear stress distribution because the wall shear stress is one of the most important parameters due to the direct effect on the driving force that is required to withdraw the plate. Following Sakiadis, Tsou et al. has performed an experiment on the Sakiadis flow and has proved that the solution of Sakiadis flow is physically realizable (Tsou et al. 1967). After an intensive investigation made in the literature no simulation has been found concerning the Sakiadis flow which is fundamental flow in fluid mechanics, and this is the goal of this work.

2 FORMULATION OF THE PROBLEM

The governing equation of motions for Blasius and Sakiadis flows can be formulated with equations (1) – (3) to consider the problem of hydrodynamic and thermal boundary layers over a flat plate.

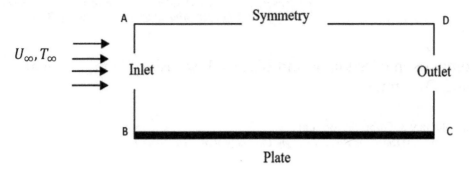

Figure 1. Blasius flow configuration.

The governing equations concerning the Blasius flow (see Figure 1) are as follows

$$\frac{\partial u}{\partial x} + \frac{\partial v}{\partial y} = 0, \tag{1}$$

$$u\frac{\partial u}{\partial x} + v\frac{\partial u}{\partial y} = \upsilon\frac{\partial^2 u}{\partial y^2}, \tag{2}$$

$$u\frac{\partial T}{\partial x} + v\frac{\partial T}{\partial y} = \alpha\frac{\partial^2 T}{\partial y^2}, \tag{3}$$

where
  u and v : components of the velocity in x and y directions, respectively,
  U_∞ : free stream velocity,
  T_∞ : temperature of free stream,
  α : thermal diffusivity of the fluid,
  υ : kinematic viscosity of the fluid.

For the Blasius flow the motion is produced by free stream with constant velocity $U_\infty = 1\text{m/s}$ and temperature $T_f = 300\text{K}$ (see Figure 1). The impermeable plate is stationary with $T_w = 400$ K.
  The equations (1)-(3) are subjected to the following boundary conditions:

at the plate as: $y = 0,$ $u = U_w = 0,$ $v = 0,$ $T = T_w,$ (4)

and far from the plate as: $y \to \infty,$ $u = U_\infty,$ $v = 0,$ $T = T_\infty.$ (5)

Figure 2 shows the Sakiadis flow, where the plate moves with constant velocity $U_w = 1\text{m/s}$ and $T_w = 400$ K. The fluid is quiescent ($U_\infty = 0m/s$) with temperature $T_\infty = 300$ K, and the boundary conditions are as follows:

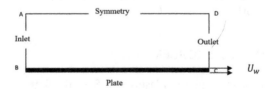

Figure 2. Sakiadis flow configuration.

130

at the plate as:
$$y = 0, \quad u = U_w, \quad v = 0, \quad T = T_w, \quad (6)$$

and far from the plate as:
$$y \to \infty, \quad u = U_\infty = 0, \quad v = 0, \quad T = T_\infty. \quad (7)$$

The Blasius and Sakiadis flow systems (1)-(2) can be solved using similarity method applying stream function $\psi = \sqrt{2vxU_\infty}\, f(\eta)$, where f is the non-dimensional stream function and $\eta = y\sqrt{\frac{U_\infty}{2vx}}$ is the similarity variable. Then, system (1)-(2) can be transformed into the well-known Blasius equation with $u = \frac{\partial \psi}{\partial y}$ and $v = -\frac{\partial \psi}{\partial x}$:

$$f''' + ff'' = 0, \quad (8)$$

with boundary conditions:

$$f(0) = 0, f'(0) = 0, f'(\eta_\infty) = 1. \quad (9)$$

It is known that for the Sakiadis flow the equation for the similarity function is the same as (8) while the boundary conditions are different:

$$f(0) = 0, f'(0) = 1, f'(\eta_\infty) = 0. \quad (10)$$

For the energy equation (3), the similarity transformation

$$\phi = \frac{T - T_\infty}{T_w - T_\infty}$$

is used. Then, equation (3) can be rewritten in the form:

$$\frac{\alpha}{v}\phi'' + \phi'f = 0, \quad (11)$$

with boundary conditions
$$\phi(0) = 1, \phi(\eta_\infty) = 0$$

for a suitably chosen finite value of η_∞. Applying Pr $= \frac{v}{\alpha}$ one gets from (11) that

$$\phi'' + \text{Pr}\ \phi'f = 0.$$

One can express the temperature and velocity components with similarity function f and ϕ as follows:

$$T = \phi(T_w - T_\infty) + T_\infty, \quad u = U_\infty f'(\eta), \text{ and } v = \sqrt{\frac{vU_\infty}{2x}}(\eta f' - f).$$

The analytical solution for the wall shear stress (τ_w), and drag coefficient (C_D), can be given by the expressions below (Schlichting et al. 2000):

$$\tau_w = \mu f''(0), \quad (12)$$

The drag coefficient is defined as:

$$C_D = \frac{F_x}{\frac{1}{2}\rho U_\infty^2 L}. \quad (13)$$

The flow is governed by non-dimensional parameters, that is, the Reynolds number, which is given by the following relationship (Schlichting et al. 2000).

$$\mathrm{Re} = \frac{\rho U_\infty L}{\mu}. \qquad (14)$$

These equations have been applied to find the analytical solution for Blasius and Sakiadis flows. The analytical solution has been used extensively in the literature, both for Blasius and Sakiadis flows (see for example Refs. (Sakiadis. 1961) (Tsou et al. 1967) (Bataller. 2010). Analytical solution is compared with the experimental one and the result of CFD simulation in the model validation.

3 COMPUTATIONAL FLUID DYNAMICS METHOD (CFD)

Our goal is to study the fluid flow properties of these two flows with CFD simulation using ANSYS Fluent R18.1. The geometry has been created using Design Modeler. After creating the two-dimensional geometry ANSYS fluent mesh is used to create the meshing and to create the name section of the model. The applied boundary conditions are the following according to the ANSYS fluent code: the boundary AB is defined as velocity inlet where, the horizontal velocity is constant, and the vertical velocity is zero. The boundary CD is defined as pressure outlet, where the static pressure is placed equal to ambient pressure and all other flow quantities are extrapolated from the interior. The plate BC is defined as wall, where both the horizontal and vertical velocities are zero, and the far field AD is considered as symmetry, where the velocity gradients in the vertical direction are forced to be zero. For the Sakiadis flow (see Figure 2) the boundary BC is defined as moving wall. The simulation is two-dimensional and laminar solver was used for solving the governing equation with second order unwind scheme. The pressure was set to standard and the solution was monitored using a residual monitor with convergence criteria 10^{-8}. Table 1 shows the fluid properties of the air.

Table 1. The physical properties of the air.

Name	Fluid properties	Value
density	ρ	1.225
dynamic viscosity	μ	1.7894×10^{-5}
length	L	1
ambient fluid velocity	U_∞	1.5
ambient fluid temperature	T_∞	300
wall temperature	T_w	400
thermal conductivity of the fluid	k	0.0242
Prandtl number	Pr	0.744176

4 MODEL VALIDATION

In Table 2 we present the values of the wall shear stress for the two flow cases. The results are compared with the analytical and experimental results. In the analytical solution, it was found that the absolute value of the wall shear stress is higher in Sakiadis flow than in Blasius flow by 33.63%, where the authors compared the value of $f''(0)$ which define the magnitude of the wall shear stress at the plate (see equation 12). The analytical solution shows a good agreement with experiment, where the wall shear stress was higher in Sakiadis flow about 34% and about 32.35% in the present simulation. After the above validation tests, next the result for the effect of Reynolds

Table 2. Comparison of the wall shear stress for the two flows (validation test).

Method	Wall shear stress	Difference [%]
Analytical (Bataller. 2010)	Higher in Sakiadis	33.63%
Present work (CFD)	Higher in Sakiadis	32.35%
Experiment (Tsou et al. 1967)	Higher in Sakiadis	34%

number between 2000-10000 on drag coefficient, average Nusselt number and average convection heat transfer will be presented.

5 RESULTS AND DISCUSSION

5.1 *Blasius flow*

The velocity distribution for Blasius flow over the plate is presented in Figure 3. The velocity at the plate is the lowest and it increases till became uniform in the inviscid area, which located above the boundary layer edge. The velocity vectors along the plate show that the velocity of fluid particles at the plate is zero due to the no-slip condition and this occurred due to the viscous effect, where the fluid particles stuck to the plate.

The comparison of velocities at various locations for Blasius flow is presented in Figure 4. it can be seen that maximum value of the velocity decreases from the left side to the right side far from the inlet (leading edge), the thickness of the boundary layer increases while the velocity is decreasing as far from the leading edge we move (see Figure 4).

5.2 *Sakiadis flow*

Figure 5 presents the velocity profiles obtained by CFD at a given x-location along the plate for Sakiadis flow. The velocity of the fluid near the plate is equal to the velocity of the wall and as we move above the plate the velocity decreases until becoming zero and this variation due to the domination of the viscous forces in the flow.

The influence of the distance x from the leading edge along the plate is shown in Figure 6, where the velocity change in the boundary layer is opposite to the Blasius flow.

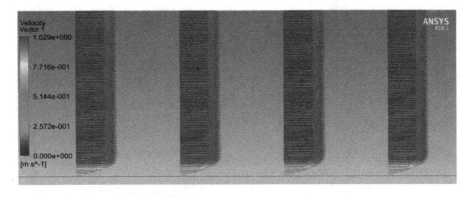

Figure 3. The velocity direction of Blasius flow.

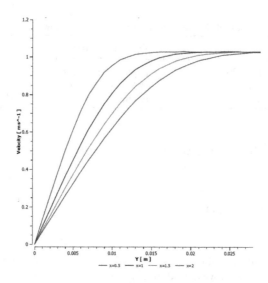

Figure 4. Velocity at different distances from the leading edge (Blasius flow).

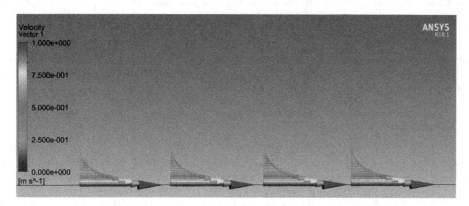

Figure 5. The velocity direction of Sakiadis flow.

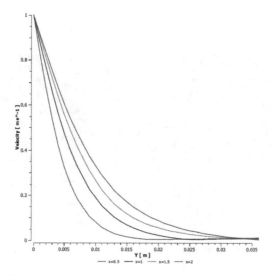

Figure 6. Velocity at different distances from the leading edge (Sakiadis flow).

Table 3.   The absolute value of the wall shear stress for the Blasius and Sakiadis flows.

X[m]	Sakiadis $[\tau_w]$	Blasius $[\tau_w]$	Difference [%]
0.1	0.007979154	0.00609105	30.99801
0.2	0.005117607	0.00362213	41.28707
0.3	0.004028016	0.00295360	36.3764
0.4	0.003438466	0.00256227	34.19577
0.5	0.003049742	0.00229598	32.82953
0.6	0.002769098	0.00210021	31.84819
0.7	0.002552552	0.00194816	31.02374
0.8	0.002382217	0.00182561	30.48861
0.9	0.002239298	0.00172409	29.88282
1	0.002119074	0.00163820	29.35365

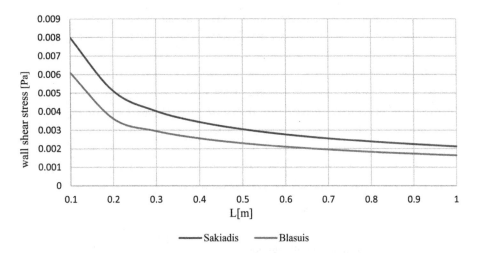

Figure 7.   Comparison of the wall shear stress.

5.3   *Wall shear stress*

The analysis of the wall shear stress for both fluid-flows is discussed here. Table 3 shows the result of the wall shear stress at different locations along the plate from the leading edge. The comparison between both flows shows that the wall shear stress in the Sakiadis flow is higher than in the Blasius flow (see Figure 7). The trend, where the magnitude of the shear stress is smaller in comparison to situation of moving flat plate was already predicted by Sakiadis theoretically (Sakiadis. 1961), and Tsou experimentally with this increase and he obtained an increasing about 34% in the wall shear comparison to Blasius (Tsou et al. 1967). In the present study, we found that the wall shear stress may increase by 32.35%.

5.4   *The effect of Reynolds number on drag coefficient*

The effect of Reynolds number on F_x and C_D has been analysed for different values of Reynolds number. Table 4 shows the effect of changing Reynolds number for both flows on the force acting in the x direction. It can be seen that as the Reynolds number increases the force that acting on the plate in the x direction increase and this occurred in both flows, however, the value of the force in the Sakiadis flow is much greater than in the Blasius flow. As a result, the drag coefficient which has been calculated using equation (13) is higher in case of the Sakiadis flow (see Figure 8). Table 5 represents the drag coefficient for both flows with variation of Reynolds number.

Table 4.   Comparison of the F_x for both flows for different Reynolds number.

Re	F_x (Sakiadis)	F_x (Blasius)
2000	-0.00024943911	0.0002217277
4000	-0.00041907231	0.0003644280
6000	-0.00056741406	0.0004880396
8000	-0.00070283638	0.0005999706
10000	-0.00082944862	0.0007051837

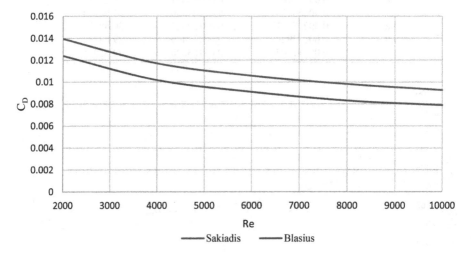

Figure 8.   Effect of Reynolds number on the drag coefficient.

Table 5.   Comparison of the drag coefficient for Sakiadis and Blasius flows.

Re	C_D (Sakiadis)	C_D (Blasius)
2000	-0.0139	0.0124
4000	-0.0117	0.0101
6000	-0.0105	0.0090
8000	-0.0098	0.0083
10000	-0.0092	0.0078

5.5   *Average Nusselt number* $\overline{Nu}_L$

The comparison of the value for the average Nusselt number shows that in the Sakiadis flow the value of $\overline{Nu}_L$ is higher than the value obtained for Blasius flow for the same fluid properties. Table 6 presents the results that obtained to analyse the effect of Reynolds number on the average Nusselt number, and it can be seen as Reynolds number increases the $\overline{Nu}_L$ increase. The comparison between both flows shows that the number is higher in Sakiadis flow for all the Reynolds number. Figure 9 represents the variation of the average Nusselt number with Reynolds number.

5.6   *Average convection heat transfer* $(\bar{h})$

Last, we are interested in the relation of the results of the average convection heat transfer with that of Blasius flow. For the uniform wall temperature, the respective values of the $\bar{h}$ for the same Reynolds number in the two flows cases show that the value $\bar{h}$ for continuous-surface

Table 6. The comparison of $\overline{Nu}_L$ for the two cases.

Re	$\overline{Nu}_L$ for Sakiadis	$\overline{Nu}_L$ for Blasius
2000	45.9120	40.9054
4000	54.2696	48.3072
6000	60.3136	53.2645
8000	64.5780	57.0709
10000	68.3271	60.2337

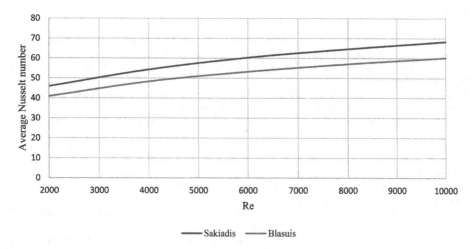

Figure 9. The effect of Reynolds number on the average Nusselt number.

Table 7. The comparison of the average convection heat transfer for the two cases.

Re	$(\bar{h})$ for Sakiadis	$(\bar{h})$ for Blasius	h
2000	1.111071	0.9899111	0.9362
4000	1.313325	1.169034	1.11372
6000	1.459591	1.289001	1.2325
8000	1.562788	1.381117	1.3244
10000	1.653516	1.457658	1.40

boundary layer exceeds that for the flat plate boundary layer. Average convection heat transfer of both flows is represented in Table 7 at different Reynolds number. It can be seen that the value of $\bar{h}$ in both cases increase when Reynolds number increase and the comparison between the two cases shows that the values of the $\bar{h}$ in Sakiadis flow is higher compared to that the value in Blasius flow.

6   CONCLUSIONS

In summary, numerical approach has been used to give suitable solutions to two classical boundary layer problems in fluid mechanics. Our computations verify that the proposed procedure offers an effective tool for solving this nonlinear problem in fluid mechanics using CFD method. The main findings of this study are summarized as below:

I.   The boundary layer thickness increases while wall shear stress decreases with increasing distance along the plate for both flows.
II. The velocity rises in Sakiadis flow and decreases in Blasius flow with increasing plate length.
III. The wall shear stress is higher in Sakiadis flow than in Blasius flow.
IV. Increasing Reynolds number impacts a decrease in the drag coefficient and an increase in the average convection heat transfer for both flows.

REFERENCES

Abbasbandy, S. 2007. A numerical solution of Blasius equation by Adomian's decomposition method and comparison with homotopy perturbation method, Chaos, *Solitons and Fractals*. 31. 257–260.

Agassant, F.Avens, P. Sergent, J., and Carreau, P.J. 1991. Polymer Processing: *Principles and Modeling*, Hanser Publishers, Munich.

Asaithambi, A. 2005. Solution of the Falkner – Skan equation by recursive evaluation of Taylor coefficients, *Computational and Applied Mathematics*. 176(2005), 203–214.

Bataller, R.C. 2010. Numerical Comparisons of Blasius and Sakiadis Flows, *MATEMATIKA*, 26, 187–196.

Blasius, H. 1908. Grenzschichten in Flüssigkeiten mit kleiner Reibung, *Z. Math. Phys*. 56, 1–37.

Howarth, L. 1938. On the solution of the laminar boundary layer equations, *The Royal Soc. London*, A 164(1938), 547–579.

Najafi, M., Khoramishad, H., Massah, H., and Moghimi, M. 2005. A study of Blasius viscous flow: an ADM Analytical Solution, Applied *Mathematics*. 9, 57–61.

Sakiadis, B. 1961. Boundary layer behavior on continuous solid surfaces: Boundary layer equations for two-dimensional and axisymmetric flow, *AIChE* J. 7, 26.

Schlichting, H. Gersten, K. 2000. *Boundary layer theory*, Springer Verlag, Berlin Heidelberg New York.

Tsou, E.S. and Goldstein, R, Flow and Heat Transfer in the Boundary Layer on a Continuous Moving Surface. *Int. J. Heat Mass Transfer*, 1967. 10: 219–235.

Wang, L. 2004. A new algorithm for solving classical Blasius equation, *Applied Mathematics and Computation*. 157, 1–9.

Zhang, J. and Chen. B. 2009. An iterative method for solving the Falkner–Skan equation, *Appl. Math. Comput*. 210, 215–222.

Solutions for Sustainable Development – Szita, Jármai & Voith (eds.)
© 2020 Taylor & Francis Group, London, ISBN 978-0-367-42425-1

Techniques for evaluation of mixing efficiency in an anaerobic digester

Buta Singh, Zoltán Szamosi & Zoltán Siménfalvi
Institute of Energy Engineering and Chemical Machinery, University of Miskolc, Hungary

ABSTRACT: Mixing is a very crucial factor which has a significant effect on the efficiency of anaerobic digester. Mixing in digester is affected by various internal and external factors such as hydrodynamics of digester and impellers, viscosity of slurry, temperature and total solid content. In this study, various methods and techniques have been described that can be used to evaluate the effectiveness of any design employed for mixing in an anaerobic digester. The efficacy of different methods has been analyzed. Various numerical and empirical models have been developed in the past. Being a chemical process, mixing efficiency in anaerobic digester cannot be only described by numerical methods because the actual effect of shear stresses on the organisms can be determined by experimental methods. With a better knowledge of the relationship between mixing and biogas production rates, the overall process efficiency can be improved and optimized.

Keywords: biogas, mixing efficiency, evaluation, slurry rheology

1 INTRODUCTION

Waste to energy is one of the trending sources of renewable energy. Anaerobic digestion is one of the technologies of converting organic waste to energy. Efficiency of anaerobic digester depends on various factors such as chemical and physical properties of the substrate, pH, temperature, OLR, HRT and mixing. All of the above mixing is very necessary to enhance the efficiency of biogas plant. Advantages of mixing include the homogenous distribution of nutrients throughout the active volume of digester, avoiding dead zones, floating layers and temperature gradient. Effect of mixing is very significant for the digesters with higher solid content and lower HRT.

Mixing in an anaerobic digester has been intensively studied by many researchers in the past few years but still, it is debatable subject (Singh, Szamosi 2019). From the literature, it can be concluded that intermittent mixing is best in terms of biogas production and power consumption. But mixing time and intensity is still to be optimized because these two parameters directly depend on the design of the setup and the rheological properties of slurry in the digester. Due to this large number of studies can be found on analysis of mixing but the results of every study vary due to variation in the above given factors. Many numerical, empirical and experimental methods have been adopted by the researchers.

Anaerobic digestion is a very complex biological process which involves many steps before the biogas is produced. Every research corresponds to a different approach to evaluate mixing. Cfd modelling is one of the trending criteria to determine the flow patterns used by many researchers (Mohammadrezaei, Zareei, and Behroozi-Khazaei 2018)(Sindall, Bridgeman, and Carliell-Marquet 2013). In many case biogas production rates is the only parameter to access the effect of mixing (Rico et al. 2011)(Bello-Mendoza and Sharratt 1998)which is not the best approach because without knowing the exact geometry of the digester it is not possible to determine the flow pattern and the shear stress exerted by impellers on the micro-organisms.

Moreover, in case of evaluation of mixing it is very hard to reach the precise results because of the involvement of various biological, chemical, mechanical and hydro-dynamic aspects. For instance, flow patterns and impeller mixing can be analyzed by a mechanical or hydrodynamic expert but the effect of various shear stresses on micro-organisms can be assessed by a micro biologist. So, it requires a multidisciplinary col-laboration of experts from different fields and different approaches to evaluating the effect of mixing. This article is representing the detailed criteria of analysis of the evaluation of mixing in an anaerobic digester at both lab-scale and large-scale biogas plants.

2 MIXING IN AN ANAEROBIC DIGESTER

Mixing in an anaerobic digester can be achieved by three methods: mechanical mixing, biogas recirculation and slurry recirculation refer to Table 1. Mechanical mixing refers to the use of impeller submerged in the slurry propelled by an external motor to mix the slurry. In case of biogas recirculation, the biogas from the head of digester is compressed by a pump and then diffused throughout the digester by nozzles. Similarly, the slurry is drawn from the center of digester by the external centrifugal pump and diffused through nozzles throughout the active volume of digester.

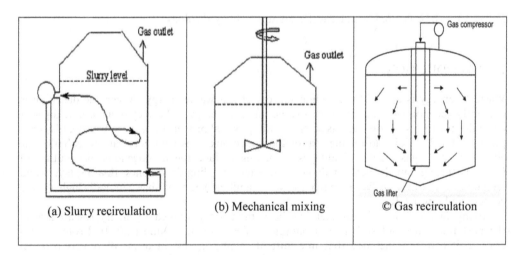

Figure 1.   Various types of mixing techniques for mixing in an anaerobic digester.

3 EVALUATION OF MIXING

As mixing is a very complex process in case of anaerobic digester many different types of evaluation methods can be applied. These methods include tracer methods, CFD simulation and modelling and lab experiments. The fluid flow in the digester at various mixing intensities and varying hydrodynamic designs can be found by the computer modelling but the actual impact of shear rates and shear stress can be only known by the lab experiments. So, it is always recommended to combine both numerical and experimental results to find the opti-mum mixing regime for the anaerobic digester.

3.1 Theoretical considerations

Rheological study of slurry for anaerobic digestion process is a very important aspect to design the digester, mixing and transport equipment. From the literature data, it is confirmed that if TS>2.5% then the sludge possesses non-Newtonian shear thinning behavior. In the laminar regime (approximately < 10-100). For this instance, the power law model can be proposed to calculate the apparent viscosity and shear rate.

$$\mu_a = K \cdot \gamma_a'^{(n-1)} \tag{1}$$

For a non-Newtonian shear thinning the value of n is always less than 1. For this instance, the rheological data for the waste water sludge is taken from the literature presented in Table 3 (Cao 2016). The average shear rate inside the vessel can be calculated as per the equation

$$\gamma_a' = k_s \cdot N \tag{2}$$

Here k_s is Otto-Metzer constant which is directly associated with the impeller geometry. Further, shear stress can be calculated as per equation

$$\tau = K \cdot \gamma_a'^{(n)} \tag{3}$$

$$R_e = \frac{\rho \cdot N \cdot D^2}{\mu_a} \tag{4}$$

The relationship between speed of impeller, rheological characteristics and Reynolds number can be expressed by the following equation.

$$N_P = \frac{P}{\rho \cdot N^3 \cdot D^5} = \frac{K_p}{R_e} \tag{5}$$

$$P = 2\pi \cdot N \cdot C \tag{6}$$

K_p constant depends on the design of impeller.

3.2 CFD simulations

Computer aided experiments are always a better option as compared before fabrication of prototype as it can save time and initial investment for the designing of any industrial equipment. Computational fluid dynamics is the science of prediction of fluid flow, heat and mass transfer and related phenomena. This section addresses the use of CFD, mass balances and kinetic models in evaluating and improving mixing. Large number of studies can be found implementing CFD models of lab-scale as well as large scale anaerobic digesters to evaluate the performance of different designs of mixers and mixing intensity. CFD can be used to determine the flow patterns, velocity and viscosity distribution, turbulence, particle trajectories, movement of dissolved components and dead zones. Validation of CFD model against the experimental data is very important to get more accurate and precise results.

For the CFD modelling first the geometry of the experiment setup is constructed in a designing software. The second step is to fit the entire geometry with the mesh and divide the entire volume into smaller cells by providing boundary conditions for inlet, outlet and walls. The properties of different phases are defined and depending on whether the problem is single phase or multiphase, different solvers and turbulence models are selected to calculate how the phase/phases are affected by the geometry and boundary conditions in each individual cell defined by the mesh.

Table 1. Different models used by researchers in CFD simulations to analyze effect of design and intensity on particle velocity and flow patterns.

Study	Model	Rheology of slurry
H. Caillet et al.(Caillet, …, and 2018 2018)	LES- turbulence model	Newtonian and non-Newtonian
Binxin Wu et al.(Wu 2010)	k–ε turbulence model	Non-Newtonian
J. Ding et al. (Ding et al. 2010)	k–ε turbulence model	Non-Newtonian
R. Meroney et al.(Meroney and Colorado 2009)	k–ε turbulence model	-
J. Bridgeman (Bridgeman 2012)	Standard k–ε (S k–ε), Realizable k–ε (Rk–ε),	Non-Newtonian
Vesvikar et al.(Mehul S. Vesvikar Muthanna Al-Dahhan 2005)	k–ε turbulence model	-

Many different CFD software has been used by researchers such as ANSYS Fluent, SPH-flow etc (Lebranchu et al. 2017). Many studies show that flow predictions are particularly sensitive to the turbulence model implemented Thereby, the choice of the turbulence model is important in the simulations. Mainly Reynolds- average Navier-Stokes (RANS) simulations were conducted on digesters. The SST k-omega model the standard k-epsilon model and the realizable k-epsilon model are recommended by many authors among the RANS models. Furthermore, other authors claim that the Reynold stress model (RSM) is the most suitable model to predict the behavior of this bioreactor.

3.3 *Power consumption*

Efficiency on a biogas plant rests in terms of conversion of waste of biogas and the overall power consumption by a plant itself. Significant amount of energy is utilized by a biogas plant for various purposes such as transportation of slurry, heating, mixing and feeding. 29-54% of overall power need of biogas plant is utilized by the mixing equipment (Dachs G 2006) (Naegele et al. 2012). Mixing intensity is defined as power used per unit volume (P/V). So, it is always a major challenge to homogeneous mix the slurry in the digester with minimum power consumption. This can be achieved by optimizing the digester design and geometry of mixing equipment. According to the United States EPA 5.8 W/m^3 of digester volume is recommended as power input for mixing in anaerobic digester but it is still conflicting subject (Kariyama, Zhai, and Wu 2018). The below give equations can be referred to determine the power consumption by different mixing methods.

Impeller mixing (Conti et al. 2019)

$$P_{shaft} = \frac{2\pi N}{60}\left(T_{substrate} - T_{air}\right) \tag{7}$$

Biogas recirculation

$$\frac{P}{V} = \frac{\lambda G_r P_2}{\lambda - 1}\left[\left(\frac{P_1}{P_2}\right)^{(\lambda-1)/\lambda} - 1\right] \tag{8}$$

Slurry recirculation

$$P = \rho g H Q \tag{9}$$

On the other side, the specific electricity consumption by stirring is also an interesting aspect to determine the mixing efficiency in a biogas plant. Specific stirring electric power consumption can be evaluated by two methods (DACHS, G. and REHM 2006):

i. Comparing power consumption to the active digester volume and mixing time

$$E_{speci.V} = \frac{\sum E_{stirrer,d}}{V_{active\ digester}} \quad (10)$$

i. Comparing power consumption to daily added feedstock

$$E_{spec.FM} = \frac{\sum E_{stirrer,d}}{\dot{m}_{FM,d}} \quad (11)$$

3.4 Calculation of dead zones

The degree of agitation depends on the preservation of homogeneity within the digester. Usually, dead zone volume is used to analyze mixing, but it depends how dead zone is interpreted on the basis of local velocity gradient, local velocity data and RTD. According to Vesvikar et al. (Mehul S. Vesvikar Muthanna Al-Dahhan 2005) dead zone is a zone in which velocity of particles remain lower than 5% of the highest velocity within the digester. By inadequate mixing of slurry in digester the stagnant zones can prevail in the reactor which can result in ineffective decomposition of slurry and there can loss of digestion process. A perfect digester should have approach of 100% mixed volume with zero dead zones (Tenney, M. W. & Budzin 1972). Results of few studies on dead zones while mixing in an anaerobic digester with different impeller configurations are discussed in this section.

Wolf and Resnick (1963) proposed a general washout equation to determine dead zones by calculating the fractional volume of cells with low liquid velocities.

$$\frac{C_{so}(t)}{C_o} = \exp\left(-\frac{1-f}{ar(1-d)T_{HRT}}\left(t - L - \frac{p(1-f)rT_{HRT}}{1-f} + \beta ar(1-d)T_{HRT}\right)\right) \quad (12)$$

3.5 Biogas production rates

The final product of anaerobic digestion is biogas. During the evaluation of different mixing techniques, the biogas production rates are measured continuously on a daily basis in every experiment in the literature. Cumulative analysis of biogas production rates is done from the day of starting until the failure of anaerobic digester by liquid displacement method. This method is one of the simplest and accurate methods to determine the volume of biogas produced at lab-scale experiments. Methane content is generally determined by gas chromatography.

The volume of gas collected, and the gas laws can be used to calculate the number of moles of gas collected. The experimental arrangement demonstrates one minor complication. The gas pressure inside the cylinder 1 is the sum of biogas pressure and water vapours. The water in the cylinder will reach an equilibrium state where the number of molecules leaving the surface will be the same as the number returning. According to Dalton's law of partial pressure:

$$P_T = P_{bio} + P_{H_2O} \quad (12)$$

From the above equation, P_{bio} can be easily calculated as

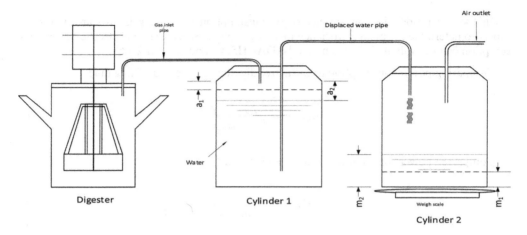

Figure 2.   Schematic diagram of biogas volume measurement apparatus.

Table 2.   Representation of data referring to the various approaches adopted by researchers to evaluate the effect of mixing in an anaerobic digester.

Reference	Digester scale	V	Numerical Appr.	Empirical Approach	Digester Geometry	Mixer Geometry	Microbial analysis	Biogas yield analysis	CH$_4$ yield analysis
(Sindall, Bridgeman, and Carliell-Marquet 2013)	Lab-S	6 l	CFD	♦	♦	♦	♦	♦	o
(Luo and Angelidaki 2013)	Lab-S	1 l	o	♦	o	o	♦	♦	♦
(Rico et al. 2011)	PS	1.5 m^3	o	♦	o	o	o	♦	♦
(Tian et al. 2014)	Lab-S	5 L	o	♦	o	♦	♦	o	♦
(Bello-Mendoza and Sharratt 1998)	n.a.	n.a.	MATLAB	♦	o	o	o	o	o
(Stalin et al. 2007)	PS	168 l	o	♦	♦	n.a.	o	♦	o
(Andrew G. Hashimoto 1982)	Lab-S	4 l	o	♦	o	♦	o	♦	♦
(Ong, Greenfield, and Pullammanappallil 2002)	Lab-S	10 l	o	♦	o	o	o	♦	♦
(K. C. Lin and 1991)	Lab-S	7 l	o	♦	♦	♦	o	♦	♦
(Zhai, Kariyama, and Wu 2018)	PS	1.6 m^3	CFD	♦	♦	♦	o	o	♦
(Noorpoor and Dabiri 2018)	LS	30 m^3	CFD	o	♦	♦	o	o	o
(Vavilin 2007)	Lab-S	1 l	CFD	♦	o	o	♦	o	♦
(Sulaiman et al. 2009)	PS	500m3	o	♦	o	o	♦	♦	♦

♦: data available; o: missing elements; n.a.: data not available

$$P_{H_2O} = 6.1121 \ \exp\left\{ \left(18.678 - \frac{T_c}{234.5} \right) \times \frac{T_c}{257.14 + T_c} \right\} \qquad (13)$$

Equation of gas volume calculation by weighing displaced water from cylinder 2

$$V_o = \frac{T_o \cdot (m_b - m_a)}{T \cdot P_o \cdot \rho} \left[P_T - P_{H_2O} + \rho \cdot g \left(a_1 + a_2 + \frac{V_a}{A} \right) \right] \qquad (14)$$

Here V_o is volume of gas produced at standard temperature and pressure, T_o is standard temperature, a and b represents the height of gas and liquid respectively, m is mass of liquid measured, 1 and 2 represent initial and final height, ρ represents density of water, A is area of cross section, g is acceleration due to gravity.

4 CONCLUSION

Mixing is very crucial to enhance the biogas plant efficiency. It can be concluded that evaluation of mixing in an anaerobic digester is a very complex subject due to a variety of fields involved. It requires the interdisciplinary collaboration of engineers, biologists, chemists and hydrodynamic experts to clearly define the statement for optimum mixing in an anaerobic digester. Experimentation requires huge setup for a parallel set of experiments to analyze mixing with variation of speed, design and other operational parameters. In this article different approaches by researches have been analyzed. It is concluded that the only numerical methods and computer modelling are not enough to evaluate mixing because the effect of shear stresses and flow patterns can be only determined by the actual experiments.

NOMENCLATURE

ρ	Density [kg m^{-3}]
γ'_a	Average shear rate [s^{-1}]
R_e	Reynolds number
K	Consistency index
k_s	Otto-Metzner constant
D	Diameter of impeller [m]
N_P	Power number
AD	Anaerobic digestion
HRT	Hydraulic retention time [days]
$E_{stirrer,d}$	Daily total stirring electric energy consumption [kWh$_{el}$/d]
$V_{Active\ digester}$	Active digester volume [m^3]
$E_{spec.FM}$	$E_{speci.}$ per tonne added feedstock [kWh$_{el}$/t$_{FM}$]
P_T	Total pressure of cylinder [Pa]
P_{bio}	Pressure of biogas [Pa]
T_c	Temperature of gas [°C]
P	Power [W]
τ	Shear stress [Pa]
μ_a	Apparent viscosity [Pa s^{-1}]
n	Power number
N	Revolution per min [rpm]
C	Experimental torque [Nm]
CFD	Computational fluid dynamics
OLR	Organic loading rate
$\dot{m}_{FM,d}$	Daily added feedstock
$E_{speci.}$	Specific stirring electric energy consumption (kWh$_{el}$)

$E_{speci.V}$ $E_{speci.}$ per 100 m^3 active digester volume [kWh$_{el}$/100m$^3_{active\ digester.d}$]

$E_{stirrer,d}$ Daily total stirring electric energy consumption [kWh$_{el}$/d]

P_{H_2O} Pressure of water vapours [Pa]

REFERENCES

Andrew G. Hashimoto. 1982. "Effect of Mixing Duration and Vacuum on Methane Production Rate from Beef Cattle Waste." *Biotechnology and Bioengineering* 24 (1): 9–23. doi:10.1002/bit.260240103.

Bello-Mendoza, Ricardo, and Paul N Sharratt. 1998. "Modelling the Effects of Imperfect Mixing on the Performance of Anaerobic Reactors for Sewage Sludge Treatment." *Journal of Chemical Technology & Biotechnology* 71 (2): 121–130. doi:10.1002/(SICI)1097-4660(199802)71:2<121::AID-JCTB836>3.0.CO;2-7.

Bridgeman, J. 2012. "Computational Fluid Dynamics Modelling of Sewage Sludge Mixing in an Anaerobic Digester." *Advances in Engineering Software* 44 (1): 54–62. doi:10.1016/j.advengsoft.2011.05.037.

Buta, Singh; Zoltán, Szamosi; Zoltán, Siménfalvi; 2019. "State of the Art on Mixing in an Anaerobic Digester: A Review." *Renewable Energy* 141(C): 922–936. doi:10.1016/j.renene.2019.04.072.

Caillet, H, ... A Bastide - ... on Engeneering for, and Undefined 2018. 2018. "Anaerobic Digestion of Vinasse and Cfd Modelling Approach." *Hal.Archives-Ouvertes.Fr*, no. February. https://hal.archives-ouvertes.fr/hal-01857234/.

Cao, Xiuqin. 2016. "Rheological Properties of Municipal Sewage Sludge: Dependency on Solid Concnetration and Temperature." *Procedia Environmental Science*, 113–121. doi:10.1016/j.proenv.2016.02.016.

Conti, Fosca, Leonhard Wiedemann, Matthias Sonnleitner, Abdessamad Saidi, and Markus Goldbrunner. 2019. "Monitoring the Mixing of an Artificial Model Substrate in a Scale-down Laboratory Digester." *Renewable Energy* 132: 351–362. doi:/10.1016/j.renene.2018.08.013.

DACHS, G. and REHM, W. 2006. "Eigenstromverbrauch von Biogasanlagen Und Potenziale Zu Dessen Reduzierung." Solarenergieförderverein Bayern. 2006. http://www.sev-bayern.de/content/bio-eigen.pdf.

Dachs G, Rehm W. 2006. "Der Eigenstromverbrauch von Biogasanlagen Und Potenziale Zu Dessen Reduzierung." In *Solarenergieförderverein Bayern*. Munich.

David wolf and william Resnick. 1963. "Residence Time Distribution in Real Systems." *Industrial & Engineering Chemistry Fundamentals* 2 (4): 287–293.

Ding, Jie, Xu Wang, Xue-Fei Zhou, Nan-Qi Ren, and Wan-Qian Guo. 2010. "CFD Optimization of Continuous Stirred-Tank (CSTR) Reactor for Biohydrogen Production." *Bioresource Technology* 101 (18): 7005–7013. doi:10.1016/J.BIORTECH.2010.03.146.

K.C. Lin and, M.E.J. Pearce. 1991. "Effects of Mixing on Anaerobic Treatment of Potato-Processing Wastewater." *Canadian Journal of Civil Engineering* 18 (3): 504–514. doi:10.1139/l91-061.

Kariyama, Ibrahim Denka, Xiaodong Zhai, and Binxin Wu. 2018. "Influence of Mixing on Anaerobic Digestion Efficiency in Stirred Tank Digesters: A Review." *Water Research* 143: 503–517. doi:10.1016/j.watres.2018.06.065.

Lebranchu, Aline, Stéphane Delaunay, Philippe Marchal, Fabrice Blanchard, Stéphane Pacaud, Michel Fick, and Eric Olmos. 2017. "Impact of Shear Stress and Impeller Design on the Production of Biogas in Anaerobic Digesters." *Bioresource Technology* 245 (June): 1139–1147. doi:10.1016/j.biortech.2017.07.113.

Luo, Gang, and Irini Angelidaki. 2013. "Co-Digestion of Manure and Whey for in Situ Biogas Upgrading by the Addition of H2: Process Performance and Microbial Insights." *Applied Microbiology and Biotechnology* 97 (3): 1373–1381. doi:10.1007/s00253-012-4547-5.

Mehul S. Vesvikar Muthanna Al-Dahhan. 2005. "Flow Pattern Visualization in a Mimic Anaerobic Digester Using CFD." *Biotechnology and Bioengineering* 89 (6): 719–732. doi:10.1002/bit.20388.

Meroney, Robert N., and P.E. Colorado. 2009. "CFD Simulation of Mechanical Draft Tube Mixing in Anaerobic Digester Tanks." *Water Research* 43 (4): 1040–1050. doi:10.1016/j.watres.2008.11.035.

Mohammadrezaei, Rashed, Samira Zareei, and Nasser Behroozi-Khazaei. 2018. "Optimum Mixing Rate in Biogas Reactors: Energy Balance Calculations and Computational Fluid Dynamics Simulation." *Energy* 159: 54–60. doi:10.1016/j.energy.2018.06.132.

Naegele, Hans Joachim, Andreas Lemmer, Hans Oechsner, and Thomas Jungbluth. 2012. "Electric Energy Consumption of the Full Scale Research Biogas Plant 'Unterer Lindenhof': Results of Long-term and Full Detail Measurements." *Energies* 5 (12): 5198–5214. doi:10.3390/en5125198.

Noorpoor, Alireza, and Soroush Dabiri. 2018. "Sedimentation and Mixing Analysis in Cattle Manure Feedstock in a Stirred Tank of Anaerobic Digestion Sedimentation and Mixing Analysis in Cattle Manure Feedstock in a Stirred Tank of Anaerobic Digestion," no. March.

Ong, H.K., P.F. Greenfield, and P.C. Pullammanappallil. 2002. "Effect of Mixing on Biomethanation of Cattle-Manure Slurry." *Environmental Technology (United Kingdom)* 23 (10): 1081–1090. doi:10.1080/09593332308618330.

Rico, S, José Luis Rico, Noelia Muñoz, Beatriz Gómez, and Iñaki Tejero. 2011. "Effect of Mixing on Biogas Production during Mesophilic Anaerobic Digestion of Screened Dairy Manure in a Pilot Plant." *Engineering in Life Sciences* 11 (5): 476–481. doi:10.1002/elsc.201100010.

Sindall, R., J. Bridgeman, and C. Carliell-Marquet. 2013. "Velocity Gradient as a Tool to Characterise the Link between Mixing and Biogas Production in Anaerobic Waste Digesters." *Water Science and Technology* 67 (12): 2800–2806. doi:10.2166/wst.2013.206.

Stalin, N, HJ Prabhu, S.V.R. Kumar, S.S. Nagaraju, N.T. Binh, N.C. Thanh, C.S. Rao, P.S. Raju, G.A. E.S. Kumar, and others. 2007. "Performance Evaluation of Partial Mixing Anaerobic Digester." *ARPN J. Eng. Appl. Sci* 2 (3): 1–6. http://arpnjournals.com/jeas/research_papers/rp_2007/jeas_0607_43.pdf.

Sulaiman, Alawi, Mohd Ali Hassan, Yoshihito Shirai, and Suraini Abd-aziz. 2009. "The Effect of Mixing on Methane Production in a Semi-Commercial Closed Digester Tank Treating Palm Oil Mill Effluent." *Australian Journal of Basic and Applied Sciences* 3 (3): 1577–1583.

Tenney, M.W. & Budzin, G.J. 1972. "No Title." *Water Wastes Eng.*, 9 (5), 57-.

Tian, Zhuoli, Léa Cabrol, Gonzalo Ruiz-Filippi, and Pratap Pullammanappallil. 2014. "Microbial Ecology in Anaerobic Digestion at Agitated and Non-Agitated Conditions." *PLoS ONE* 9 (10). doi:10.1371/journal.pone.0109769.

Vavilin, V.A. 2007. "Anaerobic Digestion of Solid Material: Multidimensional Modeling of Continuous-Flow Reactor With Non-Uniform Influent Concentration Distributions." *Biotechnology and Bioengineering* 97 (Ii): 354–366. doi:10.1002/bit.

Wu, B. 2010. "CFD Simulation of Mixing in Egg-Shaped Anaerobic Digesters." *Water Research* 44 (5),: 1507–1519.

Zhai, Xiaodong, Ibrahim Denka Kariyama, and Binxin Wu. 2018. "Investigation of the Effect of Intermittent Minimal Mixing Intensity on Methane Production during Anaerobic Digestion of Dairy Manure." *Computers and Electronics in Agriculture* 155 (September): 121–129. doi:10.1016/j.compag.2018.10.002.

Part C: Waste Management and Reverse Logistics

Solutions for Sustainable Development – Szita, Jármai & Voith (eds.)
© *2020 Taylor & Francis Group, London, ISBN 978-0-367-42425-1*

Vehicle routing in drone-based package delivery services

A. Agárdi & L. Kovács
Department of Information Technology, University of Miskolc, Miskolc, Hungary

T. Bányai
Institute of Logistics, University of Miskolc, Miskolc, Hungary

ABSTRACT: The Industry 4.0 enables the operation of new logistics equipment and new logistic solutions based on cyber-physical systems. The usage of new transport devices is discussed in this article in case of the Vehicle Routing Problem. In basic Vehicle Routing Problems, the demands of customers must be served. This article discusses a new Vehicle Routing Problem where drones are also used to improve the flexibility, reliability and cost efficiency of the whole last-mile solution. During the problem, a drone is assigned to the truck performing package delivery services, and some customers are visited by the truck while the drone is sent to certain customers located in the near environment of the truck. The drone has a capacity limit, so it is able to visit only a limited number of customers. After visiting the customers, the drone recovers the goods from the truck and serves the demands of new customers. In this article, the above mentioned problem is solved with construction and improvement algorithms. The construction algorithms are the Nearest Neighbour Algorithm and Arbitrary Insertion Algorithm, the improvement algorithms are the Genetic Algorithm and the Hill Climbing Algorithm.

Keywords: Vehicle Routing Problem, drone, construction heuristics, improvement heuristics, last mile logistics

1 INTRODUCTION

The Industry 4.0 enabled the use of new logistic equipment. Thus, not only the in-plant and out-of-plant material flow processes are varied, but also the entire supply chain has altered. The use of these logistic equipment requires new mathematical models of vehicle routing problems and algorithms to solve them. Such new equipment are for example the drones. The drones can be applied primary in out-plant material handling. This article describes a vehicle routing where the truck has a single drone. In the problem the positions of the customers to be visited were given. We also know the demand of the customers. A single truck with a single drone serves the demands of the customers. The drone visits all customers and has capacity limit. The drone starts the route with the truck. Then, from the truck pick up some demands (it depends on the capacity limit of the truck), and serves some demand of the customers. After the drone served all the loaded products, returns to the truck, and travel with the truck to the next customer. All customer must be served exactly once. The truck and the drone returns their route to the first visited customer. This type of vehicle routing is cost-effective because not all customers are served by the truck. The drone consumes less fuel than the truck.

2 VEHICLE ROUTING PROBLEM

The Vehicle Routing Problem (VRP) (Laporte 1992) is a logistical problem that deals with the delivery and collection of goods. The position of a depot is known in the classical problem, and the positions and demands of the customers are also known. Also, the number and capacity of the vehicles are known. Vehicles deliver the goods from the depot to the customers and then return to the depots. In case of the classical problem, the goal function is the minimization of the length of the route. There are many variations of the task that adapt to industrial needs.

- In case of the Multi-Depot Vehicle Routing Problem (Cordeau 1997), the demand of the customers can be served from multiple depots.
- In case of Homogenous fleet Vehicle Routing Problem (Dondo et. al. 2007), one type of vehicles are used, while in case of Heterogenous fleet Vehicle Routing Problem (Gendreau et. al. 1999) several types of vehicles serve the demands of the customers.
- In case of Multi-Echelon Vehicle Routing Problem (Dondo et. al. 2011), the depot does not directly serve the customers. The products are transported from the depot to intermediate locations, called satellites. From the satellites, the products are delivered to the customers. The level of satellites can be also defined. Then the products are transported to the first level of satellites to the second level of satellites, then from the second level of satellites to the third level of satellites and so on, and last to the customers.
- In case of Fuel Consumption Vehicle Routing Problem (Xiao et. al. 2012) not the minimization of the length of the route, but the minimization of the fuel consumption is the goal function.
- In case of Vehicle Routing Problem with Time Window (Desrochers et. al. 1992), the customers must be visited within a time interval.
- In case of the Open Vehicle Routing Problem (Brandão, 2004) the vehicles start their route from the depot, but they do not return to the depot.
- In case of the Vehicle Routing Problem with Inter-Depot Routes (Crevier et. al. 2007), several depots are applied. The vehicles start their routes from one of the depots, visits the customers and then returns to one of the depots (not necessarily to the depot from which the vehicle started their route).
- In case of Vehicle Routing Problem with Pickup and Delivery (Min, 1989) the goods are transported from the depot to the customers while other types of goods are collected from the customers and transported to the depot.
- In case of Fuzzy Vehicle Routing Problem (Cao et. al. 2010), some parameters are not discrete numbers, the parameters are fuzzy numbers. Fuzzy numbers can be the demand of the customers, the time windows of the customers, the time of the route between two customers and so on.
- In case of Stochastic Vehicle Routing Problem (Gendreau et. al. 1996) some parameters are not known in advance, only the distribution of the parameters are known. These parameters can be the demand of the customers, the time windows and so on.
- In case of Truck and Trailer Vehicle Routing Problem (Chao 2002) there are two types of vehicles. Some customers can be visited only with truck, some can be visited with trailer, and some can be visitied with both the truck and trailer.
- In case of Multi-Product Vehicle Routing Problem (Coelho 2013) the customers have multiple product demands which must be served.
- In case of Vehicle Routing with Inter-Depot Route (Crevier et. al. 2007), the vehicles must not return to the same depot from which they started their route.
- In case of Periodic Vehicle Routing Problem (Angelelli et. al. 2002) the customers' demand must be served not once, but periodically.
- In case of Cumulative Capacitated Vehicle Routing Problem (Ke et. al. 2013) not the minimization of the length of the route but the minimization of the waiting time is the goal function.
- In case of the Vehicle Routing Problem with Cross-Docking (Wen et. al. 2009) first, all of the products are picked up from the customers, then all of the products are served to the customers.
- In case of Selective Vehicle Routing Problem (Aras et. al. 2011) not all customers must be served, but only the customers which are profitable.
- In case of Vehicle Routing Problem with Perishable Food Products Delivery (Hsu et. al. 2007), the products have some expiration date which must be taken into account.

3 IMPLEMENTED ALGORITHM

3.1 *Construction algorithms*

The construction algorithms take locally the best steps, but their exclusive usage does not lead to the global optimum. The running time of these algorithms are short. In this article two construction algorithms are discussed, These are the Nearest Neighbour Algorithm (NNA) and the Arbitrary Insertion Algorithm (AIA). The pseudo code of the NNA can be seen in Figure 1. The pseudo code of the AIA algorithm can be seen in Figure 2.

3.2 *Improvement algorithms*

The improvement algorithms improve one or more solutions iteratively. The running time of these algorithms can take long base on the stop condition. In this article two improvement algorithms are discussed, these are the Genetic Algorithm and the Hill Climbing Algorithm. The genetic algorithm (Grefenstette et. al. 1985) simulates the nature processes. The algorithm operates on the population of the solutions. The pseudo code of the algorithm can be seen in Figure 3.

   The hill climbing algorithm (Yuret 1993) operates on one solution. The algorithm always takes the neighbour of the actual solution, which is better than the actual solution. The process of the algorithm can be seen in Figure 4.

4 VEHICLE ROUTING PROBLEM WITH DRONES

The article describes a new type of Vehicle Routing Problem. We extended the VRP with the delivery of drones. In case of our problem, the positions and the demands of the customers are known in advance. In case of our model, there is only one truck with only one drone. The demand of the customers can be satisfied with the drone, but some customers are also visited with the truck. The drone has some capacity limit. The task is to determine which customers are visited with only the drone, and with both the drone and truck and also determine the order of the

```
BEGIN PROCEDURE
Taking a customer randomly.
        WHILE not all customer is selected DO
                Taking the unvisited customer that is closest to the last selected customer.
        END WHILE
Taking the first customer.
Evaluation of the permutation.
END PROCEDURE
```

Figure 1.   Pseudo code of the Nearest Neighbour Algorithm (Golden et. al. 1980).

```
BEGIN PROCEDURE
Taking a customer randomly.
        WHILE not all customer is selected DO
                Taking the unvisited customer and insert to the route so, that the cost of the tour increase is
                minimal.
        END WHILE
Evaluation of the permutation.
END PROCEDURE
```

Figure 2.   Pseudo code of the Arbitrary Insertion Algorithm (Golden et. al. 1980).

```
BEGIN PROCEDURE
Creating the initial population.
Evaluation of the population.
        WHILE stopping condition is not met DO
        Certain individuals are still transferred to the next population.
                WHILE the next generation hasn't been uploaded DO
                        Choosing two individuals, these will be the parents.
                        Crossing the parents, these will be the children.
                        Mutation can be also applied to the children. Evaluation of the children.
                        Inserting the children into the next population.
                END WHILE
        END WHILE
END PROCEDURE
```

Figure 3. Pseudo code of the Genetic Algorithm.

```
BEGIN PROCEDURE
Taking a possible solution. In the beginning, this will be the actual solution.
        WHILE stopping condition is not met DO
                Producing the neighbour of the actual solution.
                Evaluation of the neighbour solution.
                If the neighbour is better than the actual solution, the neighbour solution will be
                the actual.
        END WHILE
END PROCEDURE
```

Figure 4. Pseudo code of the Hill Climbing Algorithm.

drone and truck route. The goal function is the minimization of the length of the route and also satisfying the capacity limit of the drone. Figure 6 demonstrates the problem.

In Figure 5. there are 15 customers, which demand must be served with the drone. Some customers must be also visited with the truck. Customers 2, 14 and 9 are visited with the truck and drone, and customers 4, 3, 1, 8, 7, 10, 6, 5, 11, 13, 15, 12 are visited with only the drone. In case of the problem the truck starts the route from customer 2, the drone stars their route from customer 2 then the drone visits customer 4, 3, 1, 8, 7 and returns to customer 2. Then the truck continues the route to customer 9. The drone then starts the route from customer 9, then visits customer 10, 6, 5, 11 and then return to customer 9. Then the truck continues the route with customer 14. The drone starts the route from customer 14, and then visits customer 13, 15, 12 and then returns to customer 14.

In the following, we present the general mathematical model of the problem (with many trucks and drones).

The objective function is the minimization of the total route length:

$$min \sum_{i=1}^{nc} \sum_{j=1}^{nc} \sum_{m=1}^{nt} c_{i,j} x_{i,j,m} + \sum_{i=1}^{nc} \sum_{k=1}^{nc} \sum_{l=1}^{nc} \sum_{m=1}^{nt} \sum_{n=1}^{nd} c_{k,l} x_{i,k,l,m,n} \tag{1}$$

subject to

The same number of edges go into the nodes as many as come out in the truck level:

$$\sum_{i=1}^{nc} \sum_{j=1}^{nc} x_{i,j,m} = \sum_{i=1}^{nc} \sum_{j=1}^{nc} x_{j,i,m} \ i \neq j, \forall m \in TR \tag{2}$$

The same number of edges go into the nodes as many as come out in the drone level:

154

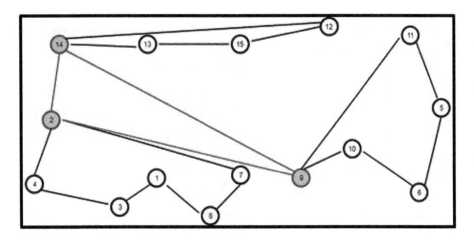

Figure 5. Vehicle routing problem with drones.

$$\sum_{i=1}^{nc}\sum_{k=1}^{nc}\sum_{l=1}^{nc} x_{i,k,l,m,n} = \sum_{i=1}^{nc}\sum_{k=1}^{nc}\sum_{l=1}^{nc} x_{i,l,k,m,n} \quad k \neq l, \forall m \in TR, \forall n \in DR \tag{3}$$

$$x_{i,j,m} \in \{0,1\} \tag{4}$$

1, if m. truck visits customer i. and j. else 0.

$$x_{i,k,l,m,n} \in \{0,1\} \tag{5}$$

1, if n. drone visits customer k. and l. else 0.
Each customer must be visited exactly once by the trucks:

$$\sum_{i=1}^{nc} x_{i,j,m} = 1 \, \forall j \in CU, i \neq j, \forall m \in TR \tag{6}$$

$$\sum_{k=1}^{nc} x_{i,k,l,m,n} = 1 \forall i \in CU, \forall l \in CU, \forall k \neq l, \forall m \in TR, \forall n \in DR \tag{7}$$

The capacity limit of the trucks must be respected:

$$\sum_{i=1}^{nc}\sum_{j=1}^{nc} de_j x_{i,j,m} + \sum_{i=1}^{nc}\sum_{k=1}^{nc}\sum_{l=1}^{nc} de_k x_{i,k,l,m,n} \leq qv_m \forall m \in TR, \forall n \in DR \tag{8}$$

The capacity limit of the drones must be respected:

$$\sum_{i=1}^{nc}\sum_{k=1}^{nc}\sum_{l=1}^{nc} de_k x_{i,k,l,m,n} \leq qd_n \forall m \in TR, \forall n \in DR \tag{9}$$

In our case the number of trucks is 1, so $nv = 1$, and the number of drones is also 1,so $nd = 1$

Table 1. Notations and their definitions.

Notation	Definition
nc	Number of customers
$CU = \{cu_1, \ldots, cu_{nc}\}$	Customers
$DE = \{de_1, \ldots, de_{nc}\}$	Demands of each customers
nt	Number of trucks
$TR = \{tr_1, \ldots, tr_{nt}\}$	Trucks
$QV = \{qv_1, \ldots, qv_{nt}\}$	Capacity of trucks
nd	Number of drones
$DR = \{dr_1, \ldots, dr_{nd}\}$	Drones
$QD = \{qd_1, \ldots, qd_{nd}\}$	Capacity of drones
i, j, k, l	Vertex indices
m	Truck index
n	Drone index
$x_{i,j,m}$	Decision variable: 1 if the m. truck travel from i. customer to j. customer directly. 0 else
$x_{i,k,l,m,n}$	Decision variable: 1 if the n. drone of m. truck begin the route from i. customer of the m. truck and travel from k. customer to l. customer directly. 0 else
$c_{i,j}$	Cost of the arc from the i. customer to the j. customer

5 IMPLEMENTED ALGORITHMS

In this article, two types of algorithms are applied. The construction algorithms create one solution. They give locally the best steps, but they exclusively usage usually does not lead to the global optimum. The running time of these algorithms is small. The other type of the algorithms is the improvement algorithms. These algorithms improve one or more solutions iteratively. In this article, we use the improvement algorithms to improve the results of construction algorithms. The algorithms use the same representation forms. In the following, we describe this representation.

The algorithms use permutation representation. The permutation contains the numbers of the costumers. The evaluation of the permutation is the following: we give the elements of the permutation until the capacity constraint of the drone is reached. These customers will belong to one route. Then we give the following elements of the permutation until the capacity constraint of the drone is not met. This will be the second route of the drone. We continue these process until not all costumers are selected. It will give the number of routes of the drone, and this also means the number of customers of the truck. In the following, we give a case of study with some explanation and figures. The permutation of the customers can be described with the permutation (2, 4, 3, 1, 8, 7, 9, 10, 6, 5, 11, 12, 15, 13, 14). The demand of the customers is also given: (20, 30, 50, 10, 20, 40, 60, 30, 10, 20, 40, 50, 50, 10, 10).

If the capacity of the drone is 200 unit, the routes of the drone are the followings: 1. route: 2-4-3-1-8-7 (170 unit); 2.route: 9-10-6-5-11 (160 unit); 3.route: 12-15-13-14 (120 unit). The drone takes Hamilton circles, so the route of the drone will be the followings: 1.route: 2-4-3-1-8-7-2; 2.route: 9-10-6-5-11-9; 3.route: 12-15-13-14-12. The result is demonstrated in Figure 6.

Then, from each of the tours, a customer must be highlighted, which is also visited by the truck. From this customer begins the route of the drone and ends. This process should be done so, that the total road increase should be minimal. The selection algorithm is the following:

1. We select the two Hamilton circle where the distance between two customers is minimal. These two customers will be also visited by the truck. This process is demonstrated in Figure 7.
2. One of the customers is taken arbitrary as last selected.

156

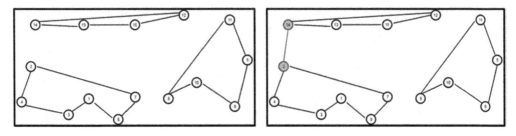

Figure 6.   The route of the drone and the first step of the selecting process of the route of the truck.

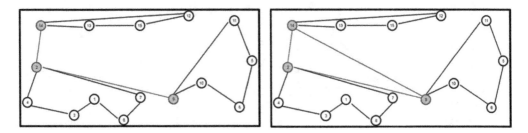

Figure 7.   The second step of the selecting process of the route of the truck and the final solution.

3. We always select the closest customer to the last selected customer that is not yet selected, belongs to the not selected Hamilton circle, and is closest to the last selected customer. This process can be seen in Figure 7.
4. If we have selected exactly one customer from each Hamilton circle, we will return to the first selected customer. Figure 7 demonstrates this method.

6   TEST RESULTS

In this section, we describe the test results of the algorithms. We used the improvement algorithms to improve the solutions of construction algorithms.

Based on the running results the genetic algorithm with the improvement of the arbitrary insertion and nearest neighbour algorithm provided the best running results for all dataset. Approximately gave 50% shorter route than the hill climbing algorithm.

Table 2.   The structure of the datasets.

Dataset	Number of customers	Drone capacity
pr01	48	500
pr02	96	500
pr03	40	500
pr04	45	500

Table 3.   The test results.

Dataset	GA+AI+NN	HC+AI	HC+NN
pr01	1432	3169	3521
pr02	3114	5049	5085
pr03	1366	2677	3076
pr04	1347	2629	3069

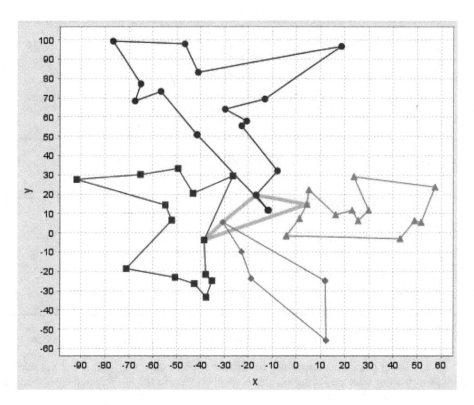

Figure 8.   Running result.

An example result can be seen in Figure 8. The purple thick line marks the path of the truck with the drone. The thin lines of different colors indicate the individual paths of the drone.

7   CONCLUSION

The Industry 4.0 has enabled the use of new logistics equipment. Such new material handling equipment are drones, which require new vehicle routing mathematical models and their algorithms. In this paper, we introduce such a new mathematical model and algorithms. During the problem the truck which has larger capacity limit or the drone which has smaller capacity limit satisfy the demands of the customers. The goal is the minimization of the length of the route respect the capacity limit of the drone and the truck. We solved the problem with construction and improvement algorithms, and the efficiency of the algorithms are also evaluated. Based on the test results the Genetic Algorithm gave the best solutions.

ACKNOWLEDGEMENT

The described article/presentation/study was carried out as part of the EFOP-3.6.1-16-00011 "Younger and Renewing University – Innovative Knowledge City – institutional development of the University of Miskolc aiming at intelligent specialisation" project implemented in the framework of the Szechenyi 2020 program. The realization of this project is supported by the European Union, co-financed by the European Social Fund.

REFERENCES

Laporte, G. 1992. The vehicle routing problem: An overview of exact and approximate algorithms. *European journal of operational research*, 59(3), 345–358.

Cordeau, J.F., Gendreau, M., & Laporte, G. 1997. A tabu search heuristic for periodic and multi-depot vehicle routing problems. *Networks: An International Journal*, 30(2), 105–119.

Dondo, R., & Cerdá, J. 2007. A cluster-based optimization approach for the multi-depot heterogeneous fleet vehicle routing problem with time windows. *European Journal of Operational Research*, 176(3), 1478–1507.

Gendreau, M., Laporte, G., Musaraganyi, C., & Taillard, É.D. 1999. A tabu search heuristic for the heterogeneous fleet vehicle routing problem. *Computers & Operations Research*, 26(12), 1153–1173.

Dondo, R., Méndez, C.A., & Cerdá, J. 2011. The multi-echelon vehicle routing problem with cross docking in supply chain management. *Computers & Chemical Engineering*, 35(12), 3002–3024.

Xiao, Y., Zhao, Q., Kaku, I., & Xu, Y. 2012. Development of a fuel consumption optimization model for the capacitated vehicle routing problem. *Computers & Operations Research*, 39(7), 1419–1431.

Desrochers, M., Desrosiers, J., & Solomon, M. 1992. A new optimization algorithm for the vehicle routing problem with time windows. *Operations research*, 40(2), 342–354.

Brandão, J. (2004). A tabu search algorithm for the open vehicle routing problem. *European Journal of Operational Research*, 157(3), 552–564.

Crevier, B., Cordeau, J.F., & Laporte, G. 2007. The multi-depot vehicle routing problem with inter-depot routes. *European Journal of Operational Research*, 176(2), 756–773.

Min, H. 1989. The multiple vehicle routing problem with simultaneous delivery and pick-up points. *Transportation Research Part A: General*, 23(5), 377–386.

Cao, E., & Lai, M. 2010. The open vehicle routing problem with fuzzy demands. *Expert Systems with Applications*, 37(3), 2405–2411.

Gendreau, M., Laporte, G., & Séguin, R. 1996. Stochastic vehicle routing. European Journal of Operational Research, 88(1), 3–12.

Chao, I.M. 2002. A tabu search method for the truck and trailer routing problem. *Computers & Operations Research*, 29(1), 33–51.

Coelho, L.C., & Laporte, G. 2013. A branch-and-cut algorithm for the multi-product multi-vehicle inventory-routing problem. *International Journal of Production Research*, 51(23–24), 7156–7169.

Crevier, B., Cordeau, J.F., & Laporte, G. 2007. The multi-depot vehicle routing problem with inter-depot routes. *European Journal of Operational Research*, 176(2), 756–773.

Angelelli, E., & Speranza, M.G. 2002. The periodic vehicle routing problem with intermediate facilities. *European journal of Operational research*, 137(2), 233–247.

Ke, L., & Feng, Z. 2013. A two-phase metaheuristic for the cumulative capacitated vehicle routing problem. *Computers & Operations Research*, 40(2),633–638.

Wen, M., Larsen, J., Clausen, J., Cordeau, J.F., & Laporte, G. 2009. Vehicle routing with cross-docking. *Journal of the Operational Research Society*, 60(12),1708–1718.

Aras, N., Aksen, D., & Tekin, M.T. 2011. Selective multi-depot vehicle routing problem with pricing. *Transportation Research Part C: Emerging Technologies*, 19(5), 866–884.

Hsu, C.I., Hung, S.F., & Li, H.C. 2007. Vehicle routing problem with time-windows for perishable food delivery. *Journal of food engineering*, 80(2), 465–475.

Golden, B., Bodin, L., Doyle, T., & Stewart Jr, W. 1980. Approximate traveling salesman algorithms. *Operations research*, 28(3-part-ii), 694–711.

Grefenstette, J., Gopal, R., Rosmaita, B., & Van Gucht, D. 1985. Genetic algorithms for the traveling salesman problem. *In Proceedings of the first International Conference on Genetic Algorithms and their Applications* 160 (168)160–168. Lawrence Erlbaum.

Yuret, D., & De La Maza, M. 1993. Dynamic hill climbing: Overcoming the limitations of optimization techniques. *In The Second Turkish Symposium on Artificial Intelligence and Neural Networks* 208–212.

Solutions for Sustainable Development – Szita, Jármai & Voith (eds.)
© 2020 Taylor & Francis Group, London, ISBN 978-0-367-42425-1

Cyber-physical waste collection system: A logistics approach

M.Z. Akkad & T. Bányai
Institute of Logistics, University of Miskolc, Hungary

ABSTRACT: The increased population has a great impact on the volume of household waste. The conventional waste management systems can be divided into two main parts. The first part is represented by recycling technologies, including incineration, processing, reusing and shredding. The second part includes the processing related logistics and material handling operations. The collection of household waste is performed in a wide geographical area which means that collection represents a significant part of the whole costs. Waste management systems need up-to-date technical, technological and logistics solutions to increase the efficiency, the reliability and the flexibility. The application of Industry 4.0 technologies offers a good opportunity to transfer conventional waste collection and processing systems into a cyber-physical system. Within the frame of this paper, the authors show the concept of smart waste collection as a cyber-physical system. After a systematic literature review, this paper introduces the concept of smart waste collection including smart bins based on RFID technology in order to support waste volume prediction, cloud and fog computation based decision making in order to optimise the allocation and utilization of technological and logistics resources.

1 INTRODUCTION

The term industrial revolution represents a quantum leap in the industry, which means raising the quantity, quality or both of them in the industry and adopting innovative industrial methods through new technologies (Roblek et al. 2016). So far, there have been three industrial revolutions. We are now in the midst of the Fourth Industrial Revolution, or briefly called "Industry 4.0", which is now being developed and dominated by the different industrial sectors comprehensively (Liao et al. 2017). The most prominent feature of Industry 4.0 is the adoption of intelligent technologies that rely on the Internet of things and remove the lines that separate the physical, digital and biological areas (Illés et al. 2018). Industry 4.0 applications include the most recent technology, especially in telecommunications, internet and nanotechnology, which allowed us to use small devices with great efficiency. This combination of advanced technologies allowed us to obtain various applications that have revolutionized the world of industry and changed the traditional concept of communication between machine and human into having the concept of communication between machine and machine (Dobos et al. 2018; Zhong et al. 2017). It is easy to observe the rapid pace of development of the industry, which makes it imperative for us to follow up the new applications of Industry 4.0 with eagerly so we can keep abreast of this development and benefit from it in our field of specialization. These applications have moved the logistics field to a new level (Glistau & Machado 2018).

The pace of industrial development is constantly increasing. The results of the technological revolution that we are living, in addition to the intelligent technologies built on the internet and resulted from Industry 4.0 make it imperative to pursue these techniques in different fields from CAD modelling (Felhő & Kundrák 2014) to digital twinning solutions (Lu 2017).

This study provides a model for the adoption of these technologies in the field of collection and treatment of municipal waste in an optimal manner. Giving the opportunity to start developing logistics systems and building remote management is a vital area that exists in every city in the world. In order to reach the required system that applies these technologies in the waste treatment field with its characteristics and benefits. The system is interconnected in several areas, linking the field of information technology, artificial intelligence, logistics systems and environmental engineering to each other.

This paper is organized as follows. Section 2 presents a literature review of waste processing solutions. Section 3 describes the structure of the proposed waste collection system, while section 4 focuses on the processes and working mechanism. Section 5 presents the system characteristics Conclusions and future research directions are discussed in Section 6.

2 WASTE PROCESSING

Waste is any excess material that is undesirable and it can mean rubbish or trash. In biology, waste means excess substances or toxins that come out of living organisms. Waste collection is a main part of the waste management processes. It is the process of solid waste transferring from the specified disposal site after use to the site of treatment or landfill. The curbside collection of recyclable materials can also be considered waste collection even that technically they are not waste. The collection is done by municipal services, similar institutions, public or private companies, specialized institutions or the general government. Waste treatment refers to the needed processes to ensure that waste has the least possible effect on the environment. Although the importance of waste treatment, these methods vary from a country to another one. It is possible that some governments consider waste treatment is just a necessary and indispensable cost should be paid to obtain a clean environment that is not harmful to the health of inhabitants. On the other hand, other governments give great importance to waste treatment because it might save many of the resources we need. There are many developed countries that implemented successfully waste treatment projects to get benefit from waste like recycling. When countries do, resources would increase besides having the streets and roads clean of waste. With the many methods of waste treatment that are available, the main methods have been classified in a waste management hierarchy according to each one's preference (Bagchi 2004) as follows:

1. Waste Minimization. The most effective option even it is very simple. Much of waste can be dispensed. Waste reduction or prevention can minimize generated waste in the first place. This means that finding a better option is the best solution instead of dealing with it. For example, plastic bags in supermarkets can be replaced with reusable carry bags, preferably made of organic cotton (Mallak et al. 2018).
2. Waste Reuse. It is the next most desirable option. Reuse is using again of the material without any structural changes in it. Waste reusing often requires collection but relatively little or no processing (Gebrezgabher et al. 2019). For example, many clothes or furniture can be reused.
3. Waste Recycling. It means turning waste into new materials and objects. It is an alternative to traditional waste disposal that can save raw materials and help reduce greenhouse gas emissions. Recycling can stop the dissipation of potentially useful materials and reduce consumption of new raw materials; moreover, it reduces energy use and air pollution from incineration or water pollution from landfills (Xu et al. 2017).
4. Incineration (with energy recovery). It is the technique of waste disposal by burning organic compounds and other materials. The process of incineration is called heat treatment. Incineration involves the transfer of waste into ash that is directed down the chimney. The heat and the released gases are used to generate electricity. By the end of using the gases to produce electric power, they are released into the air after purification from contaminated materials (He & Lin 2019).

5. Incineration (without energy recovery). It is the same as the previous option but without producing electricity. In general, it is not a good option as it increases air pollution without any benefit of energy production. However, it is a suitable option for countries that do not have enough areas. Moreover, it is an alternative to waste landfilling or waste throwing in the oceans and seas that result in long-term pollution.

6. Land Disposal. It is the disposal of waste to be decomposed by the natural. There are some differences in this method implementation, such as landfilling with/without treatment to speed up the decomposition process, surface impoundment or injection well. It is possible to get the benefit of this way in producing power by the released gases of the decomposition process. However, this method remains the worst option that should be avoided as much as possible for its longtime need, especially for inorganic materials and its impact on natural resources as soil and water (Keren 2017).

Different waste collection solutions are analysed in the literature focusing on different aspects of evaluation, like technology, logistics, human resources, policies, and social aspects (Bányai et al. 2019). The optimal structure of the waste collection system influences the performance of waste collection processes. A Portugal case study shows that strategic expansion plans of waste management companies can be supported by complex mathematical models and heuristic optimization algorithms (Gomes et al. 2011). The importance of multi-level solutions is highlighted with a three-phase hierarchical approach in the Spanish region of Galicia (Mar-Ortiz et al. 2011) and in Ankara (Demirel et al. 2016), where authors are focusing on facility location and routing problems. Waste collection systems represent a wide range of uncertainties, as shown in a Hong Kong case study, where the difficulties of the design of appropriate infrastructure for waste collection and recycling are described (Chung et al. 2011). Other case studies from Denmark (Grunow et al. 2009), Kampala City (Kinobe et al. 2015), Italy (Gamberini et al. 2009) and Taiwan (Yu & Wu 2010) demonstrate the importance of new technologies in municipal waste collection systems.

3 THE SYSTEM STRUCTURE

A new municipal waste collection system based on advanced Industry 4.0 technologies is demonstrated in this section. Municipal waste means all the kinds of garbage, which result of normal life in residential communities such as houses, apartments and villas, or places attached to population groups such as supermarkets, shops, grocery stores and similar places. In another expression, all solid waste related to humans as long as they have no chemical, biological or potentially hazardous effects on humans are considered as municipal waste. The waste, which results from the demolition and construction process, is also a municipal waste, but it is not included in this system because it does not exist in inhabited communities or it only exists as temporary work and the resulted waste should be transferred by special trucks directly to the landfill. This system includes dealing with the waste starting from the source points until the waste treatment facilities. Figure 1 shows the scheme of this system. The system management cloud is connected directly to all the other system's parts.

This system can be divided into five parts: containers, treatment facilities, collection and transfer station, trucks and system management cloud.

1. Containers. Two types of containers are used. Firstly, O type, which is used for organic waste. There should be a container for each building with a different capacity up to 1000 litres depending on the size of the building. This container has a sensor to measure the size of the waste inside it. This sensor can give three different colours as notifications, depending on the amount of waste inside the container. A yellow notification, which means it contains at least 50%, an orange notification, which means it contains at least 75% and a red notification, which means it contains more than 90%. An example of different O type containers can be seen in Figure 2.

Secondly, M type, which is used for inorganic waste (mixed). There should be a container of this type for each group of buildings where the citizens can throw the inorganic waste

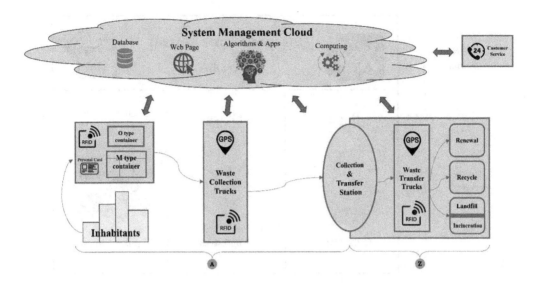

Figure 1. The system scheme.

Figure 2. O type containers (Oneplus 2019).

directly in them without the need to separate them. The person who wants to throw the waste needs to use his own specific ID card. Each user's data is stored on the system's server automatically with the amount of trash he/she has thrown out and the time. Therefore, people who do not have an ID card cannot use this container, to avoid any damage that may result from the dumping of organic garbage or stones for example. The capacity of this container is 10000 litres and it is divided into four main sections.

The M container has two parts, Figure 3. The first part is 1000 litres capacity and it is above the ground, which is used by the people to throw the waste inside it directly after using the ID card. The second part is divided into three sections, 3000 litres for each one. The first one is for paper and carton, the second one is for glass and the third one is for electronics and the other waste types. After throwing the waste into the first part, the waste is sorted automatically into the suitable section in the second part. Weight measuring, size measuring and X-ray are used in the sorting process. The second part is underground and cannot be reached without using special work ID by the workers so the container is emptied into the waste collection vehicle by using hydraulic lifting equipment. As O type container, each section in the second part has a sensor to measure the size of the waste inside it. This sensor can

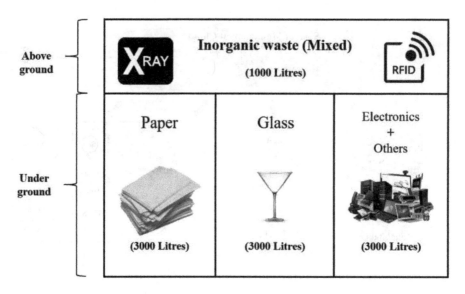

Figure 3. An illustration of the structure of M type container.

give three different colours as notifications, depending on the amount of waste inside the container. A yellow notification, which means it contains at least 50%, an orange notification, which means it contains at least 75% and a red notification, which means it contains more than 90%.

Both types of containers have active RFID to send their information continuously to the system. A notification is also sent to the system every time the containers are emptied, by using the worker's ID card.

2. Treatment facilities. They are the final stage, which the waste is transported to be treated or used in the best manner in. These facilities are divided according to the type of waste they deal with into four sections.

Firstly, renewal. In this section, waste is reused or disassembled for useful parts. The most targeted waste here are clothes, electrical and mechanical equipment. After completion of the dismantling and evaluation phase, any excess material is transferred to one of the other sections when there is a sufficient amount to fill a full transport truck.

Secondly, recycle. In this section, the waste's raw materials are obtained for reuse, which means getting raw materials resources that can be used again by saving the waste amount that needs to be disposed of. The targeted waste here is paper and glass.

Thirdly, incineration. In this section, unusable and unrecyclable materials are collected to be burned; the obtained heat is used to produce energy.

Fourthly, landfill. In this section, the remaining waste is buried after treatment to have faster biodegradation. The gases produced by the biodegradation of organic waste after burying can be collected and utilized.

3. Collection and transfer station. Waste, which is coming from containers, is collected at this station depending on the type. An additional sorting is done within this station to avoid any mistake in the type of waste, which might happen in the containers and when there is a problem with the type of waste in one shipment, the records can be checked to identify the person who randomly dumps the waste to give him/her a warning or punishment. Information is collected from gentelligent technology devices in this station.

This station is close to the city to speed up the process of transporting waste and there is no need to be a very large area because the amount of transported waste is calculated. The purpose of this station is to organize the waste sort and transport operations. On the other hand, large trucks are used to transport the waste to the treatment facilities as they are relatively far from the city.

4. Trucks. These trucks are dedicated to transport waste and handle loading & unloading waste easily. Two types of trucks are used in this system. Firstly, waste collection trucks, to move the waste from the containers to the collection and transfer station. The size of these trucks is suitable to be used for containers unloading and for moving within the city. Secondly, waste transfer trucks to move the waste from the collection and transfer station to the treatment facilities. Their size is bigger than the first type to be suitable for transporting waste outside the city faster

5. System management cloud. All the above-mentioned parts are directly connected to the system management using the internet. Cloud computing is used to store data and deal directly with all the system parts. It also allows administrators to access their accounts for monitoring and guidance, according to their permissions. All data about the transportation, delivery of waste, collection trucks and waste quantities in each part of the system, as well as the records of surveillance cameras are saved. Programs with special algorithms are used to create routes of waste collection trucks according to the waste type and quantities within the containers. In addition, there is available customer service for complaints and remarks at any time, connected to the system management.

4 THE SYSTEM WORK MECHANISM

The storage capacity of each of the treatment facilities should be determined and stored in sufficient quantity for the next day only within this storage. The amount of waste should not be more than the daily capacity, nor less. At the beginning of the day, this waste quantity is moved into the treatment facility, allowing in the following hours for the waste trucks the possibility to transfer the daily quantity of waste from the collection and transfer station.

Collection and transfer station capacity should be at least five times the total daily capacity of the treatment facilities to avoid any bottlenecks problems. In addition, the waste in collection and transfer station should not be less than 40% of its total capacity and preferably less than 80%. The system management is responsible about avoiding any possible stop problem in the transporting to the following facilities. Especially that the work within this station is not only collection, but there are also additional sorting process and information collecting.

The algorithms that create the routes of waste collection trucks target the containers that give an orange notification until it is enough to fill the truck capacity. If additional containers are needed, containers that give yellow notification are added to the route. If there is a container with a red notification, it means an emergency condition and this container should be emptied within 12 hours at most. It should be noted that the amount of waste in the collection and transfer station should be between 40% and 80% of its capacity, which means having the flexibility to create the routes depending on the needed waste amount and there is no need to have the same routes every day.

For regulatory reasons, this system is divided into two parts A and Z. A part represents the physical parts within the city. In this part, the waste is transferred until the collection and transfer station. Z part represents the physical parts outside the city. In this part, the waste is transferred from the collection and transfer station until the treatment facilities. Different administrators can be assigned to each one, depending on the size of the system, but all the parts are always directly connected to the system management.

5 THE SYSTEM CHARACTERISTICS

This system has many advantages that encourage adopting it. These advantages can be summarized in the following points:

1. Modern structured system. This system controls the process of waste transporting in an ideal way. There will be no excess waste in containers without collection nor the need to store waste above the required limit. In addition, data storage & automation permit

knowing the performance and efficiency of the system easily, allowing logistic management to fully regulate the operations of the collection system, which give absolute flexibility in controlling the system. This flexibility is due to the connection of all parts of this system to a single information processing centre within cloud computing, which allows the complete management with any modifications easily. Containers unloading functions can be modified as needed. For example, the system can be modified by stopping the containers unloading for one day or more. Instead of that, any container contains more than 50% on the previous day would be emptied. This flexibility is not affected by the size of the city or the population because of management automation.

2. Reliability and connectivity. This system provides high reliability of IT side by using a private server to save the information related to the system, accounts of users and administrators and their permissions, or for security side by using surveillance cameras and electronic ID for workers or citizens. The connectivity is through IoT, which is applied to all the parts of the system. This connectivity is continuous twenty-four hours per day and seven days per week, through an IP address to each part. Therefore, all the parts are connected and allowed to send and receive signals, orders and notifications. In addition, it is automation. Digital twinning and artificial intelligence, which are used in this system, allow automation and programming of operations within the system in accordance with the given conditions. Moreover, all the processes are saved under each person's name did it with the exact time.

3. Transform waste processing into an investment project with an economic return. This system focuses on both the waste collection part and the processing part, which means eventually obtaining the maximum amounts of reusable materials, recycled materials and raw materials. In addition to the energy that can be produced from the process of treatment. This means economic abundance covering the cost of the project.

4. Environment protecting, by applying the most modern methods of waste transfer and treatment. In addition to applying JIT (Just In Time) philosophy, which means transferring of the needed amount of waste according to the capacity of the treatment centres exactly, without the need for storage centres to retain them, which offers protection of possible biological environment pollution.

5. Conducting studies. The collected data through this system give details about the types of produced waste and the exact quantities continuously, allowing studies to be conducted through them. For instance, the types of waste and its increasing can be identified throughout the year. In addition to the possibility of saving the number of the population associated with each container, which means studying the changes in waste quantities over time and expecting the resulting increase from the increase in population. This means it is possible to reach accurate results on the required areas and energies for treatment facilities to manage waste in the future.

6. Effectiveness. The direct connection of all parts to the management means that it is easy to identify any problem that may occur immediately. Over time, a problem or malfunction would occur in any part of the system. The system can identify this problem automatically with finding the best solution and send it directly to the specialist team depending on the problem type without the need of the human intervene, which means saving time.

7. Although all the system parts are linked to each other directly, it is very easy to adopt new technologies or develop one or more of the system parts in order to replace them with ones that are more modern. Because the linking in this system depends on the internet and smart technologies that do not require many physical connections so, the system developing does not require big changes in general.

This system has two disadvantages that should be noticed. The first one is that this system requires a specific limit for the city size. Therefore, this system cannot be applied in any city. The second one is that applying this system requires a big financial budget at the beginning because the M type container building is expensive; next to establishing the system management cloud. Noticing that creating the same system in another city will not be the same expensive because there is no need to create another system management.

6 SUMMARY

This study provided details of a cyber-physical municipal waste collection system based on the latest technologies developed. To reach a system that serves humanity in an optimal manner while preserving time & effort and reducing environmental damage. Details of each part of this system were provided with illustrations of the system structure and its interconnection. The main course of action of this system was clarified as well as the reasons for its creation and its advantages. This study is a clear example of applying modern technologies in the field of waste management logistics. The system's structure offers the possibility to modify this system and suit the size of the city. In large cities, more than one system can be applied to suit the required size, such as making more than one collection and transfer station or divide the city into two, three or more sections with individually responsible systems that are connected on the management level only. Which means centralized and decentralization management at the same time in order to achieve greater flexibility.

ACKNOWLEDGEMENTS

This project has received funding from the EFOP-3.6.1-16-00011 "Younger and Renewing University – Innovative Knowledge City – institutional development of the University of Miskolc aiming at intelligent specialization" project implemented in the framework of the Szechenyi 2020 program.

REFERENCES

Bagchi, A. 2004. *Design of Landfills and Integrated Solid Waste Management*. Wiley.

Bányai, T., Tamás, P., Illés, B., Stankevičiūtė, Z., & Bányai, Á. 2019. Optimization of Municipal Waste Collection Routing: Impact of Industry 4.0 Technologies on Environmental Awareness and Sustainability. *International Journal of Environmental Research and Public Health* 16(4): 634.

Chung, S.S., Lau, K.Y. & Zhang, C. 2011. Generation of and control measures for e-waste in Hong Kong. Waste Management 31: 544–554.

Demirel, E., Demirel, N. & Gokcen, H. 2016. A mixed integer linear programming model to optimize reverse logistics activities of end-of-life vehicles in Turkey. *Journal of Clean Production* 112: 2101–2113.

Dobos, P., Tamás, P., Illés, B. & Balogh, R. 2018. Application possibilities of the Big Data concept in Industry 4.0. *IOP Conf. Series: Materials Science and Engineering* 448: 012011.

Felhő, Cs., Kundrák, J. 2014. CAD-based modelling of surface roughness in face milling. *World Academy of Science Engineering and Technology* 8(5): 71–75.

Gamberini, R., Gebennini, E. & Rimini, B. 2009. An innovative container for WEEE collection and transport: Details and effects following the adoption. *Waste Management* 29: 2846–2858.

Gebrezgabher, S., Taron, A. & Amewu, S. 2019. Investment climate indicators for waste reuse enterprises in developing countries: Application of analytical hierarchy process and goal programming model. *Resources Conservation and Recycling* 144: 223–232.

Glistau, E. & Machado, N.I.C. 2018 Industry 4.0, Logistics 4.0 and Materials - Chances and Solutions. *Materials Science Forum* 919: 307–314.

Gomes, M.I., Barbosa-Povoa, A.P. & Novais, A.Q. 2011. Modelling a recovery network for WEEE: A case study in Portugal. *Waste Management* 31: 1645–1660.

Grunow, M., Gobbi, C. & Alting, L. 2009. Designing the reverse network for WEEE in Denmark. *CIRP Annals* 58: 391–394.

He, J.X. & Lin, B.Q. 2019. Assessment of waste incineration power with considerations of subsidies and emissions in China. *Energy Policy* 126: 190–199.

Illés, B., Varga, K.A. & Czap, L. 2018 Logistics and Digitization. *Lecture Notes in Mechanical Engineering* 2018: 220–225.

Keren, Y., Borisover, M., Schaumann, G.E., Diehl, D., Tamimi, N. & Bukhanovsky, N. 2017. Land disposal of olive mill wastewater enhances ability of soil to sorb diuron: Temporal persistence, and the effects of soil depth and application season. *Agriculture Ecosystems & Environment* 236: 43–51.

Kinobe, J.R., Bosona, T., Gebresenbet, G., Niwagaba, C.B. & Vinneras, B. 2015. Optimization of waste collection and disposal in Kampala city. *Habitat International* 49: 126–137.

Liao, Y.X., Deschamps, F., Loures, E.D.R. & Ramos, L.F.P. 2017. Past, present and future of Industry 4.0-a systematic literature review and research agenda proposal. *International Journal of Production Research* 55(12): 3609–3629.

Lu, Y. 2017. Industry 4.0: A survey on technologies, applications and open research issues. *Journal of Industrial Information Integration* 6: 1–10.

Mallak, S.K., Ishak, M.B., Mohamed, A.F. & Iranmanesh, M. 2018. Toward sustainable solid waste minimization by manufacturing firms in Malaysia: strengths and weaknesses. *Environmental Monitoring and Assessment* 190(10): 575.

Mar-Ortiz, J., Adenso-Diaz, B. & Gonzalez-Velarde, J.L. 2011. Design of a recovery network for WEEE, collection: The case of Galicia, Spain. *Journal of the Operational Research Society* 62: 1471–1484.

Oneplus 2019. Smart waste technologies: Available online: https://www.oneplussystems.com/2018/05/one plus-top-6-most-innovative-smart-waste-technology-systems-list/container-renders-using-metro-sensor/. (accessed on 18 May 2019).

Roblek, V., Mesko, M. & Krapez, A. 2016. A Complex View of Industry 4.0. *Sage Open* 6(2): 2158244016653987.

Xu, Z.T., Elomri, A., Pokharel, S., Zhang, Q., Ming, X.G. & Liu, W.J. 2017. Global reverse supply chain design for solid waste recycling under uncertainties and carbon emission constraint. *Waste Management* 64: 358–370.

Yu, M.C. & Wu, P.S. 2010. A simulation study of the factors influencing the design of a waste collection channel in Taiwan. *International Journal of Logistics Research and Applications* 13: 257–271.

Zhong, R.Y., Xu, X., Klotz, E. & Newman, S.T. 2017. Intelligent Manufacturing in the Context of Industry 4.0: A Review. *Engineering* 3(5): 616–630.

Solutions for Sustainable Development – Szita, Jármai & Voith (eds.)
© 2020 Taylor & Francis Group, London, ISBN 978-0-367-42425-1

Efficiency improvement of reverse logistics in Industry 4.0 environment

I. Hardai, B. Illés & Á. Bányai
University of Miskolc, Miskolc, Hungary

ABSTRACT: Nowadays, the increase in consumption and the shortening of the life cycle of products significantly increase the amount of waste generated, causing serious environmental problems. As a result, waste management has become one of the most important tasks of the 21st century, which puts a huge burden on the economy.

Waste management means the practical implementation of the protection of the environment from the harmful effects of waste on the entire life cycle of the waste. This is a process that involves the prevention, reduction, separate collection and utilization of waste generated, the temporary storage and disposal of non-recoverable waste without pollution. It follows from the above that logistics functions and services are playing an increasingly important role in waste management, which has contributed to the development of reverse logistics. The task of reverse logistics is, among other things, to collect, store and dispose of waste generated inside the company.

In this thesis, we have set out to develop a concept for the optimal design and operation of a reverse logistics system for a production company. Nowadays, in order to meet the dynamically changing customer needs, the development of Industry 4.0 capabilities of manufacturing companies is essential to increase their efficiency and expand their capacity. To achieve this goal, digitization and vertical and horizontal integration need to be strengthened. In this study, we will demonstrate, through the example of a sample factory with Industry 4.0 technology, how to use the simulation to determine the optimal design of a reverse logistics system.

Keywords: waste management, reverse logistics, Industry 4.0, optimization, scheduling

1 INTRODUCTION

As a result of social and economic development, much – and many types – of waste are generated, the recycling and handling of which is a global problem that affects us all. Improperly treated or stored waste can cause severe environmental damage, decrease water, soil and air quality, and can cause serious illness (Kaza et al. 2018).

Rational management of resources is one of the most important foundations of sustainable development. Selective waste collection and recycling are of the utmost importance as waste can be an important source of raw materials and energy.

Logistics (e.g. waste collection) is also a priority in waste management. Modern waste collection can be supported by various elements of the Industry 4.0 concept. By exploiting the benefits of digitalization, big data, digital twin solutions, and automatic systems with sensors can be developed. These solutions help increase efficiency and flexibility while reducing costs (Elke & Machado 2018).

We can use smart containers, various wireless communication solutions, self-guiding vehicles to collect waste, so an integrated system can be developed.

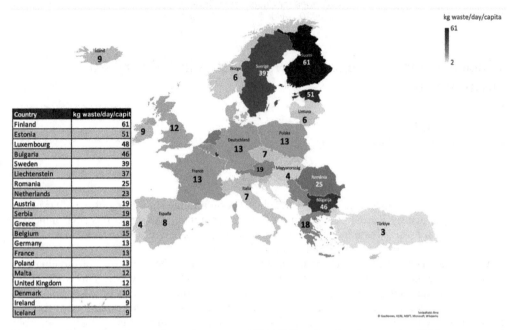

Country	kg waste/day/capit
Finland	61
Estonia	51
Luxembourg	48
Bulgaria	46
Sweden	39
Liechtenstein	37
Romania	25
Netherlands	23
Austria	19
Serbia	19
Greece	18
Belgium	15
Germany	13
France	13
Poland	13
Malta	12
United Kingdom	12
Denmark	10
Ireland	9
Iceland	9

Figure 1.   Generation of waste (hazardousness and NACE Rev. 2 activity) per capita, 2016.

2   DEFINING A RESEARCH TASK

After determining the research question, we use keyword matching as the most common method for identifying literature after determining the relevant database(s) associated with it. The main theme can be selected by reviewing article summaries and reducing the number of resources. After selecting the methodology for analyzing the articles, the less researched areas and bottlenecks can be mapped after summarizing the main scientific results (Zhang & Liu 2017).

Among the potential databases (Google Scholar, ResearchGate, Science Direct, Scopus, Web of Science, etc.) we chose Web of Science.

The focus of the search was on the terms "waste management", "reverse logistics", "Industry 4.0", "optimization" and "scheduling" and combinations of these. Figure 2 shows the number of publications published between 1975 and 2019, including the number of papers published in 2018–2019.

A significant proportion of articles (15–100%) published in the subject area were written in the last one and a half years. The number of publications published each year increases exponentially, which underlines the growing importance of the topic (Figure 3).

Based on the study of the literature, we have delimited the field of research and determined that the optimal design of a reverse logistics system using industry 4.0 technology and the waste collection system operating under these principles is the subject of our research.

3   MODELING THE WASTE COLLECTION SYSTEM

In the factory of the future, the collection of waste requires the establishment of a reliable, easily variable, flexible system. Most workflows should be adaptive, universal, and value-creating. The control and supervision must be integrable, suitable for industrial use and also efficient.

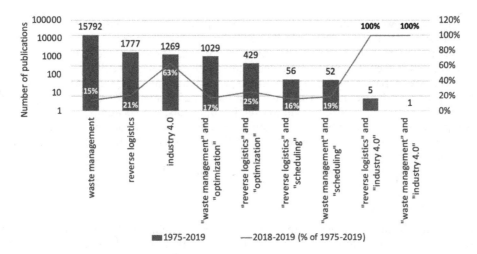

Figure 2. The number of published publications, the ratio of the latest ones.

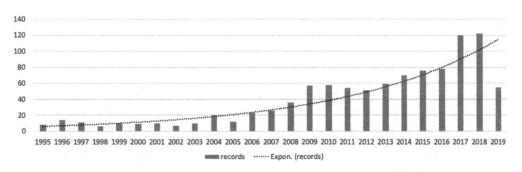

Figure 3. Number of publications in "waste management" and "optimization" broken down by year.

3.1 *Waste collection processes and abilities*

Depending on the diversity of the selectively collected waste and the quantity to be stored, the number and size of depots can be determined.

Garbage collection vehicles leave the depot in an empty state and then return to the collecting station that meets the material quality of the collected waste at the end of the collection process along the predefined route, and the vehicles are emptied there.

Containers to be emptied are transported manually or automatically.

The collected waste will be sent to the recycling site after further sorting and classification.

During the collection process, the waste containers in which the amount of waste has reached the level of signaling are emptied.

If more or other sizes of waste container or waste collection vehicle are disposed of, it must be documented in the system, and the design will follow the changed parameters. The procedure is similar when opening new depot areas or reducing the area/number of existing ones.

3.2 *Characteristics of the model used*

Characteristics of a model for waste collection based on logistics 4.0 principles:

– the number of waste depots can be determined by the number of selectively collected materials,

- the number of waste containers can be dynamically changed as needed,
- the capacity of the on-site storage facility is optional,
- the amount of waste deposited in the storage facilities can be monitored and known in real time, thanks to the use of modern data collection and sensor technologies,
- the temporal variation in the amount of waste stored in the containers can be estimated using predictive data and predictable methods using statistical methods,
- the total capacity can be freely changed (number of depots, size, location),
- Automated Guided Vehicles (AGVs), automated vehicles can support collection processes,
- the location and condition of the waste collection vehicles can be tracked and analyzed in real time,
- freely programmable logistics by linking processes and AGVs.

3.3 *Benefits of the model*

The most significant advantages of the model are the scalability of the total capacity of the system, the number of components can be increased according to the change needs. The high degree of flexibility also appears at the level of depots, vehicles and waste containers: automatic data collection, processing, and analysis.

The route of the collection process is determined automatically, according to the optimum solution, thus reducing the number of trips made to the collection vehicles, resulting in lower cost levels, less emissions.

4 SAMPLE FACTORY – SIMULATION

An example of a sample factory demonstrates how the waste generated by the production processes of the various components can be collected selectively and along with the Industry 4.0 principles in a cyber-physics system. All of this requires modern automated transport vehicles, flexible extensible system components, and the necessary infrastructure (sensors, transmitters, server, software) for data collection. This equipment, devices, and applications provide flexibility and high efficiency.

By utilizing the possibilities provided by the simulation, it is possible to visualize, improve and then define the optimal design of the logistics system (Telek 2018).

In case of a large task to be solved, the bottleneck is the runtime environment (hardware) and the time it takes (Kota 2012). While in a smaller system, taking into account all the possible cases, the best waste collection schedule and route can be determined in real time, solving a larger problem, we have to solve an NP-hard task, which can be assisted by various heuristic algorithms (Gubán & Gubán 2012).

In the first phase of software implementation, the simulation does not yet use heuristic algorithms and will be done in a later research.

The optimization process was divided into 3 main segments: optimal layout, optimal collection groups, and optimal route definition.

For optimum layout, the best location of the depots can be determined, taking into account the areas that can be used freely.

In the case of optimal collection groups, waste storage containers with the levels indicated by the sensors indicate the need for waste collection is identified. The quantity-time functions of the waste entering each storage can be defined as the input parameters of the simulation.

The optimal route is the route containing the positions of the containers containing the waste marked for collection, with the lowest time, cost, and emission.

4.1 *Optimal layout*

Different size cells can be used in our sample factory, which represents the working areas.
The rules of the Road Traffic Code apply in the area of the factory.

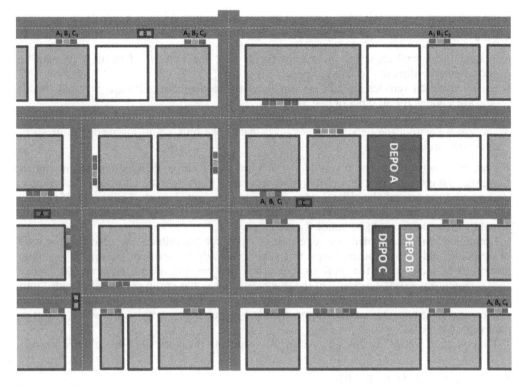

Figure 4.   Visualization of the sample factory using the simulator.

The number of selectively collected materials can be determined freely, and three types of waste (yellow, blue, green) have been considered for the sample factory in, shown in Figure 4. We must designate at least the same number of depots as the diversity of materials to be collected. By examining the possible uses of the available areas (cells) and combinations of materials to be collected in a given cell, the optimal arrangement can be determined.

The position of the occupied cells and the position of the deposited waste containers are randomly generated. The entire factory area is predefined – of which any number of empty (free) can be simulated, thus simulating the diversity of the factory area.

The optimal arrangement is determined by depot use / non-deposition of all possible cells and by individual testing of combinations of material types that can be stored in the cells used, and by comparing the results obtained for each parameter system/sample factory, taking into account the route rules.

4.2   Optimal route, optimal collection groups

When determining the task, the number, capacity, speed of the vehicles, the time needed for loading and unloading, the transport capacity of each container can be specified for the given material type.

All vehicles deliver only one type of waste at a time, and mixed transport is not allowed. The vehicle only collects and delivers as much cargo as its capacity allows.

The number of vehicles is not limited in advance, their optimal number can be determined during the simulation according to the parameters of the given task.

In this sample factory, we are counting on failures in waste collection vehicles, sensors, transmitters, system control software, hardware.

In the first phase of the simulation software, the following restrictions apply when determining the route:

- scalability of total capacity: the size of the system, the number of components can be increased according to the changed needs,
- waste collection vehicles are all the same, have two constant speeds (straight track, bend) and are suitable for all types of materials.
- the paths between and around the cells are two-lane.
- movements within the depots are not investigated, we only count an average time spent there (unloading time).

For waste storage containers, two fullness levels are used (can be specified during the simulation), the first indicates that the emptying process can already be planned when a new route is created, and the second signal level shows it has to be emptied on the next route. If there is more than one container for collecting the same material in one position, then the data collected by the sensors should be handled aggregated.

Containers may be included in a collection group where the fullness exceeded the first level of signaling, and the containers where this value is already higher than the second signaling level must be included. If the total number of selected containers and the amount of waste they represent exceeds the capacity of the transport vehicle, the group shall be subdivided into several subgroups so as to minimize the total length of the routes in the subgroups. Collection groups and subgroups are created automatically by analytical analysis of possible cases (by control software).

The collection groups are formed separately according to the material of the collected waste.

5 SIMULATION SOFTWARE

The software was created individually, specifically for the problem being investigated. It consists of three major parts, which are detailed below.

5.1 *Simulator*

The simulator visually shows the operation of the plant in real time with a layout (frame/s). Visualization helps to identify bottlenecks and correct/faulty operation solutions.

Input parameters:

- quantity-time functions of waste generated per container,
- capacity of the containers,
- cell locations and dimensions on the map (x, y),
- types and locations of containers on the map (x, y),
- node locations on the map,
- routes between nodes, cells and special points (e.g. depot)
- loading and unloading times,
- the speed of the transport vehicle is in a straight line,
- speed of the transport vehicle at corner, braking, acceleration (used within X distance before and after the point),
- capacity of transport vehicles,
- first and second signaling levels,
- the length of the simulated activity over time.

Output:

- any time of production process can be viewed in real time. In fact, the output of the planning scheduler is visualized.

5.2 Collection scheduler

The collection scheduler performs the collection according to the given order for a given layout.

Output:

- simulating collection according to a specified parameter system and recording running information over t time axis,
- "what happened" file broken down into seconds => the simulator works from it,
- times of each route, distance traveled => this can be used to calculate moving average ("search for crowded or empty periods"),
- vehicle log: which vehicle transported what and when (states: not working, loading, unloading, waiting, waiting to go) => finding solutions to avoidable waiting cases
- collection container log (start collecting, emptying, fullness levels, reaching signal levels) => from this it is possible to examine the utilization, to find the optimal emptying algorithm based on the log.

5.3 Layout variant

The layout variator sends the planning scheduler through all possible layouts.

It is advisable to choose an initial layout that is expected to give good values, so the worse can be thrown away without having to go through the entire process.

First, we try to find a solution in a general-purpose programming language (C++).

Output:

The following aggregate data on the process for getting the best N route results (length of the journey made during the entire collection process):

- recommended cell layout,
- total number of trips and times (routes per collection group)
- time and distance average, variance, minimum and maximum time and distance between two products,
- maximum transport vehicle number requirement,
- transport vehicle utilization, broken down to % of statuses,
- utilization data for collection containers in %.

If it is known that the worst result is Y from the top N route, then if a process reaches the Y distance before the time runs out, the scheduler stops and gives "not worth counting" status in response.

6 MATHEMATICAL BACKGROUND FOR THE MODEL

A, B and C indicate in general the three types of containers for selective waste disposal. A_i, B_i, C_i are the containers on position 'i'. Capacity for them are K_{Ai}, K_{Bi}, K_{Ci} $[m^3]$, current usage level are S_{Ai}, S_{Bi}, S_{Ci} measured in $[m^3]$. There are two types for signaling of used capacity: L_{1Ai}, L_{1Bi}, L_{1Ci} and L_{2Ai}, L_{2Bi}, L_{2Ci} in percent.

The following formulas define conditions, limitations and goal-function for container type A. B and C types can be defined in the same way.

When the path of disposal includes the container then $D_{Ai} = S_{Ai}$, otherwise $D_{ai} = 0$. Length of the journey of one route:

$$S_1 = \min\left\{ d_0^1 + \sum_{i=1}^{k} d_i^{i+1} + d_k^0 \right\} \tag{1}$$

where d_i^{i+1} is the distance between point 'i' and '$i+1$' in meters. The path consists of 'k' points. S_1 defines the minimal length among all available combinations in order. Point '0' is a depo.

The container 'i' will be on the next disposal journey if:

$$K_{Ai} \cdot L_{2Ai} \leq S_{Ai} \qquad (2)$$

The quantity of waste gathered by the journal 'j' at least will be:

$$G_j = \sum_{i=1}^{n} D_{Ai} \qquad (3)$$

where 'n' indicates the possibility of positions.

Number of journeys required for collections will be:

$$N_V = F\left[\frac{G_j}{K_V}\right] + 1 \qquad (4)$$

where K_V is the capacity of waste collector car [m^3]. $F[]$ is the floor function.

If $N_V \neq F\left[\frac{G_j}{K_V}\right]$, then waste can be collected from containers which have not been signed for collection yet, assuming the following formula is true:

$$K_{Ai} \cdot L_{2Ai} \leq S_{Ai} \qquad (5)$$

Total length of the journey during N_v number of routes:

$$S_{N_V} = min\left\{ \sum_{j=1}^{N_V} \left(d_0^{1j} + \sum_{i=1}^{k_j} d_i^{i+1} + d_{k_j}^0 \right) \right\} \qquad (6)$$

The total costs (C_{N_V}) and greenhouse gas emission (P_{N_V}) during N_V number of routes are:

$$C_{N_V} = S_{N_V} \cdot c \qquad (7)$$

$$P_{N_V} = S_{N_V} \cdot p \qquad (8)$$

where costs for 1km is 'c' and emission is 'p'.

Instead of extending the number of collector cars for higher collecting rate, it is well worth seeing an algorithm for searching optimal path to be able to achieve the highest available usage in capacity with the lowest car numbers.

Summarizing length of routes, the costs and emissions make the totals of their values during the monitored time frame.

7 CONCLUSIONS

Our planet covered more and more waste. It is very expensive and polluting for environment to get rid of the accumulated garbage. To prevent this trend the best and most efficient way is recycling.

The recycling process is started with good separation of the different kind of waste using the most cost-efficient method.

This thesis shows one possibility for waste collection routing.

ACKNOWLEDGEMENTS

This project has received funding from the EFOP-3.6.1-16-00011 "Younger and Renewing University – Innovative Knowledge City – institutional development of the University of Miskolc aiming at intelligent specialization" project implemented in the framework of the Szechenyi 2020 program.

REFERENCES

Glistau, E. & Machado, N.I.C. 2018 Industry 4.0, Logistics 4.0 and Materials - Chances and Solutions. *Materials Science Forum* 919: 307–314.

Gubán, M. & Gubán, Á. 2012. Production scheduling with genetic algorithm. *Advanced Logistic Systems: Theory and Practice* 6, 33–44.

Kaza et al. 2018. What a Waste 2.0: A Global Snapshot of Solid Waste Management to 2050. Urban Development;. Washington, DC: World Bank. https://openknowledge.worldbank.org/handle/10986/30317 License: CC BY 3.0 IGO.

Kota, L. 2012. Optimization of the supplier selection problem using discrete firefly algorithm. *Advanced Logistic Systems: Theory and Practice* 6, 10–20.

Telek, P. 2018. Process-based planning of material handling in manufacturing systems. *IOP Conference Series: Materials Science and Engineering* 448, 012018.

Zhang, L. Liu, W. 2017. Precision glass molding: Toward an optimal fabrication of optical lenses. *Frontiers of Mechanical Engineering*. March 2017, Volume 12, Issue 1, pp 3–17.

Solutions for Sustainable Development – Szita, Jármai & Voith (eds.)
© 2020 Taylor & Francis Group, London, ISBN 978-0-367-42425-1

Application of sigmoid curves in environmental protection

F.J. Szabó

Institute of Machine- and Product Design, University of Miskolc, Miskolc, Hungary

ABSTRACT: Sigmoid curves can describe a lot of phenomena in the world: iteration history of optimization algorithms, time history of sports world records, competition of populations, product lifetime, etc. Approximation of a sigmoid curve can give many interesting and useful information about the most important characteristics of the growth speed, expectable maximum of the phenomenon, maximum possible value achievable, steepness of the curve. The investigation of the derivative and the integral of the sigmoid curve can give more information about where is the maximum of the growth speed, and how durable the maximum speed is. These results could increase our knowledge when investigating a phenomenon (growth history, saturation, future behaviour of the phenomenon). In this paper, the quantity of plastic contamination of the world oceans is investigated. The history curve of the world's plastic production is a sigmoid curve (logistic curve), so every useful information which can be derived from the curve, from its derivative and integral can be applied to the present and to the future of the contamination quantity, because there is a strong connection between the quantity of produced plastics and of the plastic contamination quantity of the seas. The results of the investigation of the sigmoid curve of the plastic contamination will give us information about the history and the future of the contamination process. A better understanding of these characteristics of the phenomena could help to design better the future steps to stop or at least decrease this dangerous contamination form. During the approximation of the sigmoid curves, the Nelder – Mead optimization technique is applied for the solution of the least squares approximation problem.

1   INTRODUCTION

The life- curve of the plastic production of the world and of the plastic production of Europe is a sigmoid- type curve (logistic curve), just like the life curve of any other product. This curve can be approximated on the basis of the existing data concerning the production quantity. Other sources say that approximately 10% of the total plastic quantity fabricated will contaminate the water of the oceans and seas of the world. Approximating the sigmoid curves of plastic production in the world and in Europe, applying the EBSYQ (Evolutionary Based System for Qualification of Group Achievements) curve analysis method (Szabó, 2017), it is possible to show the future of the contamination, which could help specialists to prepare and foresee the quantity of the necessary measures. The analysis of the first derivative of the curves and of the integral of the curves gives also interesting details about the most important characteristics of the contamination quantity. The Lorentz- function, its derivative and integral could enlarge the scope of these results. The EBSYQ method and the curve analysis is applicable in many other fields of our life.

2   OVERVIEW OF SOME SIGMOID CURVES

A typical sigmoid curve could be the iteration curve of evolutionary type optimization algorithms. The iteration history curve of the RVA optimization algorithm for a simple demonstration optimization problem can be found in Figure 1. It can be seen from the figure, that in the beginning phase of the optimization, the improvement in the objective function is high regarding more and more generations, but this "improvement speed" is decreasing in the final phase of the optimum search. This effect can be called as "saturation" and this gives the sigmoid shape of the curve.

On the basis of the mathematical representation of the iteration history sigmoid curve of the algorithm, it will be possible to see the most important parameters determining the shape of the curve and also it is possible to see the most important characteristics of the result (maximum possible value achievable, steepness of the curve which is in connection with the improved speed of objective function, etc). Table 1 shows the curve shape, the first derivative and the integral function shape of some different sigmoid type functions, in order to see and compare the most important characteristics of the sigmoid shape curves.

One can draw some conclusions from the Table 1:

Two different types of curves are possible: in the beginning phase with a curvature (e.g. Pearl- Reed function) or without beginning curvature. Maybe not the curve itself has the sigmoid shape but its integral (e.g. Life- curve). Derivative of the curves also can have different shapes (e.g. Törnquist curve (Törnquist, 1936, 1981), Mitscherlich curve, Life- curve or Pearl-Reed curve).

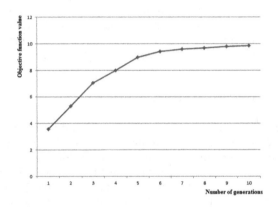

Figure 1.　Iteration history of RVA algorithm.

Table 1.　Some sigmoid curves.

curve	derivative	integral
Pearl–Reed		

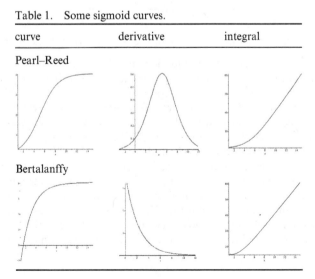

| Bertalanffy | | |

179

For further investigations the following curves will be selected: Pearl- Reed curves because of its beginning curvature, one curve without beginning curvature (Bertalanffy), and the Life-curve because of its very special shape.

Equations of the curves and their derivatives or integrals are as follows:

a.) Pearl–Reed (logistic) curve: (Pearl & Reed 1920) equation of the curve:

$y(x) = \frac{K}{1+ce^{-rx}}$, first derivative: $\frac{dy(x)}{dx} = \frac{Kcre^{-rx}}{(1+ce^{-rx})^2}$,

integral:

$$\int y(x)dx = -\frac{K}{r}\ln(e^{-rx}) + \frac{K}{r}\ln(1 + ce^{-rx}) \tag{1}$$

b.) Bertalanffy-growth curve: (Bertalanffy 1960) equation of the curve:

$y(x) = K(1 - ce^{-rx})$, first derivative: $\frac{dy(x)}{dx} = Krce^{-rx}$,

integral:

$$\int y(x)dx = Kx + \frac{Kc}{r}e^{-rx} \tag{2}$$

In order to analyse the plastic production of the world and Europe, it is enough to select two curves (Bertalanffy and Pearl–Reed) from the possible sigmoid curves of Table 1. Analysis of the life- curve can give special results (eigenvalue of the algorithm, Lorentz-profile, spreading characteristics), so this curve will be studied separately in the future.

3   APPROXIMATION OF THE CURVES

The production curve is approximated by the Pearl- Reed and Bertalanffy curve, by using the method of least squares, determining the parameter values of K, r, c in the equation of the curves which give the best approximation to the iteration history curve.

During the method of least squares, it is necessary to approach the given discrete values (x_i, y_i), $i = 1, 2, 3, \ldots, n$, by a function $y^* = f(x)$, while the parameters of the curve should give the minimum possible value of the sum of the squares of the differences. This means that regarding the function values $f(x_i) = y^*_i$, we have to find:

$$H = \sum_{i=1}^{n} \left(y_i - y_i^*\right)^2 = min. \tag{3}$$

The minimum is possible if the first derivative of the function H is 0, therefore:

$\frac{\partial H}{\partial K} = 0$, $\frac{\partial H}{\partial r} = 0$, $\frac{\partial H}{\partial c} = 0$, this gives three equations for the three unknowns K, r and c, so it is possible to find the parameters for the best approximation. Another possible way to find the minimum of H as a function of the three parameters is to solve the problem as an uncon-strained minimization task of H using the three parameters as design variables. In this paper, this method of optimization is selected for the calculation of the best curve-parameters during the approximations. For the numerical solution of this optimization task, the Nelder–Mead „simplex" algorithm (Nelder & Mead 1965) is used.

The linear regression coefficient will be used to check the quality of the approximation. Thus it is necessary to calculate the regression coefficient for both of the curves. Since the two selected curves are non-linear, before the analysis of the regression it is necessary to transform the equations of the curves into linear form. The regression coefficient calculated for these resulting linear functions will show which curve has the better correlation with the discrete data, so the conclusions derived from that curve will be stronger, i.e. more realistic.

The value of the regression coefficient is always between -1 and +1. If it has a value of 0, that means there is no relationship between the curve and the discrete values. The closer the regression coefficient's absolute value to 1, the better the correlation between the data and the approximation curve. If the regression coefficient is negative, it shows a decreasing tendency, while positive value shows an increase. This means that the conclusions derived from a curve having „weak" regression coefficient will be not „true", not „strong" or not accurate enough, but the conclusions derived on the basis of a curve having good correlation will be true and adequate, or „strong".

For calculation of the regression coefficient, the curve equations need to be transformed into linear form for both of the selected functions. Linear transformation of the Bertalanffy- function:

$$y(x) = K(1 - ce^{-rx}), \ ce^{-rx} = \frac{K - y(x)}{K}, \ lnc + lne^{-rx} = ln\left(\frac{K - y(x)}{K}\right), \tag{4}$$

therefore the linear function for the Bertalanffy- curve is:

$y^* = a + bx$, where $a = \ln c$, $b = -r$.

The linear transformation of the Pearl- Reed function can be done in a similar way:

$$y(x) = \frac{K}{1 + ce^{-rx}}, \ \frac{K - y(x)}{y(x)} = ce^{-rx}, \ lnc + lne^{-rx} = ln\frac{K - y(x)}{y(x)}, \ y^* = a + bx \tag{5}$$

The regression coefficient can be calculated as:

$$R_{lin} = \frac{A_{xy} - \frac{B_{xy}}{n}}{\sqrt{\left(C_x - \frac{D_x}{n}\right)\left(C_y - \frac{D_y}{n}\right)}} \tag{6}$$

where:

$$A_{xy} = \sum_{i=1}^{n} x_i y_i, \ B_{xy} = \sum_{i=1}^{n} x_i \sum_{i=1}^{n} y_i, \ C_x = \sum_{i=1}^{n} x_i^2,$$

$$D_x = \left(\sum_{i=1}^{n} x_i\right)^2 \text{and} \ C_y = \sum_{i=1}^{n} y_i^2, \ D_y = \left(\sum_{i=1}^{n} y_i\right)^2. \tag{7}$$

In Equation (6) one can calculate the linear regression coefficient of the y* transformed function determined in equation (4) or (5), but for simplicity, we returned back to the y notation.

4 ANALYSIS OF PLASTIC PRODUCTION QUANTITY

The quantity of the plastic production in the world and Europe can be seen in Figure 2 in million tones per year, between 1950 and 2010. (Source of the image: internet homepage, plasticseurope.org, 2012).

The shape of the curve is sigmoid curve, which is not surprising because normally the life-curve of any product (and plastics are also products) is a logistic curve (Pearl- Reed curve), which is sigmoid type curve. Another possible curve is the Bertalanffy curve, which is also a sigmoid type curve. Figure 3 shows the approximation of the world's curve, using Pearl-Reed function and the Bertalanffy function. The approximated shape of the function can be extrapolated over 2010 until even 2050, showing the possible future of the plastic production quantity of the world. Comparing the two curves, the Bertalanffy curve is a "pessimistic" guess of the plastic production, showing higher quantity in the future, while Pearl- Reed

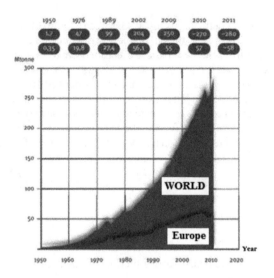

Figure 2. Plastic production quantity of the world and Europe from 1950 until 2010.

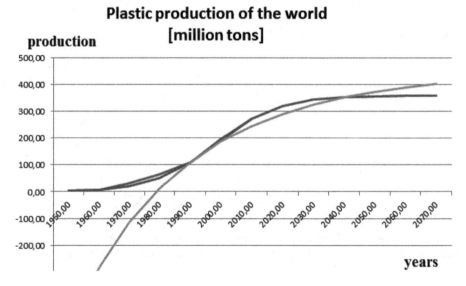

Figure 3. Approximation of the plastic production curve. Blue: real curve, red: Pearl- Reed, green: Bertalanffy.

seems to be an "optimistic" guess or more realistic because at the end of the real curve it is possible to discover some saturation- like behavior or some decrease.

The approximation of the plastic production curve of Europe can be seen in Figure 4. In the real curve, the saturation behavior is a little bit stronger than in case of the world's curve, therefore the Pearl- Reed approximation curve seems to be more realistic.

Equations of the approximation curves, derivatives and integrals can be seen in Equations (1) and (2). The values of K, r, c parameters of the curves are shown in Table 2. The parameter K always shows the possible maximal value achieved at the end of the investigated period, the parameter r is in connection with the speed of increase of the functions.

Comparing the realistic and pessimistic future approximation results, it can be said that the difference is "not too high", which means that concerning, for example, the 2050 year

Plastic production of Europe [mio tonnes]

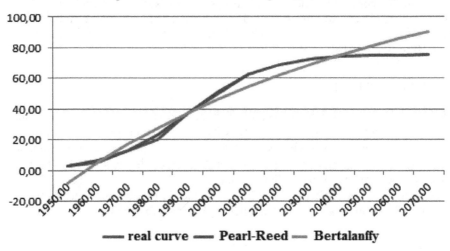

Figure 4. Approximation of Europe's plastic production quantity.

Table 2. Parameters of the approximating curves.

| Parameter | Pearl Reed | | Bertalanffy | |
	Europe	World	Europe	World
K	75,44599681	358,33	118,0199693	445,0
r	0,07931015112	0,09872031606	0,009338095157	0,02576604032
c	3,536853450e+68	4,7080616561e+85	8,564893849e+07	1,396407556e+22

foreseen production quantity, the realistic and pessimistic curve are very close to each other. If we believe the realistic curve, the increasing speed of the plastic production will decrease until 2050 and the production quantity will tend to be constant in the future. This could be also realistic because possibly the production of biodegradable plastics and the environmental- conscious plastic usage and production will do its effects on the decrease of the contamination quantity, too.

5   CONCLUSIONS

Since plastic objects are products, the life curve or logistic curve is an important characteristic of the plastic production quantity in the world and Europe.

Parameters of the curves and their derivatives, integrals can show some details of the future concerning the possible maximum of the production quantity, when it will reach its maximum, how quick is the increase of the production, etc. Since the plastic contamination of the seas is approximately 10% of the production, these results could be available for the observation of the future of plastic contamination of the seas and oceans, too.

The logistic function approximation (Pearl- Reed curve) of the sigmoid curve of plastic production could be the realistic future, because it is based on the product life behaviour of the plastics and plastic production, while the Bertalanffy curve approximation could be a pessimistic future, showing higher quantity of products and higher contamination in the future. The analysis of the approximation curves could show the future even until 2050.

The EBSYQ method and the curve analysis technique could be applied easily in other fields of our life (sport records of individuals and groups, comparison of students by their teacher or comparison of applicants to jobs, prizes, grants, comparison of different optimization algorithms or different settings of the same algorithm), so could be very useful for teachers, jury members, trainers, referees, etc.

ACKNOWLEDGEMENTS

The described study was carried out as part of the EFOP-3.6.1-16-2016-00011 "Younger and Renewing University – Innovative Knowledge City – institutional development of the University of Miskolc aiming at intelligent specialisation" project implemented in the framework of the Szechenyi 2020 program. The realization of this project is supported by the European Union, co-financed by the European Social Fund.

The realisation of this project is co-supported by the European Union and co-financed by the European Social Fund and by the EU Lifelong Learning Program.

REFERENCES

Bertalanffy, L. 1960. Principles of Theory of Growth. In: *Fundamental Aspects of Normal and Malignaent Growth*. Amsterdam. pp. 137–259.

Nelder, J.A. & Mead, R. 1965. A simple method for function minimisation. *Computer Journal 7.*: pp 308–313. doi: 10.1093/comjnl/7.4.308.

Pearl, R. & Reed, L.J. 1920. On the Rate of Growth of the Population of the United States since 1790 and its Mathematical Representation. *Proc. of the National Academy of Sciences.* Vol. 6. No 6. pp. 275–288.

Szabó, F.J. 2017. Evolutionary Based System for Qualification and Evaluation of Group Achievements (EBSYQ). *International Journal of Current Research*, ISSN: 0975-833X, Vol. 9, Issue 08, pp. 55507–55516, August, 2017. www.journalcra.com/sites/default/files/21246.pdf.

Törnquist, L. 1981. Collected scientific papers of Leo Törnquist. Research Institute of the Finnish Economy. Series A. ISBN 978-951-9205-74-8, 1981.

Törnquist, L. 1936. The Bank of Finland's Consumption Price Index. *Bank of Finland Monthly Bullettin*, 10, 1-8. http://www.plasticseurope.org/documents/document/20121120170458-final_plasticsthefacts_no v2012_en_web_resolution.pdf last visit: 2019 january.

Part D: Environmental Management and Ecodesign

Solutions for Sustainable Development – Szita, Jármai & Voith (eds.)
© 2020 Taylor & Francis Group, London, ISBN 978-0-367-42425-1

Comparative study of the influence of traditional walls with different typologies and constructive techniques on the energy performance of the traditional dwellings of the Casbah of Algiers

H.F. Arrar, A. Abdessemed Foufa & D. Kaoula
Lab ETAP, Institute of Architecture and Town-Planning, University Saad Dahlab 1, Blida, Algeria

ABSTRACT: The Casbah of Algiers is classified world heritage in 1992. Their constructive typologies and specific equipment's of walls distinguish its traditional dwellings. They are mainly distinguished by the principles of eco-design and climatic adaptability. The present work aims knowing the influence of a traditional wall on the energetic performances of a dwelling of the Casbah of Algiers through different compositions in order to choose the most suitable typology for the ecological and environmental question. The adopted methodology is structured on the characterization of materials and an evaluation, through a series of simulations, of the influence of thermal inertia on the energy performances. The results obtained allowed us to enrich the ecological framework of the materials by determining the optimal composition to the eco-design for a better revaluation of our built inheritance and a better safeguarding of environmental resources.

1 INTRODUCTION

Traditional building is a living complex that reflects the needs for which it was originally constructed. The house should always respond to its inhabitants' way of life; it is characterized by centrality, introversion of the built space and arrangement of rooms around the patio.

Reading the historical development of a monument, or an urban fabric, cannot be achieved without the thorough knowledge of the used materials and techniques, which help to identify the different phases of construction. The implementation of materials and work tools leave marks that cannot be interpreted and understood without prior knowledge of construction techniques (Mileto 2007). The identification of materials and construction techniques used locally is therefore a crucial and substantive step to guarantee the structural homogeneity and physical integrity of the building.

It is worth noting that about 17% of World Heritage buildings are earthen constructions (Pignal 2005), which proves the durability of the construction material used, just like several other natural and ecological materials capable of ensuring a good energy efficiency in terms of thermal inertia and resistance.

Indeed, a sufficient thermal inertia would allow the building to improve its stability in spite of the outdoor temperature variations; however, thermal inertia at night slightly raises the minimum temperatures (Cheng 2005).

Conventional building materials could be the basis for reflection about the design of a sustainable architecture that could connect man with his natural environment (Ghaffour 2014). In addition, earth is a biodegradable and recyclable material (Little 2001); it does not generate, or very little, construction waste and can be mixed with a number of natural materials such as plant fibers, wood, stone, etc. (Dubost 2011).

This material undergoes no polluting transformation from its extraction to its implementation. If a building is destroyed, it can be reused to erect other walls. It is infinitely recyclable (Moriset 2011).

In Algeria, a vast earthen heritage is encountered throughout the entire territory; in fact, southern Algeria is well known for its Ksours and the north for its architectural heritage, particularly in the historic cities of Cherchell, Kabylie, Tlemcen, and the Casbah of Algiers, which was declared a world heritage site of Unesco in 1992. There are constructions with mixed materials, such as rammed earth, brick and stone. It is in this historical context that the present work aims to treat the thermal properties of these materials through the assessment of the thermo-energetic behavior of four variants of a traditional house, which differ by their construction techniques and the construction materials used, namely "raw earth", "terracotta" and "rubble". In order to determine the most suitable typology with regard to the ecological and environmental issue. Furthermore, the concept of compactness is also investigated in order to address heat loss problems for the purpose of reducing energy consumption and improve the thermal comfort (Munaretto 2014).

Human thermal comfort involves two main approaches. The first one is the classical model, called the Fanger model, which, according to the ANSI/ASHRAE Standard 55-2013, considers that human comfort depends on the combined quantitative influence of six parameters, namely the metabolic rate, clothing insulation, air temperature, mean radiant temperature, air speed and relative humidity. The second one is the adaptive model that establishes the relationship between acceptable indoor design temperatures and outdoor meteorological and climatological parameters (Udrea 2015).

The indoor thermal environment is affected by internal sources and external sources. Common sources of heat include electrical equipment (lighting, computers, etc.), solar radiation, and human body heat. Common sources of cold include glazed surfaces, weakly insulated walls and thermal bridges in buildings. All these sources have an impact on the perception of the environment by human beings as well as on their level of thermal comfort (Corgnati 2011).

2 METHODOLOGY

2.1 *Simulation approach*

Buildings are huge energy consumers; therefore, important research on the energy efficiency within constructions is urgently needed. Modelling of energy equipment and physical phenomena within the building is the first essential step that can help us achieve optimal management of energy flows; this would be an important stage towards the sustainable development in our society (Hoang 2014).

The proper management of a construction's energy flows would certainly limit the energy consumption in a reasonable way. Thorough knowledge of energy flows is necessary and essential for making decisions about different tasks related to buildings (Morel 2009).

Thermal simulation models must meet the needs of the investigation; they are supposed to materialize the combined effect of thermal phenomena, such as heat exchange through buildings by the three modes of heat transfer, i.e. conduction, convection and radiation, as well as by ventilation and air movement (M'sellem 2007).

For cost and time reasons, simulation is an effective way to develop and study the thermal behavior of buildings under variable conditions.

Thermal analysis simulation is achieved in a perspective of integrating the climatic and physical parameters of the materials into the process of improving the thermal performances of buildings in view of exploring and optimizing certain decisions in order to achieve the best thermal comfort. This thermal analysis simulation also allows for the evaluation and thermal control of buildings.

The methodology adopted is structured based on the characterization of the materials used and also on a comparative assessment of the energy requirements and interior temperatures of four modelled variants of a traditional patio house in the Casbah of Algiers; these variants are differentiated by the materials used, i.e. terracotta, terracotta-rubble, raw earth, and brick-concrete. This was achieved based on modelling and a series of simulations using the Pleiades software, while taking into account the specific climatic conditions of the city of Algiers. The outdoor environment has a significant impact on the thermal comfort inside the houses. In addition, the geographical location, outdoor temperature and solar radiation are three environmental factors that also affect the thermal comfort (Noel 2018).

Using this approach, we will be able to determine transiently air temperatures, heating needs and air conditioning. This provides us with the opportunity to evaluate, compare and retain the most suitable variant with regard to both climate and energy dimensions.

2.2 Case of study

Traditional houses, included in our case study, are distinguished by their very interesting constructive typologies and specific mixed walls (Opus spicatum and Opus incertum) beside other building compositions based on local materials (cooked brick, rubble, wood, etc.), with a wall thickness between 40 to 70 cm (Abdessemed 2005). These constructions are characterized by the Ecodesign principles, as well as by an environmental symbiosis and a climatic adaptability.

More specifically, we have selected an old center house with the constructive characteristics of the traditional houses of the casbah-type (Ateliers Casbah1980) with "Patio" porticoes down to the street of the Frères Bechara, the surface of the house is 252.42 m² and consists of a ground floor and two levels, plus an accessible terrace. The plans for the modelling are of "DAR IV.8.1" Boussoura Abderrahmane. (Missoum 2003)

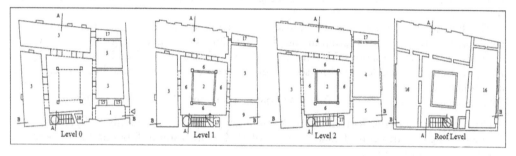

Figure 1. Plans.
Legend: 1. Entrance 2. Center of the house 3. Bedroom 4. Bedroom with bathroom 5. Bedroom 6. Gallery 7. Stairs 8. Shop 9. Kitchen 10. Toilet 11. Laundry 12. Well 13. Cistern 14. Bedroom on terrace 15 Masonry bench 16. Terrace 17. Mid-height space.

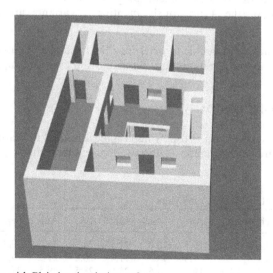

Figure 2. 3D modeling with Pleiades simulation software.

189

2.3 Specific climatic features

Table 1 Summarizes the climatic data of the city of Algiers selected during our evaluation.

Name of site = Algiers. Latitude [N] = 36°43'1" 'Longitude [E] = 3°15'0" Altitude [m] = 25

The climate of Algiers is of Mediterranean type, known for its long hot and dry summers and mild and wet winters. The rains are abundant and can be diluvial. Spades are usually from mid-July to mid-August.

2.4 Technical specificities

The study of the different variants, which are distinguished by the materials used, leads one to a comparative reading of the grids, which allows us to define their energy performance. The following summary table (Table 2) gives the details of the technical specificities of the different building compositions used.

2.5 Simulation

Computer simulation evaluation, using an efficient software program, offers a significant advantage due to its flexibility and the very interesting data it can predict. This approach is highly recommended for this kind of tests. It offers the possibility to vary different parameters relating to the building and on based the climatic data of the site to be studied.

The Pleiades V4.19 software, developed by Izuba Energies, is used in this study; it is a comprehensive software program intended for building design, and energy and environmental assessment of constructions. The first step consists of creating a "virtual" volume using the Alcyone graphical modeler, and the second one concerns the entry of the building envelope and its characteristics.

The selected scenario is the same for all 4 variants where only the materials used differ.

- The occupancy scenario corresponds to a minimum occupancy of a family of 4 persons.
- The heating scenario considers a set point of 19°C from 7 am to 8 pm and a set point of 15°C the rest of the time.
- Standard air conditioning 25°C.
- A breakdown of 0.6 vol/h.
- The lighting: the rooms: 300 Lux, and for the gallery and sanitary 50 Lux.
- The simulation covers the period of one entire year and the data for the city of Algiers are downloaded using the MétéoCalc complementary extension.
- Compositions: One of the 4 previously mentioned variants are considered in each simulation.

The software used makes it possible to determine the heating and cooling requirements for each zone; it also helps to find the minimum, maximum and average temperatures.

3 RESULTS AND DISCUSSIONS

Thermal comfort is studied from the following graphical results, indicating for the four variants the heating requirements (Figure 3), the cooling requirements (Figure 4) and the total

Table 1. The climatic data of the city of Algiers.

	Jan	Feb	Mar	Apr	May	Jun	Jul	Aug	Sep	Oct	Nov	Dec
Avg. T (°C)	11,5	12,1	13,5	15,5	18,2	21,5	24,3	25,2	23,3	19,4	15,1	12,3
Min. T (°C)	8,1	8,3	9,7	11,3	14	17,4	20,2	21	19,6	05,7	11,5	9
Max. T (°C)	14,9	15,7	17,3	19,7	22,5	25,6	28,5	29,5	27	23,2	18,8	15,6

Table 2. The details of the technical specificities of the various compositions.

	Variant 1	Variant 2	Variant 3	Variant 4
Floor	Solid brick 5.5 cm Lime mortar 4 cm Raw earth 15 cm Log of wood 10 cm Total: 34.5 cm R = 0.76 Ms = 589	Solid brick 5.5 cm Lime mortar 4 cm Raw earth 15 cm Log of wood 10 cm Total: 34.5 cm R = 0.76 Ms = 589	Solid brick 5.5 cm Lime mortar 4 cm Raw earth 15 cm Log of wood 10 cm Total: 34.5 cm R = 0.76 Ms = 589	Concrete hollow 16 cm Concrete 5 cm Cement screed under flooring: 3 cm coating 2 cm Total: 26 Cm R = 0.18 Ms = 438
External wall	2 * Lime plaster 0.7 cm on both sides 2 * Mortar 5 cm on both sides Raw earth 50 cm Total: 61.4 cm R = 0.55 Ms = 1163	2 * Lime plaster 0.7 cm on both sides 2 * Mortar 5 cm on both sides Solid brick 25 cm * 2 (Control wall with double switchgear) Total: 61.4 cm R = 0.54 Ms = 1170	2 * Lime plaster 0.7 cm on both sides 2 * Mortar 5 cm on both sides Control wall with mixed equipment (terracotta + rubble) 50 cm Total: 61.4 cm R = 0.46 Ms = 1229	2 * Cement plaster 2 cm on both sides Hollow brick 15 cm Blade of air 5 cm Hollow brick 10 cm Total: 34 cm R = 0.85 Ms = 301
Internal wall	2 * Lime plaster 0.7 cm on both sides 2 * Mortar 3 cm on both sides Raw earth 20 Cm Total: 27.4 cm R = 0.26 Ms = 517	2 * Lime plaster 0.7 cm on both sides 2 * Mortar 3 cm on both sides Solid brick 10 cm * 2 (Control wall with double switchgear) Total: 27.4 cm R = 0.25 Ms = 520	2 * Lime plaster 0.7 cm on both sides 2 * Mortar 3 cm on both sides Control wall with mixed equipment (terracotta + rubble) 20 cm Total: 27.4 cm R = 0.22 Ms = 540	2 * Cement plaster 2 cm on both sides Hollow brick 10 cm Total: 14 cm R = 0.36 Ms = 136
Roof	Lime sealing 5.5 cm Lime mortar 4 cm Raw earth 15 cm Log of wood 10 cm Total: 34.5 cm R = 0.79 Ms = 572	Lime sealing 5.5 cm Lime mortar 4 cm Raw earth 15 cm Log of wood 10 cm Total: 34.5 cm R = 0.79 Ms = 572	Lime sealing 5.5 cm Lime mortar 4 cm Raw earth 15 cm Log of wood 10 cm Total: 34.5 cm R = 0.79 Ms = 572	concrete hollow slabs 16 cm Concrete 5 cm Polystyrene 4 cm lime plaster 3 cm Heavy protection in rolled gravel 4 cm Total: 32 cm R=0.25 Ms=548

Note that: R = thermal resistance (m².K/W) and Ms = surface density (kg/m²).

energy required (Figure 5). In this section, the internal temperatures are also studied based on the data generated by different simulations (Figures 6 and 7).

The first three figures (3, 4, and 5) give a comparison of the energy needs for all four variants, based on the different simulations performed. Examining the results presented in Figure 3, which displays the heating requirements, one should note that the heating needs partly decrease as the thermal conductivity of the different materials used goes down.

The results obtained for both variants 1 and 2 (raw earth and cooked brick) are very interesting and promising; because their heating needs are estimated at 12 kWh/m²/year. However, for variant 3 (Terracotta – rubble), this is 14kWh/m²/year. The difference is resulted by the material used (rubble) whose thermal conductivity is 1.700 W/m.K, while those of the materials terracotta and raw earth, they are equal to 1.150 and 1.100 W/m.K, respectively. Regarding variant 4 (Concrete - Brick), although the wall conductivity is better than that of

191

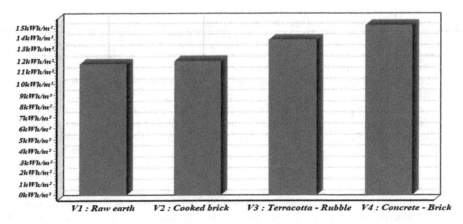

Figure 3. Heating needs.

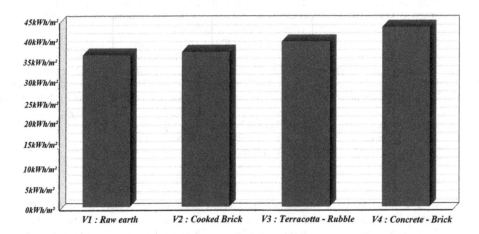

Figure 4. Air conditioning needs.

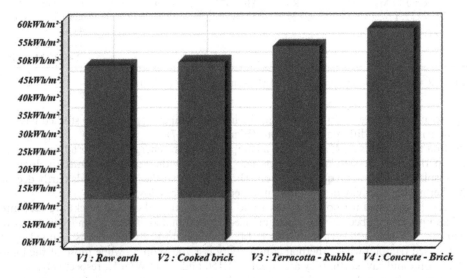

Figure 5. Energy needs.

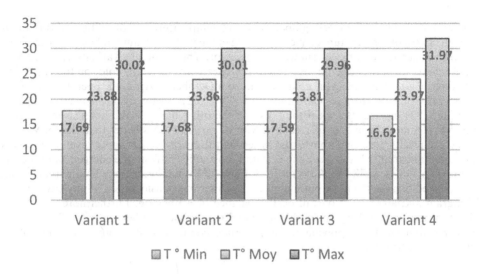

Figure 6. Internal temperatures of different variants.

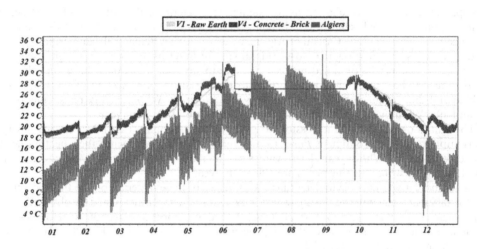

Figure 7. Opérative temperatures/Outdoor temperature.

terracotta or raw earth (0.82 W/m.K against 1.150 W/m.K and 1.100 W/m.K, but it has a smaller wall thickness, (34 and not 61.4). The floor conductivity remains very important, with 1.56 W/m.K for variant 4 (Concrete Brick) against 1.03 W/m.K for the other variants.

Similarly, for the air conditioning needs illustrated in Figure 4, it is noted that the variants with raw earth and cooked brick are those that need less air conditioning, with 36 kWh/m²/year and 37 kWh/m²/year, respectively. However, the variant with Terracotta and rubble needs 40 kWh/m²/year and that with concrete and brick demands 43 kWh/m²/year.

It should also be noted that, according to Figure 6, the internal temperatures of the different variants are satisfactory and offer good thermal comfort due in part to the type of construction of the patio house. Nevertheless, one can notice that through the variants with cooked brick (variants 2 and 3) and variant 1 with raw earth, up to 2 °C of cooling comfort in T ° Max and 1°C of heating comfort for T ° Min are gained in comparison with variant 4 (concrete - brick).

– For Maximal Text of 36 °C: Variants 1, 2, 3 (T = 30 °C) - Variant 4 (T = 31.97 °C)
– For Minimal Text of 2 °C: Variants 1, 2, 3 (T = 17.6 °C) - Variant 4 (T = 16.6 °C)

Figure 7 illustrates the highest global operative temperatures of the different variants as well as the outdoor temperature of the city of Algiers, based on an annual balance sheet. Variant 4 with concrete and brick is shown in blue color. The results for variants 1, 2 and 3 are similar and are therefore superimposed; they are represented in yellow color on the graph. It is interesting to note that during the cold season, from October to April, the operative temperatures are within the comfort standards, i.e. 16 ° < T ° < 27 °. Moreover, the energy efficiency for variant (V1) has a better yield than V4; on the other hand, in the hot season, i.e. from April to September, the overall thermal behavior of variant (V1) is better than that of V4.

From the results presented above, and based on the energy performance given in Figure 4, it appears that variants 1 and 2 are the best and most efficient from the energetic and thermal comfort points of view, 48 kWh/m²/year for V1 and 49 kWh/m²/year for V2. Note that variant 2 with cooked brick is the most optimal; it deserves serious consideration in any environmental program and also in the overall energy balance of a building. Furthermore, it is worth noting that raw earth is very vulnerable to humidity and bad weather. In spite of this, its use can be recommended by applying the appropriate coatings or incorporating adjuvants to improve its performance.

4 CONCLUSION

The present research work allowed us to examine and determine the influence of construction materials on the thermal behavior of a building, in terms of heating and cooling needs of a patio heritage house located in the Casbah of Algiers, by considering different wall compositions.

It is thus possible to know the most optimal variant in terms of energy, through a series of simulations. The results obtained make it possible to say that the cooked brick and the raw earth offer the best results in comparison with the variant with a mixture of terracotta and rubble and the variant in traditional masonry and concrete. Whose energy requirements are 48 and 49. kWh/m²/year and the heating requirements are 12 kWh/m²/year.

The values obtained make it possible to place the two variants 1 and 2 in the low energy consumption building (Bâtiment *Basse Consommation* - BBC) label since their heating needs are lower than 50 kWh/m²/year. These values can therefore be considered as close to the passive house label whose heating needs are 10kWh/m²/year.

In addition, it was possible to determine the influence of these materials on the thermal behavior for different wall compositions for which a reduction in temperature could be obtained, since for a maximum outdoor temperature of 36 °C, the indoor temperature was 30 °C. However, for a minimum outdoor temperature of 2 °C, the indoor temperature was 17.6 °C. The average annual room temperature was 24 °C.

These encouraging results urge us to optimize the use of these materials and integrate them into the wall composition of our buildings for better indoor hygrothermal comfort, without major energy consumption, in order to better enhance our built heritage and therefore protect our environmental resources.

REFERENCES

Abdessemed Foufa A. 2010. The rehabilitation manual as conservation tools in the framework of perma nentplan and enhancement of the safeguarded sector of the Casbah of Algiers. Barcelona: IEMed.
Atelier Casbah 1980. Algiers Casbah Development Project, Preliminary Development Plan. Ministry of housing and construction.
Cheng, V. & Givoni, B. 2005. Effect of envelope colour and thermal mass on indoor temperatures in hot humid climate, Freiburg: The International Solar Energy Society.
Corgnati 2011 S.P. Indoor climate quality assessment – evaluation of indoor thermal and in door air quality. Brussels: Rehva Guidebook 14. Rehva.
Dubost, O. 2011. Environment and sustainable city. Paris: Ed. Weka.
Ghaffour, W. 2014. Patrimoine architectural, entre technicité, confort et durabilité. Magister, Thesis: 49-50. Tlemcen: University of Tlemcen.

Hoang A.D. 2014. Modélisation en vue de la simulation énergétique des bâtiments: Application au pro-totypage virtuel et à la gestion optimale de PREDIS MHI. Sciences de l'ingénieur[physics]. Grenoble University.

Ioana Udreaa 2015. Thermal comfort in a Romanian Passive House. Preliminary results. Sustainable Solutions for Energy and Environment, Bucarest: Polytechnic University.

Little, B. 2001. Building with earth in Scotland: 11-12. Edinburgh: Ed. CRU.

M'sellem, H. 2007. L'évaluation du confort thermique Par les techniques d'analyse Biocli-matiques Magister Thesis. Biskra: Université of Biskra.

Noel, P. 2018. Évaluation du confort thermique à la suite d'abaissements de la température de consigne des thermostats en mode chauffage à l'aide de données mesurées in situ et de simulations - Thesis, School of Higher Technology - University of Quebec.

Morel, N. 2009. Energetic building. Lausanne:Solar energy and building physics laboratory (LESO-PB).

Moriset, S. & Misse, A. 2011. Renovate and build adobe in the Livradois-Forez Regional Nature Park.

Mileto, C. 2007. Stratigraphic analysis of architecture and its application to traditional architecture II. rehabilitation buildings. Barcelona: Ed.ApaiefladorsiArquitetesTècnis.

Munaretto, F. 2014. Study of the influence of thermal inertia on the energy performance of buildings. Paris: National School of Mines.

Pignal, B. 2005. Raw earth: Construction and restoration techniques: 10-11. Paris: Ed. Eyrolles.

Missoum, S. 2003. Algiers at the time Ottoman (sixteenth-twentieth century): The medina and the tradi-tionnel house. Aix-en-Provence: Ed. Edisud.

Solutions for Sustainable Development – Szita, Jármai & Voith (eds.)
© 2020 Taylor & Francis Group, London, ISBN 978-0-367-42425-1

Rules of environmentally friendly packaging

Á. Takács
Institute of Machine and Product Design, University of Miskolc, Miskolc-Egyetemváros, Hungary

ABSTRACT: The paper deals with environmentally friendly packaging. As the environment consists of not only the environment itself but the human and the product are also part of it, the interrelation of the product-human-environment cycle should be analyzed. It should be scanned what the designer should keep in front of the eye during the design process of an environmentally friendly package. Of course, on behalf of protecting the environment, one of the most significant questions is the material that is used for the package. It should be kept in one's mind what kind of effects have the packaged product on the package and how does this effect work backwards. If we look on the package as it is the product we should analyze the effect of the package on our green environment and on the users as well. For example, is the material of the package degrades, is it easy to open the package? The paper deals with these questions of designing environmentally friendly packaging.

1 ENVIRONMENTALLY FRIENDLY DESIGN

It is not easy to shortly define the environmentally friendly design. According to the several components it has, it is quite a complex process. According to Zilahy environmentally friendly design systematically concentrates to the environmental impacts that are potentially coming in the fore during the whole life-cycle of products and services, and to reduce or eliminate these expectable impacts still in the design process. (Zilahy, 2004)

Orbán defines DFE as a design, that minimalizes the undesired impacts for nature (DFE = design for the environment). (Orbán, 2006) DFE is the necessity of the developed product causes a less harmful impact on the environment that is an ever-growing claim of today.

Due to the literature of the field environmentally friendly design or DFE or Green design or eco design mean only the protection of nature does not pay any attention to the protection of the human that is a component of the green environment, only indirectly referring to it. It is essential to notice that the man only as a designer but also as part of the green environment appears in the machine-human-environment cycle. The elements of the cycle are interrelationship continuously, as Figure 1 shows. So, the human designs for itself and for the environment as well. Machine has the effect for the human and for the environment too. The environment also has the impact for the human and for the machine. So, the environment means not only the nature over the office, the factory, but the direct environment of the human where it works, so the workplace. As for the further researches, it would be practical to mention and analyze ergonomics as the element of the environmentally friendly design.

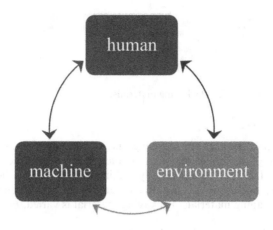

Figure 1.   Machine-human-environment cycle.

2   FRIENDLY PACKAGING

2.1   *Tasks of packaging*

The purpose of the packaging is to preserve the quantity and the quality of the product from production to its intended use. In addition, from the point of view of environmental protection, it is important that the product does not cause harm to our environment and that the packaging is as environmentally friendly as possible. The tasks of packaging:

– product protection
– rational design of handling and delivery units
– designing appropriate storage units
– designing appropriate sales units
– awareness raising, informing = product sales - marketing task.

If Figure 1 is slightly rewritten, Figure 2 can be defined. As the environment has an impact on humans and packaging, packaging and humans are also affected by the environment. It is therefore not enough to pack the product to protect it from the environment and the environment from it. We also should prescribe the packaging material and the way of packaging. This is true even if we look at this triple cycle from the perspective of man. The product should be protected from mankind (creaking), but it should also be easy to open, while the packaging has the least possible impact on the environment.

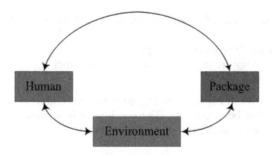

Figure 2.   Human-environment-package cycle.

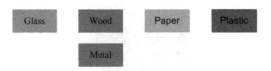

Figure 3.   Packaging materials.

2.2   *Materials for packaging*

Nowadays, environmental protection, recycling and waste reduction are playing an increasingly important role in global pollution. Waste reduction has a key role in a cleaner future. It cannot be equal to the fact that we are already picking up the waste at home, and e.g. we use yoghurt jars for different household tasks after we ate the yoghurt. So most of the household waste comes from packaging materials. Packaging material is varied, but the plastic is heavily predominant.

Figure 3 illustrates the types of materials that occur most frequently, colors show the eco-friendly value of the given material. Of these materials, priority should be given to those that are most recyclable. The glass is perfect for collecting and refilling again and again (beer glass, compote jar), but it is also easy to recycle. That is the reason for it is signed with light green in Figure 3. Wood is not only an easily recyclable organic material, but wood packaging can be used several times before it should be repaired or disposed of (pallets, fruit boxes). The metal and the paper can be very good recycled. Metal is signed with dark green in Figure 3 because it can be recycled with high efficiency. Wood is also signed with dark green: although it is not so easy to recycle, we can reuse wood packaging several times before recycling. Paper is signed with yellow, it can be reused, but not so many times as wood packaging; it can be recycled quite easy, but not so easy like metal. The problem is with plastic packaging because they are not organic, so they are signed with red in Figure 3. Although their loading capacity is very good, as they are resistant to shocks and drops, their recycling is not always possible and therefore is not a good choice for the environment. Certain plastics are also dissolved in foods, so they should be carefully chosen. When selecting the material for the packaging, if the plastic cannot be replaced, the aim should be to reduce the used amount at least (Otto & Wood 2008, Boylston 2009). From materials introduced in Figure 3 glass, wood, metal and paper are eco-friendly materials, because if we do not recycle them, they have no harmful effect on the environment. In contrast, plastics are releasing deteriorative substances whether we recycle them or not.

2.3   *Eco-friendly package*

If we start a search on the web with the keywords "eco-friendly packaging" and we are curious about the images, the first few results are shown in Figure 4. Based on the results, the following can be defined:

– These packages usually made of paper and the paper can be recycled at low environmental load.
– They may already be made of recycled paper.
– Low toner, practically they are made of entirely natural paper.
– Only one material is used.

Among packaging shown in Figure 4 the egg holder in the middle row made by Maja Szczy-pek, a Polish designer can be interesting. The egg holder is made of straw, and the straw is bonded without any adhesive under hot pressure.

2.4   *User-friendly package*

If we keep in mind that the packaged product who will be bought by it is advisable to consider the user-friendly aspects. These aspects also depend on age, gender, nationality and income,

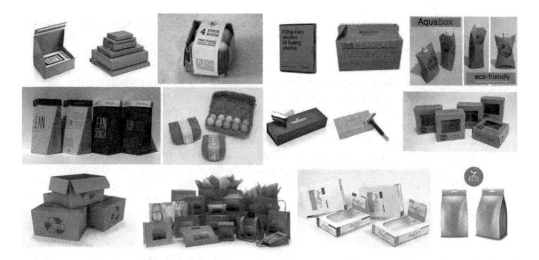

Figure 4. Eco friendly packaging, similarity can be seen. (Source of photos: Internet, Google Search Engine).

but primarily the age and gender of the user, which is the most influential in terms of human power.

It should not be mentioned that men are generally stronger than women (Figure 5). Men and middle-aged people are generally stronger than the elderly and children. The strength of the elderly and the children is usually the same, the difference is in the size of their hands. It is also important that certain disadvantages reduce physical strength (e.g. rheumatism). Normal grip force is around 200N. (Heiniö et. al.)

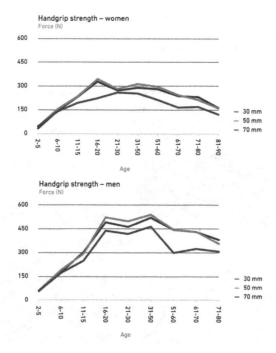

Figure 5. Changing of the handgrip strength in case of women and men (Heiniö et. al. 2008).

199

These should be taken into account if, for example, we plan to pack a medicine. If we design a vitamin box for an elderly person, it should be easy to open, but we should pay attention to prevent a child from opening it because the medicine is not candy!

Packaging of toys may be considered to be the most critical in the point of view of the environment and the users. Manufacturers prefer transparent plastic packaging because this way we can see inside the box, so the product offers itself. In addition to today's printing technology, a realistic photo of the product can be made for the packaging, the child would see what is in the box. Thus, the various fixing elements could be left out, as the product would fit easily in a smaller box and the walls of the box would not allow any movement of the product.

In the case of toys, great care must be taken to ensure that the packaging can be easily removed by one hand, as the parent may open the toy for the elder child as the younger one sits on his/her lap and must also hold the younger child. We also should pay attention not to leave any tiny piece of the package on the product. (For example, plastic pieces due to electrostatic charge or pieces of binding materials.) One of the most prominent examples for the remaining pieces of the packaging in the toy is the packaging of some dolls. In the box, the doll's head is fixed to stand under the transparent foil as in a shop window. The fastening gimp cannot be removed from the doll's head, but it can be cut between the hair. The remaining part can damage the child. Of course, there are also many excellent examples of safe packaging. One of these is the packaging of a toy tablet for babies. The packaging is mainly made of paper, open from the front so that the customer can look at the toy or try it out. Figure 6 shows that the carton of the packaging can be torn at the back so that the two plastic keys that are attaching the toy to the carton become available. By turning the plastic keys, the packaging and the plastic key can be removed from the game. The figure also shows the back of the toy with the two grooves into which the plastic key fits and in which this plastic key can be turned. This product is an excellent example of the fact that packaging does not start after the product is ready. The packaging should be designed in parallel with the design of the product.

The following requirements are recommended to consider in case of user-friendly packaging (Heiniö et. al. 2008):

– The opening mechanism should be visible.
– The opening mechanism should be easy to grip.
– The packaging needs to be easy to hold and lift.

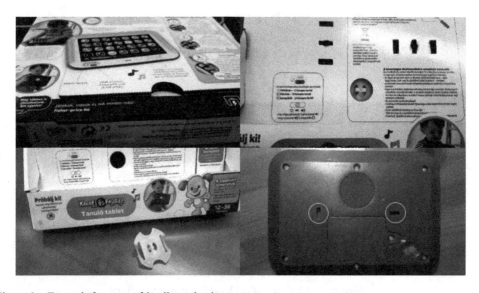

Figure 6.   Example for a user-friendly packaging.

- The force needed for opening should be low.
- Minimize the need for using both hands to open the packaging.
- Consider the stiffness and smoothness of the packaging material.
- Minimize the risk for leaks in and breakage of the packaging.
- Avoid the packaging being broken during opening.
- Minimize squirts and waste of the product when opening the packaging.

3 SUMMARY

This paper summarizes some of the possible guidelines that you should take into account when designing environmentally friendly packaging. The article says that environmentally friendly packaging design is not only about protecting the environment, but also about protecting users and the products as well. Among these basic principles, designers should concentrate better on the fact, that recyclable is not enough, reuse and biodegradable materials would be better. Packaging introduced in Figure 4 is almost fully made from paper, which was signed with yellow in Figure 3. It seems to be a dead-end if paper means the eco-friendly packaging, while there are materials that have a smaller impact on our life.

In our globalized world, a paper bag counts as a very eco-friendly choice for carrying our things. But is it really eco-friendly when we realize that this paper bag we bought here in Europe was transported by ship from China where it was produced from the wood that was extracted in India?

ACKNOWLEDGEMENT

The described article/presentation/study was carried out as part of the EFOP-3.6.1-16-2016-00011 "Younger and Renewing University – Innovative Knowledge City – institutional development of the University of Miskolc aiming at intelligent specialization" project implemented in the framework of the Szechenyi 2020 program. The realization of this project is supported by the European Union, co-financed by the European Social Fund.

REFERENCES

Boylston, S., 2009. Designing Sustainable Packaging, Laurence King Publishing.
Heiniö, R.-L., Åström, A., Antvorskov, H., Mattsson, M. & Østergaard, S., 2008. NT Technical report: Scientific background for the basis of an international standard for easy-to-open packages, ISSN 0283 7234, Nordic Innovation Centre.
Orbán F. 2006. Környezetszempontú tervezés. PTE PMMK Gépszerkezettan tsz. HEFOP-3.3.1., előadásfóliák http:\\www.pmmf.hu
Otto, K. & Wood, K., 2008. Product Design – Techniques in Reverse Engineering and New Product Development, Prentice Hall.
Zilahy Gy. 2004. A tisztább termeléstől az ipari ökológiáig - irányzatok a vállalati környezetvédelemben *267-280 in* Kerekes S. & Kiss K. (ed.) *Környezetpolitikánk európai dimenziói* Bp.: MTA Társkut. Közp., 304 p.

Part E: Circular Economy and Life Cycle Approaches

Solutions for Sustainable Development – Szita, Jármai & Voith (eds.)
© 2020 Taylor & Francis Group, London, ISBN 978-0-367-42425-1

Aluminium infinite green circular economy – theoretical carbon free infinite loop, combination of material and energy cycles

Norbert Babcsán
Innobay Hungary Ltd., Misolc, Hungary
University of Miskol, Miskolc, Hungary

ABSTRACT: Aluminium is the 3rd most abundant element of the Earth's crust. Novel tech-nologies which appeared recently decreasing the carbon consumption of the aluminium life cycle. Aluminium is widely used as structural material and an energy carrier. In this paper, an infinite double life cycle is described demonstrating, that the economy circle of aluminium as a permanent material or as an energy carrier can be combined into one infinite loop. Several recent innovations played a role in closing the loop as novel structural materials like hybrid aluminium and novel energy conversion devices as aluminium batteries. The material circle: aluminium production, fabrication, manufacturing and waste handling can be made green by the invention of novel technology of carbon free melting of the scrap. The energy cycle for aluminium production, assembly, discharging, charging can be made green by the novel carbon free smelting process thus combining the two cycles into one infinite green loop. In this context, the suggested aluminium production, utilization and recycling processes can pro-vide important global solutions for general utilization of aluminium for future applications.

Aluminium is the 3rd most abundant element of the Earth's crust. Thus, the source of the metal is nearly infinite. Aluminium is one of the most important contributors and playing the role of the main accelerator of the transition from a linear to a circular economy (Circular Economy, 2016). The current approach to the circular economy oversimplifies the classification of materials and products as renewable or non-renewable, re-usable or non-reusable or even bio-degradable or non-biodegradable (Permanent Materials, 2016). *A material is defined as permanent if its inherent properties do not change during use and through solid-liquid transformation, it can revert to its initial state. This is the case when the material consists of basic components, which are either chemical elements or robust chemical compounds, making repeated use and recycling possible without change of inherent material properties* (Conte, 2014). Permanent materials slowly degrade during recycling, but the material kept in the loop without losing its intrinsic material characteristics. Aluminium is a permanent material which is more than a re-usable material, one for which the inherent proper-ties do not change during use and following repeated recycling into new products. Of course, alu-minium once used, has to be collected and sorted properly, in order to make it available for its next use phase.

The first ton of aluminium was produced in the US in 1886 when Alcoa started his operation as the first aluminium smelter of the world. In 2018 (Bertram, 2017) the world aluminium stock in use was 1066 Mt (140 kg per capita). The total aluminium ingot was 97 Mt from which 65 Mt primary aluminium from alumina and 32 Mt secondary aluminium from scrap. The aluminium stock of the world is developed during more than a 100 years period. The quantity of the alumin-ium stock shows an exponential increase in time and will be doubled within 15 years from now. The primary aluminium production rate of the 2000's is also doubled for 2030. Assuming a 2000 USD/t price, the total market volume of aluminium production in the world is a 200 billion USD. This amount is around Toyota car manufacturer revenue or 50% of GDP of Sweden.

Players of the world aluminium industry can be classified into two major groups: aluminium raw material producers and aluminium product manufacturers (aluminium transformers). The former bauxite mining companies and the raw material production units of the large aluminium raw material production groups are merging creating huge industrial concerns (Rio Tinto and the fusion of the metallurgical part of Alcan, splitting of Alcoa into New Alcoa and Arconic). Aluminium product manufacturers 30-30% share of the transportation industry and the construction industry and the remaining 40% is distributed equally between consumer goods, power lines, packaging materials and machine parts. For example, some international companies have a significant role in certain market segments: Magna (transportation), Arconic (all three), Alucoil (construction). Both production and consumption are nearly equally distributed between China and the rest of the world. Due to the unique situation in China, the structural reorganization of the Chinese aluminium industry is driven by other processes than the rest of the world, but certain fusion can be observed between the Russian, Chinese, and Rio Tinto Group.

1 ALUMINIUM AS A MATERIAL

1.1 *Aluminium material circle*

Aluminium is used in its permanent material form as a structural or functional material. As a structural material, aluminium is used in the building industry and mobility. As a functional material, it is used in the cable and the packaging industry. The *properties of the aluminium are mostly set by alloying* and traditional aluminium transformation technologies as solidification, casting, rolling, extrusion, wire drawing etc. The raw aluminium enters into the material cycle through a high temperature metallurgical process which transforms the oxide of aluminium into aluminium metal. The aluminium material cycle contains four main steps:

Production of aluminium metal: this step is the transformation of aluminium scrap or secondary aluminium into purified and re-alloyed bulk aluminium blocks.

Fabrication of aluminium metal blocks: in this step, the aluminium block is transformed into semifinished products as aluminium foil or sheet by rolling, into aluminium bars by extrusion, or into 3D shapes by casting.

Manufacturing of aluminium is the main step of making final products, sold to customers. The main technological processes are machining, welding and other joining techniques. Finally, the surface of the aluminium is treated (anodization, painting, etc.).

Waste handling of aluminium metal contains several processing steps beside waste sorting. The combination of waste handling and production of new secondary aluminium metal are summarized as recycling technologies shown in Table 1.

Table 1. A Summary of Recycling Technologies for Cans and Automotive Scrap (Das, 2007).

Step	Description
Can Recycling Technology	
Collection	Shredding bales or briquettes and removing ferrous contaminants through an air knife.
De-Lacquering	Employing a rotary kiln with a sophisticated re-circulating system for products of combustion gases.
Alloy Separation	Screening out can lid from can body in thermo-mechanical chamber by the onset of incipient melting.
Melting, Preparation, and Casting	Transferring the melt metal from dedicated melters to on-line melting furnaces for can stock manufacturing.
Automotive Scrap Recycling Technology	
Physical Separation Methods	Using electromagnetic or eddy current methods to separate nonferrous metal from nonmetallic particles.

1.2 Novel aluminium material family, the hybrid aluminium

Aluminium is also expected to be a major part of the vehicle's materials. Very light structures can be constructed from aluminium, as aluminium has higher specific strength than steel. While the maximum strength for steels is 1500 MPa and for aluminium is 650 MPa the densities are 7.8 and 2.7 g/cm^3, and the specific strengths are 192 and 240 respectively. Developing new materials cost long time and significant amount of money. Introducing a new materials family into the market usually take 30 years (Schafrik, 2003). Hybrid aluminium is called nano-, micro- or macroscopically structured aluminium. Besides the alloying elements, the material contains solid mostly *ceramic particle phases by an addition process called mixing*. The properties are set in by the ceramic particle and the foam bubble sizes and concentration (volume). The new technology steps of this type of aluminium are the foaming and 3D printing (additive manufacturing). Depending on the complexity of the structure, we are talking about single, double and triple hybrid aluminium (Figure 1.).

- 0x hybrid aluminium = base material: conventional aluminium alloy or aluminium waste.
- 1x hybrid aluminium: also known as *aluminium matrix composite (AMC) or aluminium MMC*, which, in addition to aluminium alloy, also contains nano- or micron-size particles (e.g. ceramic) which remains solid after melting the aluminium. *The most promising reinforcing particle is the own oxide of aluminium (alumina or Al$_2$O$_3$) which can be made totally recyclable as hybrid aluminium.* The 1x hybrid aluminium can also be foamable. Aluminium matrix composites may have other advantageous properties, for example as higher temperature resistance materials, compared to conventional aluminium alloys. In addition, casting properties can also be improved using 1x hybrid aluminium.
- 2x hybrid aluminium: such as stabilized *aluminium foams*, in this case, the starting material is the 1xybrid aluminium, i.e. the aluminium matrix composite. Aluminium foam is a new material that is so light that it floats on water. Currently, as a semi-finished product, it is used as building cladding, protection material against blast and cash, and part of vehicle bodies (especially in electric car weight and safety technology as battery holder) and dampers of machine vibration and sound reduction.
- 3x hybrid aluminium: semi-finished or finished products of macroscopically alternating 0,1,2x hybrid materials containing aluminium parts. For example, metallurgically bounded aluminium foam core *sandwich panel* materials and weight-reduced aluminium castings.

Some of the collected secondary aluminium raw material (aluminium waste) has high silicon and iron content. The use of these wastes is limited and can only be reused when mixed with primary aluminium. However, hybrid aluminium can also use highly alloyed or oxidized aluminium and produce high added value hybrid aluminium products.

Aluminium matrix composite and aluminium foam patents are available from the end of 1980's and recently created for the market by the Canadian company Cymat and the Hungarian company Aluinvent. Cymat, which is established in 1999, is already on the market using the Rio Tinto Alcan company made MMC. Aluinvent is a new company established in 2012 making their own MMC and aluminium foam. But the breakthrough is expected from the cooperation of Alucoil and Cymat joint venture, making the first large aluminium foam core sandwich façade production line in Spain with the aim to capture a significant portion of the $ 6.3 billion aluminium composite panel market (Cymat, 2018). The hybrid aluminium material class developed into

Composite/MMC

Foam

Sandwich

Figure 1. Hybrid aluminium classes.

a serial product on the market in the last 30 years. Moreover, there is still even greater potential in the hybrid aluminium since the nanoparticle reinforced hybrid aluminium can also bring steel strength aluminium into the stage. In addition, the hybrid aluminium can also be an important raw material for aluminium casting making the heat treatment process unnecessary.

2 ALUMINIUM AS AN ENERGY CARRIER

2.1 *Energy mix of aluminium smelting*

The production of aluminium from the ore (oxide) requires a significant amount of energy, and therefore aluminium smelting process is currently associated with low-energy price countries. Alcoa, Rio Tinto and Century Aluminium use hydroelectric power and geothermal in Iceland, Quatar Aluminum using oil in Quatar, Norsk Hydro and Rio Tinto using hydroelectric power in Norway and Canada respectively, Chinese aluminium industry operates on coal etc., which determines the place of production of primary aluminium. Aluminium metal stores significant amounts of energy (energy bank). It is a very dense energy carrier with comparable energy density as gasoline. For the time being, primary aluminium production requires 15-13.5 kWh/kg of energy, and when it burns, raise its energy content by 32 MJs (8.9 kWhs) per kg of Al or 85 MJs (23.6 kWh) per L. Therefore, its role can be decisive in the future. Melting of aluminium or secondary recycling of aluminium requires only 5% of the energy needed for the primary production or fully oxidization of aluminium.

The primary aluminium production of the world relies on renewable resources. 25% of recent (2015) global aluminium industry power mix is a renewable hydro energy (World Aluminium, 2017) but there are other projects based on solar (Vorrath, 2016), wind and geothermal (Alcoa Iceland, 2007) energies as well. Smelting aluminium by renewable energies increased Iceland's GDP by 30%. The abundance of renewable resources combined with advanced technologies resulted in effective, inexpensive methods to produce aluminium. This has attracted aluminium producers from abroad. By harnessing the abundant hydroelectric and geothermal power sources, Iceland's renewable energy industry provides nearly 85% of all the nation's primary energy – proportionally more than any other country – with 99.9% of Iceland's electricity being generated from renewables. In addition, Iceland's national power company Landsvirkjun negotiated power deals with aluminium companies that were particularly favorable, long of term, and tied to the LME price of aluminium – the three smelters, one each owned by Rio Tinto Alcan, Century, and Alcoa, paid an average total of 24 US $/MWh in 2014. Operating an aluminium smelter in Iceland is over one-third cheaper than the global average, which is roughly 10 US $/MWh above the average price paid by smelters on the island. The industry has taken full advantage of this fact, as the three aluminium smelters in Iceland consume over seventy percent of the electricity produced each year.

2.2 *Aluminium batteries*

The Israeli company Phinergy (Phinergy, 2019) is working on a new primary nonrechargeable aluminium battery pack (the aluminium battery is removed from the car and replaced with a new one) together with Alcoa (Arconic) but the production line will be established finally in China. Moreover, the current rechargeable aluminium-ion batteries can reach more than 2.5 times (1 kWh/kg) the lithium-ion battery theoretical maximum (300 Wh/kg) which can drastically change the future battery industry (Elia, 2016). The recent Al-ion batteries are already capable to store 400 Wh/kg energy (Alion, 2019). The Al-air battery theoretical energy density is 8.1 kWh/kg (Loveday, 2014). The value is similar than the density of the coil, the energy density of the natural gas or the gasoline, but three times as much per volume (Conversion Factors, 2012). Unfortunately, primary aluminium production requires still 2 times the energy content of the theoretical Al-air battery. However, the best current Li-ion batteries have an energy density of only 0.25 kWh/kg (Lithium-ion Battery, 2019). The increased energy content of the batteries will revolutionize first the electric car industry and later at 400 Wh/kg density the manned electric aircraft industry (Harrop, 2019). At the moment, the Tesla Model S (Tesla, 2019) is capable of running 539 km with a 100 kWh rechargeable battery pack (0.207 kWh/kg energy density, 0.19 kWh/km, 0.90 kg/km, 483 kg Li-ion battery). The new Alcoa-Phinergy electric car was able to drive 1750 km with the current 0.3 kWh/kg Al-air battery, which can easily be upgraded to 1 kWh/kg. Tesla Model S fueled with the impoved Phinergy air battery technology would use only 100 kg of Al-air battery, which means that it will be enough to refuel with 100 kg Al for such distance. If the theoretical energy capacity of 8.1 kWh/kg Al can be utilized with the current electric power train, only 23 kg of aluminium should be refueled, which is only 8.5 liters of aluminium. To compare, this amounts at least 32 kg and 40 liters of petrol.

3 INFINITE LOOP OF THE ALUMINIUM CIRCULAR ECONOMY

3.1 Infinite aluminium circular economy diagram

Summarizing, structuring and coupling the information listed previously an infinite aluminium circular economy diagram can be drawn (Figure 2.).

3.2 Novel green technologies introduced into the infinite circles

Primary aluminium production consumes 2% of the global electricity supply, and one third of the total energy consumption in primary aluminium production comes from coal-generated electricity. Air pollution from primary smelting and the production of the necessary electrical power includes hundreds of thousands of tonnes of carbon dioxide, nitrogen oxide, hydrogen fluoride, and particulates. Reducing these levels can be achieved by maximizing the use of state-of-the-art environmental control systems and environmentally friendly practices such as recycling. Recycling aluminium scrap results in the production of only 5% of the carbon dioxide produced in making new primary metal. Thus, the energy savings of recycling aluminium also translates into reduced environmental emissions. In addition, new aluminium metallurgical technology has emerged, which will significantly improve the environmental friendliness and cost-effectiveness of primary

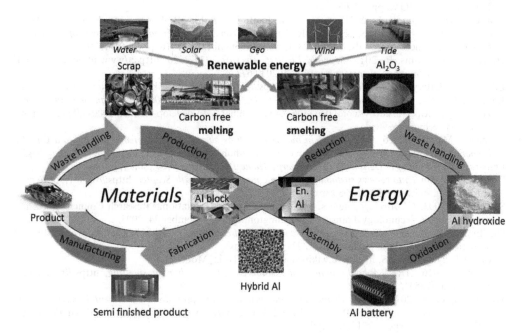

Figure 2.   Infinite aluminium circular economy diagram.

aluminium production, as O_2 is a by-product instead of CO_2 (Smelting, 2018). In addition, technological advances in the field of secondary raw material manufacturers have produced environmentally-friendly salt-free waste recycling technology (Scepter, 2018).

4 CONCLUSIONS

Aluminium is a promising candidate for a new sustainable CO_2 free material-based economy to change oil-based society for a renewable economy called aluminium economy. Although, there is a long way from the theoretical infinite loop towards the practical establishment. One of the future directions of the aluminium industry is likely to be the new type of hybrid aluminium. Novel technologies are existing and introduction into the infinite loop the aluminium can have the potential to replace oil as an energy carrier and steel as a structural material.

REFERENCES

Alcoa Iceland, 2007, *Alcoa, Iceland group to develop geothermal power*, https://www.reuters.com/article/alu minum-alcoa-geothermal/alcoa-iceland-group-to-develop-geothermal-power-idUSN1145903220070911.

Alion, 2019, *Alion: A Low-Cost Aluminium-ion Battery Project Home Page*, http://alionproject.eu/project/.

Bertram, M. & Ramkumar, S. & Rechberger, H. & Rombach, G. & Bayliss, C. & Martchek, K.J. & Müller, D.B. & Liu, G. 2017, *A regionally-linked, dynamic material flow modelling tool for rolled, extruded and cast aluminium products*, Resources, Conservation and Recycling Volume 125, October 2017, Pages 48–69., http://www.world-aluminium.org/statistics/massflow/.

Circular Economy, 2016, *Recommendations on the Circular Economy Package*, https://european-aluminium.eu/media/1495/2016-04-european-aluminium_position-paper-on-circular-economy.pdf.

Conte, F. & Dinkel, F. & Kägi, T & Heim, T. 2014, *Permanent materials*, https://carbotech.ch/cms/wp-content/uploads/Final_PeM_Report_Carbotech.pdf.

Conversion Factors, 2012, *Energy related conversion factors*, https://deepresource.wordpress.com/2012/04/23/energy-related-conversion-factors/.

Cymat, 2018, *CYMAT Provides Sandwich Panel Development Update*, https://investingnews.com/daily/tech-investing/cymat-provides-sandwich-panel-development-update/.

Das, S.K. & Yin, W. 2007, *The Worldwide Aluminum Economy: The Current State of the Industry*, 2007 November, JOM, pp. 57–63.

Elia, G.A. & Marquardt, K. & Hoeppner, K. & Fantini, S. & Lin, R. & Knipping, E. & Peters, W. & Drillet, J.-F. & Passerini, S. & Hahn, R. 2016, *An Overview and Future Perspectives of Aluminium Batteries*, Advanced Materials, Volume28, Issue35, September 21, 2016, pp. 7564–7579.

Harrop, P. & Collins, R. 2019, *Manned Electric Aircraft 2020-2030*, https://www.idtechex.com/en/research-report/manned-electric-aircraft-2020-2030/672.

Lithium-ion Battery, 2019, *Lithium-ion battery*, https://en.wikipedia.org/wiki/Lithium-ion_battery.

Loveday, E. 2014, *Alcoa Teams With Phinergy to Develop Claimed 1,000-Mile Aluminum Air Battery Technology*, https://insideevs.com/alcoa-teams-with-phinergy-to-develop-claimed-1000-mile-aluminum-air-battery-technology-wvideo.

Permanent Materials, 2016, *Why permanent materials such as aluminium are perfectly suitable to a Circular Economy*, https://www.european-aluminium.eu/resource-hub/permanent-material-and-multiple-recycling/.

Phinergy, 2019, *Phinergy: Delivering Clean Energy Home Page*, http://www.phinergy.com/.

Scepter, 2018, *Rio Tinto renews sustainable aluminium partnership with Scepter*, https://www.riotinto.com/media/media-releases-237_25708.aspx.

Schafrik, R. 2003, *GE Aircraft Engines*, briefing presented at the National Research Council Workshop on Accelerating Technology Transition, Washington, D.C., November 24, 2003.

Smelting, 2018, *Rio Tinto and Alcoa announce world's first carbon-free aluminum smelting process*, https://www.riotinto.com/media/media-releases-237_25362.aspx

Tesla, 2019, *Tesla Model S*, https://en.wikipedia.org/wiki/Tesla_Model_S.

Vorrath, S. 2016, *ARENA backs plan to use solar energy for alumina smelting*, https://reweconomy.com.au/73541/.

World Aluminium, 2017, *World Aluminium: Life cycle inventory data and environmental metrics for the primary aluminium industry*, 2015 data (June 2017), http://www.world-aluminium.org/media/filer_pub lic/2017/06/28/lca_report_2015_final.pdf.

Solutions for Sustainable Development – Szita, Jármai & Voith (eds.)
© *2020 Taylor & Francis Group, London, ISBN 978-0-367-42425-1*

Life Cycle approach of a new Industrial Symbiosis alternative

B.S. Gál, R. Bodnárné Sándor & Zs. István
Bay Zoltán Nonprofit Ltd. for Applied Research Division of Intelligent Systems (BAY-SMART) Hungary

ABSTRACT: LCA Center Association on behalf of IFKA Public Benefit Non-Profit Ltd. made this research for the Development of Industry in connection with CIRCE2020 project. This project is supported by European Union INTERREG Central Europe Program. In the frame of CIRCE2020 the participants searching new technologies in the field of waste treatment, in order to close the loop of material flows in case of different companies. During this project had to search these possibilities on the industrial areas of Central Europe. As partners form Hungary, the IFKA and Bay Zoltán Nonprofit Ltd. are working on this project. This study contains MFA, LCA and LCC result from the possible waste streams. The MFA outcomes will be an important part the further life cycle analysis. During the analysis had to compare the currently and one (or more) possible (developed) way of waste handling. This research can help achieve the concept of industrial symbiosis and circular economy on the pilot areas.

1 INDUSTRIAL SYMBIOSIS (IS)

1.1 Term of IS

The industrial symbiosis is one subpart of the industrial ecology, tool of the industrial production. The industrial ecology based on that concept that said that the industrial systems able to function as the principle of operation of natural cycles, where everything recycled (Bárány 2012).

The term industrial symbiosis is used when traditionally separate companies and industries work together in a collective approach to physically exchange energy or materials with a mutual competitive advantage (Schüch 2017). Domenech et al. (2018) determining three primary opportunities for resource exchange:

- By-product reuse, surplus energy use or the exchange of enterprise specific materials between two or more parties for use as substitutes for commercial products, raw materials or fossil fuels. The materials exchange component has also been referred to as a by-product exchange, by-product synergy, or waste exchange and may also be referred to as an industrial recycling network.
- Utility/infrastructure sharing the pooled use and management of commonly used resources such as energy, water and wastewater.
- Joint provision of services meeting common needs across firms for ancillary activities such as fire suppression, transportation and food provision.

1.2 System approach

The IS contributes to closing the loop in case of industrial processes by (Domenech et al. 2018):

- Increasing the time the material/substance remains in the anthroposphere, before it becomes waste or is discharged to ecosystems
- Reducing the volume of waste sent to landfill or disposed of in nature, such as CO2 emissions
- Increasing energy and material efficiency through further reuse and recycling of materials/substances/energy
- Creating jobs and business opportunities linked to alternative uses of existing waste streams

1.3 EXAMPLES on IS PROGRAMS FROM THE WORLD

1.3.1 National Industrial Symbiosis Programme (NISP)
That was the world's first facilitated national industrial symbiosis programme and has received recognition for its achievements from bodies including the UN, European Commission, and WWF Innovation is part of NISP's that it applies holistically addressing not just material 'waste' but energy, a systems approach, water, logistics, capacity, expertise – indeed any assets that are under-utilized (International Synergies Limited 2019).

1.3.2 Transitional Regions towards Industrial Symbiosis (TRIS)
Main goals of TRIS are identify facilitating elements, obstacles and embed them, in (or remove them from) the appropriate policy instruments. Reach out and engage with the actors that can drive and/or be impacted by the change and involve them in structured local networks. Improving the competitiveness of SME sector or rather the affecting policy tool, through the applying of IS (E. Tanka 2019).

1.3.3 CIRCE 2020
The Circe2020 project was funded within the Interreg Central Europe Programme. Aim of the CIRCE2020 is, expanding the Circular Economy concept in local productive districts of Central Europe (CE). Further project aims:

- to facilitate a larger uptake of integrated environmental management approach in specific Central European industrial areas
- changing patterns from single and sporadic company recycling interventions to an integrated redesign of industrial interactions based on the concept of circular economy.
- transition the circular economy towards.
- use of primary natural resources in production stages for Central European industrial areas, due to the processing, packaging and transportation.
- increasing the recycling rates
- initiatives to reuse of by-products by companies.

The project should also provide robust evidence about environmental and economic benefits from shifting to enhanced industrial symbiosis (Interreg CENTRAL EUROPE 2019).

2 TOOLS OF CICRE 2020

The CIRCE 2020 is an ongoing project. The current goal of CIRCE2020 is realization of a set of pilot actions based on the results of local surveys, in order to identify the unevaluated waste streams by using innovative instruments. To try to achieve this aim, we will analyse the alternative waste disposal possibilities trough the view of donor organisations, in case of those companies, what is want working together in industrial symbiosis. These observations derived from the MFA (Material Flow Analysis), LCC (Life Cycle Costing) and LCA (Life Cycle Analysis) (Interreg CENTRAL EUROPE 2019).

2.1 MFA tool

According to *Brunner and Recberger (2005)* "the material flow analysis (MFA) is a systematic assessment of the flows and stocks of materials within a system defined in space and time. It connects the sources, the pathways, and the intermediate and final sinks of a material. Because of the law of the conservation of matter, the results of an MFA can be controlled by a simple material balance comparing all inputs, stocks, and outputs of a process. It is this distinct characteristic of MFA that makes the method attractive as a decision-support tool in resource management, waste management, and environmental management."

In the case of the CIRCE2020, the project team elaborated an MFA. This methodology has applied in every county and every examined material flows. The MFA is the main contributor of the life cycle assessment and cost analysis besides other facts. The final result of MFA helped to collect out the most interesting material flows from pilot areas.

2.2 Life cycle assessment

The life cycle measurements are environmental centred methods, what able to contain the whole life cycle of products or service from the production of raw material to the waste handling or recycling. During the project, all participants have applied the LC assessment with the same method and frame for the analysis, in case all type of waste stream. The characterization was based on the rules of EF recommendation. Characterised results per life cycle stage and impact category (all 16 EF impact categories) were calculated. There are some optional step possibilities in the assessment.

According to ISO 14044 (ISO 2006), normalisation, in the context of Life Cycle Assessment (LCA), is an optional step of Life Cycle Impact Assessment (LCIA) which allows the practitioner to express results after the characterisation step using a common reference impact (Lorenzo et al. 2014).

Weighting is the optional fourth step in Life Cycle Impact Assessment (LCIA). According to Meijer (2014) the weighting is the final and is the most debated step. Weighting entails multiplying the normalised results of each of the impact categories with a weighting factor that expresses the relative importance of the impact category. The weighted results have the same unit and can be added up to create a single score for the environmental impact of a product or scenario. Simply put, weighting means applying a value judgement to the LCA results.

The chosen weighting factors able to influence the conclusions of LCA. It is a controversial step.

2.2.1 Analysed scenarios

According to the principle of the CIRCE2020 measuring methodology for the assessment, had applied two types of scenario, in case of each selected waste streams (tire and plastic). The list below containing the similarities and differences between the scenario types

Business as usual – (BaU)

- The waste is coming from production.
- The chosen waste stream goes to the conventional handling (incineration) in the BaU scenario.
- During the incineration, electricity is generated.

Circular Economy solution (CE)

- The waste comes from production.
- The waste goes through a multi levelled recovery process.
- During this process, the unusable fractions are extracted from waste.
- The result of the recovery is recycled material (That can use in production processes again).

2.2.2 *Waste flows*

Thanks for the IS ambition, the waste streams and its solutions had tried to search in the central Europe industrial areas. In Hungary, two several types of waste had chosen the tyre and the ostomy bag production.

2.2.2.1 WASTE OF TYRE PRODUCTION

One of the chosen waste streams from Hungary had connected to a tyre factory, in Hungary. From the tyre production, the main waste stream is the by-products and remaining materials of the production (other rubber waste). Currently, this waste amount had delivery to waste incineration. The pilot tries to focus on the possible CE ways and measure the differences between the original and alternative waste handling in environmental and economic aspect. This alternative scenario may give better results than the present.

2.2.2.2 WASTE OF OSTOMY BAG PRODUCTION

The other chosen waste type is the remaining part of ostomy bag production in Hungary. The content of this waste stream is really complex (e.g. PE, PP, PVC). The ostomy pouching system is a prosthetic medical device that provides a means for the collection of waste from a surgically diverted biological system (colon, ileum and bladder) and the creation of a stoma. But the focus waste is clear material from the production stage. The present waste incineration system will be changed to an alternative solution. This new system can use the waste – as a raw material – for the new product production. Here the industrial symbiosis and the circular economy get on in same time.

Besides the high and always grooving quality of the products, the company struggles to minimize the amount of waste of the manufacturing. In this modern technology, this amount of waste has decreased significantly but is still considerable. The companies are always searching for those possibilities, where the waste is utilized by sustainable methods.

2.2.3 *Aim of the life cycle assessment*

The main goals of the EF-based studies can be summarized as follows:

- quantify the potential environmental benefits of the identified CE solution compared to the current waste management practice;
- must to identify the hotspots of the CE solutions, to be used as an indication for a further improvement of the technology solution and its implementation at pilot scale.

As a reference, the functional unit (FU) has to be used as a quantified performance of a product system. Important elements of the LC studies the FUs and related reference flows. According to the definition "A functional unit is a quantified description of the performance of the product systems, for use as a reference unit" (Wenzel 2004). The comparison of the BaU and CE solutions able to accomplished only with same FU. In case of LCC and LCA studies to guarantee the full consistency also must be applied the same FU. Therefore, the functional unit was defined from the waste donor perspective, namely the treatment of 1kg waste from the production.

For the inventory analysis, primary data from waste handlers have been collected. The data of BaU scenario are coming from an incineration plant, near to the tyre production. These data have been projected to the handling of 1 kg tyre waste. In case of CE scenario, primary data has been used from the waste handler corporation (and calculated to one kilogram of waste).

The LCC data collection results a Cost Breakdown Structure (CBS) including each life cycle stage and the different cost types. During the data collection, our partners of BaU and CE versions were very cooperative in this study, so we received in most of the case very exact data directly from the industry.

3 RESULT OF TYRE ANALYSIS

Although all of the scenarios were analysed by the environmental footprint impact assessment method, but the final results were defined by the normalized, weighted values. The next diagrams (Figure 1) show these environmental impact values.

Trough LCA approach, the results show that the CE solution has less environmental impact with almost 15% comparing with the BaU scenario.

In case of BaU and CE scenarios also, the waste handling (incineration or the alternative) solution part of the LCA, has the biggest impact. The incineration process and the substitution (CE) process have the most impact on the environment. The raw material production and transportation groups have negligible roles.

From LCC point of view, the comparison is very obvious between BaU and CE version. The deficit is 7,1% in case of BaU and the profit is 1,1% for CE version from the view of waste recipients. But the situation is more complex than the first sensation.

Overall the CE scenario is a better choice than the BaU scenario in LCA and LCC point of views.

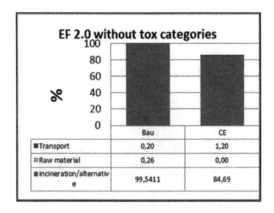

Figure 1. Comparison of LCA results of BaU and CE systems in case of tyre (IFKA 2019a).

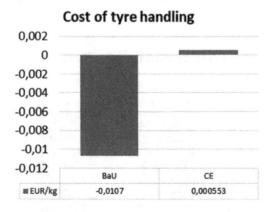

Figure 2. Comparison of LCC results of BaU and CE systems in case of tyre (IFKA 2019b).

4 RESULT OF PLASTIC ANALYSIS

Here the concept and the methodology were the same than the tire waste. The BaU and the CE scenarios were compared by LCA and LCC. The diagrams show the normalized and weighted results in case of LCA and the deficit/profit in case of LCC.

Figure 3 shows the results of the assessment of plastic. The BaU scenario has a much bigger environmental impact (60%) than the CE solution. The environmental impacts in plastic BaU scenario originate, from the whole life cycle. In case of CE (against the BaU) the environmental impact are coming from incineration with more than 90%.

The biggest impact comes from the material production (59,87%). Another important source of the environmental impacts is incineration phase. In case of BaU, this life cycle group is responsible for 38% of environmental impact, and also in the CE scenario, the incineration dominates the impacts with more than 90%.

In case of LCC assessment (Figure 4) the comparison is not so obvious between BaU and CE version. The value of profits are nearly the same in both cases: it is 2,6% profit in case of BaU and 2,5% profit. The LCC result of CE version presumably is not the best option for the future. Therefore necessary to investigating another CE version later where the technology has higher added value than this one. The profitability of this suggested technology is too slight.

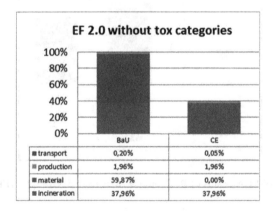

Figure 3. LCA results of BaU and CE systems in case plastic waste (IFKA 2019a).

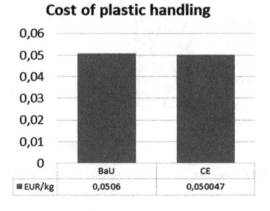

Figure 4. LCC results of BaU and CE systems in case plastic waste (IFKA 2019b).

5 CONCLUSION

In this study, the LCA and LCC assessment had fulfilled for the Hungarian pilots (the waste of the tyre and plastic production) of CIRCE2020 project. The analysed waste streams have been chosen by MFA, from the available waste sources of pilot areas. The LCA results confirm that predictions what are assumed the better environmental impacts are achievable by the CE solution. In case of the chosen waste streams (plastic and tyre), the CE scenarios have (more or less) better results. In case of economical analysis also the CE scenario shows better outcomes, but not by to high differences. In case of tyre waste, the BaU scenario show deficit and the CE is profitable. In case of plastic waste, both of the scenarios are profit makers with close to the same rate. In the future may there be an opportunity for analysis of other CE solutions. This research will continue until 30/06/2020 in the frame of CIRCE2020 project.

NOTE

"This research was made on behalf of IFKA Public Benefit Non-Profit Ltd. for the Development of Industry in connection with CIRCE2020 project. The CIRCE2020 project is supported by European Union INTERREG Central Europe Program, under the European Regional Development Fund. www.interreg-central.eu/circe2020".

REFERENCES

Bárány, Z. 2012. Másoljuk le a természetet – A Nemzeti Ipari Szimbiózis Program Magyarországon; *Ipari Ökológia*; 1(1): 135–140.

Benini, L. & Mancini, L. & Sala, S. & Manfredi, S. & Schau, E. M. & Pant, R. 2014. Normalisation method and data for Environmental Footprints; *JRC TECHNICAL REPORT*; DOI 10.2788/16415; Publications Office of the European Union.

Brunner P.H. & Rechberger H. 2005. Practical Handbook of material flow analysis; *Lewis Publishers is an imprint of CRC Press LLC*; ISBN 00-203-50720-7.

Domenech, T. & Doranova, A. & Roman, L. & Smith, M. & Artola, I. 2018. Cooperation fostering industrial symbiosis market potential, good practice and policy actions; *Final report; EUROPEAN COMMISSION*; Sources: https://www.technopolis-group.com/wp-content/uploads/2018/08/1_IS-Cooperation-Study_Final-Report.pdf (2019.05.28.).

International Synergies Ltd 2019. Industrial Symbiosis Industrial Ecologies Solutions https://www.international-synergies.com/about-us/industrial-symbiosis-ltd-not-for-profit/.

IFKA 2019a. LCA Report of Hungarian pilot, version 20 May-2019, made by LCA Center on behalf of IFKA, CIRCE2020, Manuscript.

IFKA 2019b. LCC Report of Hungarian Pilot, version 20 May 2019, made by LCA Center on behalf of IFKA, CIRCE2020, Manuscript.

Interreg CENTRAL EUROPE 2019. WHAT IS CIRCE2020? Sources: https://www.interreg-central.eu/Content.Node/CIRCE2020.html (2019.05.28).

Meijer, E.B. 2014. Weighting: Applying a Value Judgement to LCA Results; *Pre-sustainability*; Sources: https://www.pre-sustainability.com/news/weighting-applying-a-value-judgement-to-lca-results/.

Schüch, A. & Hänel, A. & Thapper, C. & Nakrosiene A. & Paulauskas, M. & Flink P., Lund, M. 2017. WP3 Technical report on industrial symbiosis; *UBIS - Urban Baltic Industrial Symbiosis*.

Tanka E. 2019. Régiók úton az ipari szimbiózis felé: *TRIS; Interreg Europe*; Sources: http://www.hermanottointezet.hu/sites/default/files/1.%20Tanka%20Eszter_TRIS_projekt_bemutato.pdf (2019.05.28).

Wenzel, H. & Petersen, C. & Hansen K. 2004. The Product, Functional Unit and Reference Flows in LCA; *Environmental News* 70.

Possibilities for adopting the circular economy principles in the EU steel industry

Á. Kádárné Horváth, M. Kis-Orloczki & A. Takácsné Papp
University of Miskolc, Miskolc, Hungary

ABSTRACT: The global economy has to face several challenges due to the scarce resources and environmental impact of economic activities. The introduction of circular economy as a business model can reduce environmental damage. In our study, we introduce the potentials the steel economy has in the transition to circular economy. Steel is the most commonly used raw material, it is essential in all fields of life. Based on the strategic documents, workpapers and reports of relevant organizations of the sector we introduce in our model the advantages and possibilities the steel industry has to engage to the circular approach. We found that the sector can be one of the drivers in transition. The raw material and energy consumption can be reduced, besides the GHG emissions, while cost efficiency and competitiveness increases. The developments have spillover effects in other sectors using steel of higher quality or through the by-product reuse and have an impact on social well-being as well.

1 INTRODUCTION

Despite of the rapid economic development, the past and present trends of using our material resources lead to the scarcity of environmental resources, volatile prices, high level of pollution and acceleration of climate change. The well-known definition of the Brundtland Report says that the sustainable development is a "development which meets the needs of the present without compromising the ability of future generations to meet their own needs" (WCED 1987, p.41), but today's patterns of production and consumption risk the fair access to scarce resources. According to Li et al. (2011), a 1% increase in real GDP per capita results in a 0.48-0.50% increase in energy consumption and therefore increases carbon dioxide emissions by about 0.41-0.43% (Li et al. 2011). Circular economy as a strategy of sustainable development receives increasing attention both in political, strategic and scientific platforms as it involves the efficient use of resources and minimization of waste. It contributes to a sustainable, low carbon and more competitive economy, while increases the level of employment by creating local jobs at all skills levels. Due to the transition to circular economy in Europe resource productivity could grow by 3%, resulting in a primary resource benefit of 600 million EUR per year by 2030 and could generate 1200 million EUR non-resource and externality benefits. The GDP increase could reach about 7 percentage points relative to the current development scenario. (Ellen MacArthur Foundation 2015).

Concerning the European Union, it can be connected to the main strategic and operational EU policy aims and priorities like jobs and growth, investments, climate and energy, innovation and sustainable development. The Action Plan "Closing the Loop" focuses on the use of finite, abiotic resources, through five priority areas, such as plastics, food waste, critical raw materials, construction and demolition waste and biomass and bio-based products. Also contains recommendations on innovation and investment, and on monitoring the process. (European Commission 2015) The European Union provides the economic actors with a wide range of financial support to accelerate the transition to the circular economy (e.g. Horizon 2020, Cohesion Policy, LIFE Program), the total sum exceeds 10 billion Euros over the 2016–2020 period. (European Commission 2019) Although the circular economy concept gains growing attention today, implementation is limited, our world is only 9% circular, meaning 9% of minerals, fossil fuels, metals and biomass that enter the economy are re-used. (PACE 2019).

The first aim of this study is to give a brief summary on what circular economy means. Afterwards, we introduce what possibilities an energy-intensive industry, the steel industry, has in transition to circular economy.

2 THE CIRCULAR ECONOMY

2.1 *The concept*

It is often emphasized that in a circular economy, there is no waste, as products of today can serve as resources for the future. However it is more than just recycling, it is rather a new model of production and consumption overarching the supply chain and sectors, being able to increase the efficiency of resource use and competitiveness, to decrease costs, to drive innovation and to contribute to sustainable development.

The concept itself is not a new one. According to the latest studies (Murray et al. 2017, Kirchherr et al. 2018) basic idea of circular economy was first mentioned in 1848, in 1966 by Boulding and was further discussed in the last 40 years by e.g. Stahel and Ready-Mulvey, Pearce and Turner. Currently, more than 100 definitions are identified with a wide range of meanings. (Kirchherr et al. 2017) As the circular economy gained importance in Europe since in 2015 the European Commission announced its package to support the EU's transition to a circular economy, we accept the definition of the European Parliament and the European Commission. Circular economy is "a model of production and consumption, which involves sharing, leasing, reusing, repairing, refurbishing and recycling existing materials and products as long as possible" (European Parliament) "and the generation of waste minimized" (European Commission 2015).

2.2 *Benefits and enabling factors*

The circular economy compared to the linear uses less input and natural resources, while the share of renewable and recyclable resources increases. The value of products, components and materials are kept as long in the economy as possible, resulting in reduced emissions and fewer material losses. (EEA 2016) Benefits are connected to the following four areas, including the 3 pillars of sustainable development:

- Resource benefits: demand for primary raw materials decreases, so does the dependency on imports
- Environmental benefits: environmental impact of economic activities decreases due to the higher resource-efficiency, thus externality costs decrease.
- Economic benefits: it fosters innovation and economic growth.
- Social benefits: creates jobs and through a more sustainable consumer behavior it can contribute to human health and safety. (EEA 2016)

Circular economy is a systemic shift that requires changes in current production and consumption patterns. Firstly new, innovative technologies must be developed and introduced (eco-innovation and eco-design). Besides the technological innovation, social and organizational innovation is inevitable. Secondly, the actors must change their interplay by giving repair, refurbishment, remanufacture and recycling higher importance. All these changes cannot be done if not supported by regulation (such as rethinking incentives, providing a suitable set of international environmental rules) and cultural shift in changing the manner of both consumers and producers. (Kirchherr et al. 2018, EEA 2016) Among the enabling factors, we would like to highlight the industrial symbiosis as a new business model, a local or global partnership where companies of different sectors collaborate to make one's waste or by-product a resource for another. By providing, sharing and reusing resources industrial symbiosis can create loops of technical or biological materials and minimize waste. (PwC 2018, EEA 2016).

In the next part of the study, we examine what potential the steel industry has in the transition to circular economy. As an explanation of our choice we cite the Director General of the World Steel Association Dr Edwin Basson:

"As steel is everywhere in our lives and is at the heart of our sustainable future, our industry is an integral part of the global circular economy." (World Steel Association 2015, 3.).

3 THE STEEL INDUSTRY IN THE CIRCULAR ECONOMY

3.1 *The status of the European steel industry concerning the EU climate policy*

3.1.1 *Steel industry of the European Union – overview of the current situation*
Steel is the second most commonly used raw material, its significance is unquestionable in our daily lives. The global steel consumption excluding net indirect exports in 2016 was 1,425,732 thousand tonnes in finished steel equivalent and has increased steadily in recent years. China ranks first in the world in terms of true steel use with 43 percent of total. It is followed by the EU with about 10.5 percent, and by the US with about 8 percent. (World Steel Association 2018a, 115–116.).

Steel production is a material- and energy-intensive process with high level of CO2 emissions. Thus, the sector is very sensitive to the changes in the commodity market and in the energy sector (like primarily access to resources or price volatility) as well as to the stringer environmental regulations.

In addition, the EU steel industry had to face further challenges over the past decades. The majority of the steel industry's products are investment goods which results in high level of cyclical sensitivity. (Barta and Poszmik 1997) After a significant (42 percent) fall in demand in 2008 as a result of the global economic crisis, steel use in the EU increased to three quarters of the pre-crisis level by 2017. (World Steel Association 2018a, 115–116.) Crude steel production followed this trend too when from 2008 to 2009 fell by 30 percent. By 2017, crude steel production increased by about 21 percent compared to the year 2009, but the pre-crisis level was far not achieved. The global crude steel production was 1,690,479 thousand tonnes in 2017 of which China accounts for 49 percent, the EU ranked second with a share of about 10 percent (168,305 thousand tonnes), followed by Japan, India and CIS with a share of around 6 percent. The share of the US share in global production is about 4.8 percent. (World Steel Association 2018a, 1–2.).

The EU steel industry is facing difficulties in maintaining its position in the international competition dominated by the growing production of emerging countries, mainly China, India, CIS and Brazil. While in 2008, the EU ruled 14.8 percent of global production, its share fell below 10 percent by 2017. (World Steel Association 2018a, 1–2.) The situation is further worsened by the overcapacity of the worldwide steelmakers. This encourages the export of steel products, increasing competition for EU steel companies and thus the likelihood of market loss within the EU. The commercial policy measures applied by the competitors (such as the introduction of subsidies for their steel sector or import duties) also have a negative impact on the competitiveness of the European steel industry.

Continuous innovation can be an effective solution to these challenges. The introduction of low-carbon, less energy-intensive technologies, the development of high-quality, tailor-made products and the exploitation of the potentials of industrial symbiosis are key to the competitiveness of the EU steel industry. In this regard, the circular economy concept can provide an effective solution for the industry.

3.1.2 *Characteristics of the steel production*
When presenting the characteristics of the steel production, we place the emphasis on the technological solutions of steel production, raw materials used in production, energy use, and CO2 emissions, as well as the effects of related environmental regulations. Although we acknowledge the importance of those factors, we do not cover the analysis of labor intensity and employment of the sector.

Two main technologies exist in steel production. In the basic oxygen furnace (BOF) within an integrated steel mill, molten iron from the blast furnace is changed into liquid steel. (Ecorys 2008) This is also known as oxygen-blown converter (OBC) technology. The main raw materials in this case are iron ore, coke, limestone and little amount of steel scrap. As energy resource predominantly coal is used. Annual capacity of over 2 million tonnes can make production viable. Introduction of the so called EAF route is a technological improvement. An electric arc furnace (EAF) is a furnace that heats charged material by means of an electric arc. (Ecorys 2008) In this technology, the raw material is primarily steel scrap, while it mainly uses electricity as energy resource. The technology applied highly depends on the type of the product and on the quality requirements. The BOF route compared to the EAF technology is characterized by higher cost of capital, higher fixed costs as well as higher energy cost. Overall, the EAF technology is a more expensive process (Ecorys 2008, 62.) Commonly the most important cost factors are cost of raw materials, energy costs and transport costs. Connected to the development of technology, the introduction of continuous casting should be mentioned as revolutionary. Improving productivity and cost-efficiency, nowadays it is predominant, 96.3% of global production applies continuous casting. (World Steel Association 2018a, 5-6.) The implementation of DRI and HBI as scrap substitutes also was of high importance. Due to the continuous technological innovation, the sector achieved lower cost of capital, economic viability at small scale, lower operational costs, higher flexibility in raw material and significant environmental benefits. (Ecorys 2008)

Access to raw materials and long-term insurance of that must receive strategic attention from the EU. Although there is no shortage of raw materials in the long run, rising demand from emerging countries puts pressure on the supply of raw materials. Thus raw material prices are rising while import dependency of the EU is high and growing steadily.

The steel industry is one of the energy-intensive sectors, its emission is constantly in the center of attention. Energy consumption of the EU steel industry in 2016 was 49 Mtoe. It is important to emphasize that this represents a 40.7 percent reduction compared to 1990 energy consumption. According to 2016 data, 47.2 percent of energy was from coal, 30.2 percent from natural gas, and 20.2 percent from electricity. The rest is made up of other energy resources. The share of electricity in the 1990s shows an increase of about 7.5 percentage points due to the spread of EAF technology. Energy consumption per tonnes of crude steel in 2016 was 0.3021 toe/t. Compared to 1990, this represents a decrease of 27.2 percent. Energy intensity of primary metals, at purchasing power parities calculated as Final consumption/Value added) at 0.7353 koe/€in2010. The value of the indicator also showed a significant decrease of about 42 percent compared to the 1995 value. (1.27675). (Enerdata-Odyssee database)

The sector is very sensitive to changes in energy prices due to its energy demand. Energy prices today increases constantly and dramatically concerning electricity, natural gas, coal and oil. Significant territorial differences exist in the development of prices, which means prices show a significant variation within the EU and the situation is even more unfavorable compared to non-EU states. For example, the advantage of the US concerning energy prices has been significant in recent years. One of the reasons for the large differences in the price was the shale gas revolution in the US, but the effects caused by the diversity of taxation of energy products are significant as well. The EU is applying stricter environmental rules, such as higher tax burdens, resulting in higher energy prices. This is certainly disadvantageous for energy-intensive industries.

In 2015, CO2 emissions from the iron and steel industry were 102.2 Mt in the EU. This represents a 41.5 percent decrease compared to 1990 levels. CO2 emission per tonne of crude steel in 2015 was 0.6172 tCO2/t, representing a 30.3 percent reduction compared to 1990 levels. (Enerdata-Odyssee database). EU environmental legislation put a significant burden on steel industry. The EU has introduced much stricter rules (e.g. EU ETS, IPPC standards) than most non-EU countries. This is a major competitive disadvantage for EU steel companies. In the period up to 2030, the rules for producers were further tightened. A 40 percent reduction in CO2 emissions and a further 27 percent improvement in energy efficiency must be reached. High expectations will enhance carbon trading, resulting in higher carbon credit prices. In 2018, the price of carbon credit increased by 4-5 times, which contributed significantly to the rise in electricity prices. According to Róbert Móger, Director of the Hungarian Iron and Steel Association, "the expectations set by the EU are not realistic, because there is currently no technology that can be operated economically on an industrial scale to ensure the achievement of these goals." (Viland 2018, 2.).

3.2 *Steel industry as an engine of circular economy*

Steel production plays a major role in the global concept of the circular economy. All the elements of the 4R framework (Reduce, Reuse, Remanufacture, Recycle) of the circular economy model can be identified. Life cycle thinking is essential to the successful transition to the circular economy. The raw material and energy consumption, emissions and waste generated must be taken into consideration at all stages of the product's life cycle from design, production of raw materials, production and use to reuse or disposal. The total impact of a product on the environment can be determined only if a full life cycle approach is applied. (World Steel Association 2015).

The extremely favorable properties of steel can make it one of the drivers of the circular economy and gives several advantages over other materials (e.g. aluminum, magnesium, plastic). Steel can be recycled 100% and many times while maintaining its original properties. Through recycling, a wide range of new steels can be produced from any type of steel waste. For example, a lower value steel scrap can be used to produce higher value steel using appropriate technology. Due to its magnetic properties, steel products can be recovered from waste streams. The high value of steel scrap makes the recycling economically viable. On one hand the excellent durability of steel results in less steel is needed to maintain everyday life, while on the other hand it means that steel scrap is available in limited amount.

Material efficiency index of the steel industry (meaning percent of materials converted to products and by-products) is around 97 percent (World Steel Association 2018b), which is quite high compared to other sectors. It means 66 percent of the raw materials and steel scrap used in production becomes new steel, 31percent is by-products and only 3 percent waste is produced. Thus, it can be said that steel production is an almost closed-loop system, the future goal is to reach 100 percent. (Eurofer 2015).

The integration of the steel industry into the global circular economy is illustrated by our model.

As Figure 1 shows involvement in the circular economy can be realized through potentials within the steel industry and by exploiting the potential of industrial symbiosis. As a result, they can reduce raw material and energy consumption, reduce CO2 emissions, increase cost

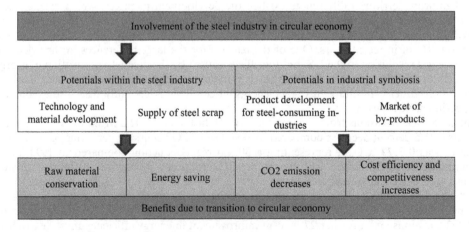

Figure 1. Model of the involvement of steel industry into the circular economy.
Source: own elaboration.

efficiency, and thus competitiveness. And most importantly, these positive effects do not only occur in the steel industry, but also in sectors that use steel or its by-products, and ultimately at social level.

3.2.1 *Potentials within the steel industry*

3.2.1.1 TECHNOLOGY AND MATERIAL DEVELOPMENT

In the previous chapter, we introduced the main features of the 2 main steel production routes. According to data from 2017, the BOF accounts for 71.6 percent of the global crude steel production, while 28 percent is produced by EAF. (The remaining 0.4 percent means other technological solutions, such as the open-hearth furnace technology.) However, the share of technologies of steel production varies significantly across regions. In the EU, the share of the BOF route is only 59.7 percent, while the more advanced EAF technology has a 40.3 percent share. In the US, the focus was even more shifted to the EAF technology (68.4%), while in China, BOF technology accounts for 90.7 percent of total production. (World Steel Association 2018a, 17-18.) The growth of electro-steel production is projected in the medium term, e.g. China has already begun to transform several oxygen-blown steel plant to change to the electro steel production process. (Krause 2019) Besides the technological milestones described, currently, carbon capture and storage technologies are also developed. In addition to reducing CO2 emissions, the reduction of other greenhouse gases and harmful substances (NO_x, SO_x, dust) is among the objectives.

3.2.1.2 SUPPLY OF STEEL SCRAP

There are 3 main sources of steel scraps. The internal by-products from the steel mills that are the industry's own circular scrap from steelworks and foundaries. Due to the constant improvements of technology, this type of scraps shows a downward trend. The second source is the new scrap from steel-processing industries that is together with the previous type is also referred to as pre-consumer steel scrap. It is 100 percent collected and recycled in steel production. The third source is the so-called post-consumer steel scrap, like used cars and other end products. The latter is the most common type of waste. (Ecorys 2008) While in BOF technology steel scrap is used up to 35 percent, in the case of the EAF this rate can be 100 percent. The supply of steel scrap is limited by the durability of steel. We distinguish products with short (e.g. small boxes), average (e.g. caring machines, cars) and long (e.g. buildings, bridges) service life. The latter serves up to 100 years. (Eurofer 2015) The 75 percent of the steel ever produced is still in use. Therefore, the development of scrap substitutes is also a key issue. Steel is the most frequently recycled material in the world, about 650 million tonnes of steel scrap is recycled annually. The recovery rate is not the same in different sectors, is 98% for industrial and commercial buildings and over 90% for machine manufacturing, while 50% for smaller electronic equipment. About 20 percent of steel scrap of the EU is exported, thus the EU is a net exporter in the international steel scrap market.

3.2.2 *Potentials in the industrial symbiosis*

3.2.2.1 PRODUCT DEVELOPMENT FOR STEEL-CONSUMING INDUSTRIES

Analysis of the share of total finished steel demand in the EU shows that the largest steel-consuming sector is construction (35%), automotive (19%), mechanical engineering (15%), metalware (14%) and tubes (11%) are also significant, while domestic appliances, other transport and the miscellaneous both have a 2% share. (Eurofer 2018) Technological development has also led to improvements in the quality of steel. High quality steel is a prerequisite for manufacturing high value added products. The EU's competitiveness in the production of high-quality, tailor-made steels is the highest.

The sector is characterized by permanent material and product innovation, as it has to meet the constantly changing needs of the main steel-consuming industries. In all sectors, steel meets different basic requirements, e.g. in the automotive industry the development of advanced high-strength steels (AHSS) is needed to be lightweight but also rigid components for the transport sector. It contributes to a decrease in weight of vehicles, while safety and quality increases, fuel consumption and emission decreases. (Ecorys 2008) According to a research cited by the World Steel Association 2015 study, vehicles manufactured using AHSS were slightly worse in the use cycle in emissions reduction than the aluminum, magnesium, plastic and steel based Super Light Cars (SLC). However, applying the lifecycle thinking, including the raw material production and the automotive industry GHG emissions as well, the use of AHSS was less harmful. Stronger and lighter steel is also demanded in the energy sector, e.g. the production of wind turbines from lighter steel significantly reduced CO_2 emissions and emissions from its transport and assembly. Using high grade steel in the construction industry, the construction of higher buildings is more efficient, the amount of steel used reduces, so does transportation cost. The construction time of plants has also decreased, and the thinner buildings allow for space utilization. Modular reinforced concrete elements can be reused, that provides with cheaper solutions than producing new items from raw materials. Reducing the thickness of steel used in the packaging industry, such as food and beverage cans, has resulted in positive environmental impacts concerning the whole life cycle. Despite of the increase in energy use of production, the total energy use, GHG emission and transportation cost decreased as less steel is needed.

3.2.2.2 MARKET OF BY-PRODUCTS

In the circular economy framework finding market to the by-products is inevitable. The steel industry is outstanding in it. One of the main by-products of iron and steelmaking is slags. In 2016 in Europe a total of 41 million tonnes slags was used, out of it 46.8 percent in the cement industry. This significantly reduces the environmental impact of cement production. (up to 59 percent CO_2 and 42 percent energy savings). The 29.8 percent of the slag was used as aggregates in road construction, metallurgical use was 6.8 percent but it could be used in crop fertilizer production (1.2%) The 5.7 percent of the slag was placed in landfill sites. Another important by-product are the process gases, which are utilized in electricity and heat production. Dust and sludge are reused in alloys, the petrochemicals are input materials for the chemical industry. (World Steel Association 2018c).

4 CONCLUSIONS

The world's environmental resources are scarce, while the amount used in production is increasing. Business models need to find sustainable solutions for preserving raw materials and decreasing the environmental impact of economic activities. The circular economy model is a new approach in practice, where maximum value is created with minimum waste by closing the loops. Transition to the circular economy became a priority in the European Union to gain competitive advantage. It is very important to take steps towards transition to all actors of the economy in all sectors and it is supported financially.

The steel industry has made considerable efforts and shows significant progress over the past decades to reduce raw material and energy consumption and greenhouse gas emission. Technology development is in the center of attention, although the efforts are partly driven by the stricter environmental regulations the steel industry and the steel-consuming sectors have to face.

The steel industry fit into the circular approach taking advantage of potentials within the sector and potentials in the industrial symbiosis. Thanks to the transition to the circular economy steel industry can reduce raw material and energy consumption, reduce GHG emissions, increase cost efficiency, while it becomes more competitive. The positive effects do not only occur in the steel industry, but also in sectors that use steel or its by-products, and at social level as well. While emphasizing the environmental and economic advantages we cannot go along the fact that innovative technologies need skilled and medium skilled workforce. Besides its sectoral advantages, the transition to circular economy can contribute to better well-being of people.

ACKNOWLEDGEMENTS

"This research was supported by the project nr. EFOP-3.6.2-16-2017-00007, titled Aspects on the development of intelligent, sustainable and inclusive society: social, technological, innovation networks in employment and digital economy. The project has been supported by the European Union, co-financed by the European Social Fund and the budget of Hungary."

REFERENCES

Barta, Gy. & Poszmik, E. 1997. A vas- és acélipar versenyképességét befolyásoló tényezők. "Versenyben a világgal" kutatási program, műhelytanulmány - Budapesti Közgazdaságtudományi Egyetem, Budapest Available at: thttp://edok.lib.uni-corvinus.hu/211/1/MT_15_Barta_Poszmik.pdf (Retrieved: 30 May 2019).

Ecorys 2008: Study on the Competitiveness of the European Steel Sector. Within the Framework Contract of Sectoral Competitiveness Studies – ENTR/06/054, Final report, August 2008. ECORYS Nederland BV, Rotterdam Available at: https://ec.europa.eu/growth/content/study-competitiveness-european-steel-sector-0_en (Retrieved: 30 May 2019).

EEA – European Environment Agency 2016. Circular economy in Europe Developing the knowledge base. EEA Report No 2/2016 Available at: https://www.eea.europa.eu/publications/circular-economy-in-europe (Retrieved: 30 May 2019).

Ellen MacArthur Foundation 2015. Growth within: A Circular Economy Vision for a Competitive Europe. Available at: https://www.ellenmacarthurfoundation.org/publications/growth-within-a-circular-economy-vision-for-a-competitive-europe (Retrieved: 30 May 2019).

Enerdata-Odyssee database.

Eurofer 2015. Steel and the Circular Economy. The European Steel Association (EUROFER), Brussels. Available at: http://www.eurofer.eu/News&Events/PublicationsLinksList/20151016_CircularEconomyA4.pdf (Retrieved: 30 May 2019).

Eurofer 2018. European Steel in Figure 2018. The European Steel Association (EUROFER) Available at: http://www.eurofer.org/News%26Events/News/European%20Steel%20in%20Figures%202018.fhtml (Retrieved: 30 May 2019).

European Commission 2015. Closing the loop -An EU action plan for the Circular Economy, COM (2015) 614 final. Available at: https://eur-lex.europa.eu/legal-content/EN/TXT/?uri=CELEX%3A52015DC0614 (Retrieved: 30 May 2019).

European Commission 2019. Report on the implementation of the Circular Economy Action Plan., COM(2019) 190 Final. Available at: https://eur-lex.europa.eu/legal-content/EN/TXT/?uri=COM:2019:190:FIN (Retrieved: 30 May 2019).

European Parliament. Circular Economy 2019. Available at: http://www.europarl.europa.eu/thinktank/infographics/circulareconomy/public/index.html Retrieved: 30 May 2019.

Kirchherr, J. et al. 2017. Conceptualizing the circular economy: an analysis of 114 definitions. *Resources, Conservations and Recycling*, Vol 127: 221–232.

Kirchherr, J. et al. 2018. Barriers to the Circular Economy: Evidence From the European Union (EU). *Ecological Economics* 150: 24–272.

Krause, G. 2019. Acél- és alumíniumhulladék-piaci trendek. Főnixmadárként születik újjá az értékes nyersanyag az olvadékból. Available at: http://www.autogyar.hu/auto-vilag-piac/acel-es-aluminiumhulladek-piaci-trendek-20190222 (Retrieved: 30 May 2019).

Li et al. 2011. Energy Consumption-Economic Growth Relationship and Carbon Dioxide Emissions in China. *Energy Policy* 39(2): 568–574.

Murray, A. et al. 2017. The circular economy: an interdisciplinary exploration of the concept and application in a global context. *Journal of Business Ethics* Vol 140: 369–380.

PACE – Platform for Accelerating the Circular Economy 2019. The Circularity Gap Report. Available at: (Retrieved: 30 May 2019).

PwC 2018. Ha a kör bezárul – a körforgásos gazdaság jelentősége és lehetőségei. Available at: https://www.pwc.com/hu/hu/sajtoszoba/2018/korforgasos_gazdasag.html (Retrieved: 30 May 2019).

Viland, G. 2018. Az acélipar fejlődése számos más üzleti területre is hatással van. *Magyar Hírlap*. 2018. október15. Available at: https://www.magyarhirlap.hu/gazdasag/Az_acelipar_fejlodese_szamos_mas_uzleti_teruletre_is_hatassal_van_

WCED - World Commission on Environment and Development 1987. Our Common Future; http://www.un-documents.net/our-common-future.pdf (Retrieved: 30 May 2019).

World Steel Association 2015. Steel in the Circular Economy. A life cycle perspective Available at: https://www.worldsteel.org/en/dam/jcr:00892d89-551e-42d9-ae68-abdbd3b507a1/Steel+in+the+circular+economy+-+A+life+cycle+perspective.pdf.

World Steel Association 2018a: Steel Statistical Yearbook 2018. Brussels, Belgium. Available at: https://www.worldsteel.org/en/dam/jcr:e5a8eda5-4b46-4892-856b-00908b5ab492/SSY_2018.pdf (Retrieved: 30 May 2019).

World Steel Association 2018b: Sustainability Indicators 2003–2017. Brussels, Belgium. Available at: https://www.worldsteel.org/en/dam/jcr:6315d64c-c3a9-460b-8f80-dcbeaeaac5c4/Indicator%2520data%25202003%2520to%25202017%2520and%2520relevance.pdf (Retrieved: 30 May 2019).

World Steel Association 2018c: Steel Industry Co-Products worldsteel position paper. Brussels, Belgium. Available at: https://www.worldsteel.org/media-centre/press-releases/2018/steel-industry-co-products—worldsteel-position-paper.html (Retrieved: 30 May 2019).

Solutions for Sustainable Development – Szita, Jármai & Voith (eds.)
© 2020 Taylor & Francis Group, London, ISBN 978-0-367-42425-1

Influence of urban morphology on the environmental impacts of district. Applied LCA

D. Kaoula & A. Abdessemed-Foufa
Institute of Architecture and Town-Planning, University Blida 1, Algeria

ABSTRACT: Rethinking the city and its entities in such a way as to place them in a purely ecological framework raises major concerns that seem urgent today in the face of a heightened environmental decline within cities. The district is a relevant base to mitigate this decline through urban morphology, which appears to be a major criterion for the environmental optimization of neighbourhoods through their configuration and organization. The present work aims to address the sustainability of a retirement district in France by seeking the most optimal morphology for environmental preservation through the influence of three criteria, namely, shape, compactness and density, on the generation of environmental impacts and then identify the contribution of the different sources of impacts to the overall assessment through a series of simulations based on the very rigorous Life Cycle Assessment (LCA) approach.

1 INTRODUCTION

Nowadays, cities suffer from urban anarchy which leads to the environmental degradation; so, however, urgent solutions must be engaged.

Unlike the districts 'environmental dimension- particularly in terms of impact generation-it energy consumption have already been studied by many authors (Long et al, 2018; Murshed et al. 2019; Mouzourides et al. 2019; Ken et al. 2019). It is important to note that Muniz et al. (2019) demonstrated that polycentric land-use planning can be a good strategy to reduce the ecological footprint of cities. Lausselet et al. (2018) developed an LCA model for neighbourhoods focused on greenhouse gas emissions based on a modular structure with five physical elements; the results highlighted the critical parameters that are emission intensities for electricity, heat production by waste incineration, and the daily distance travelled by residents. Kang et al. (2017) shown that to adequately address air quality concerns, mixed land use and the compact urban form must be considered more seriously in sustainable urban planning.

It can be seen that no study demonstrated the influence of urban form on the generation of environmental impacts. This contribution enhances this field through a multi-criteria comparative analysis of different morphological models. Shape, density and compactness are the three parameters chosen to reach the objective.

2 METHODOLOGY

2.1 *Methodological approach*

The assessment of environmental impacts on an urban scale is considered as a fruitful field of research, several existing methods can be classified according to three main approaches (Baynes & Wiedmann 2012):

- Consumption-Based Approaches (CBAs): they assess environmental impacts based on economic data;

- Metabolsim-Based Approaches (MBAs): they are based on metabolism, they account for the flows of materials, water, waste and energy;
- Complex systems approaches: these approaches address the interactions between the different components of the urban system.

However, the method chosen for the evaluation of the case study is the rigorous Life Cycle Analysis (LCA) method, which is considered as an important method for urban-scale analysis (Anderson et al. 2015). It is efficient for a combination of the CBA and MBA approaches mentioned above. It has also been concluded that it is the only method that can avoid the transfer of pollution from one stage of the life cycle to another, or from one impact to another and to the territories (Loiseau et al. 2012).

2.2 Simulation approach

The simulation approach is based on a software chain. We started this approach by modelling the case study via ALCYONE, then we carried out a dynamic thermal simulation using the COMFIE software. The results obtained are then transmitted to EQUER in order to be able to carry out an environmental impact assessment via Life Cycle Analysis (LCA).

2.3 Case study

The selected case study is a district located in France, in Vermand-Saint Quentinois (Figure 1) composed of 20 housing units for the elderly. Its technical specificities are summarized in Table 1.

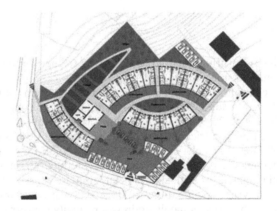

Figure 1. The case study.

Table 1. Technical specifications of the case study.

Elements	Components of the system	Thickness cm	Density kg/m^2	Thermal conductivity W/m k	Résistance m^{2}k/w
Exterior wall	Generic insulator	20	2	0.040	5.00
	Building block	20	260	1.053	0.19
Interior wall	Cinder block	20	260	1.053	0.19
Low floor	Generic insulator	15	2	0.040	3.75
	Concrete heavyweight	16	208	1.231	0.13
Intermediate floor	Concrete heavyweight	16	208	1.231	0.13
High floor	Generic insulator	25	3	0.040	6.25
	Gypsum plaster	1	12	0.420	0.02

3 RESULTS AND DISCUSSION

3.1 *Influence of form*

Keeping the same layout of the basic case study, we modelled the district in a rigid form by attributing different thermal zones according to the orientation and use of the interior spaces (Figure 2). Then we carried out a series of energy simulations and a comparative LCA (Figure 3).

The rigid form is less impact generating than the fluid form. This difference in impact generation is linked to the form, several orientations which generate a different flow of ventilation. This influences the energy consumption and generates different environmental impacts.

3.2 *Influence of compactness*

In this second case, we modelled a third variant by breaking up the buildings (Figure 4) in order to study the influence of compactness on the generation of impacts. The LCA results deduced from the energy simulation are illustrated in Figure 5.

The fragmented form generates more environmental impacts than the basic one. However, the last one generates more inert waste, resource depletion, aquatic ecotoxicity and radioactive waste. These impacts increase its cooling needs and energy consumption.

3.3 *Density influence*

We choose for the same number of dwellings, a density with two levels and another with four levels. We modelled the variants using the Alcyone software (Figure 6), then, we carried out an energy simulation and an LCA (Figure 7).

Figure 2. Modelling of the basic and rigid variants.

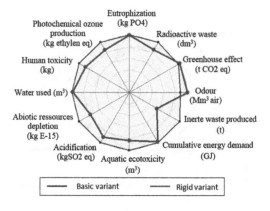

Figure 3. LCA comparison of the fluid and rigid variants.

Figure 4. LCA comparison of the fluid and fragmented variants.

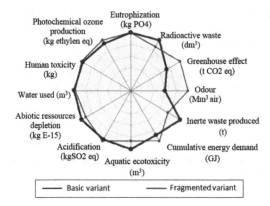

Figure 5. LCA comparison of the basic and fragmented variants.

Figure 6. Modelling of basic variant with different densities.

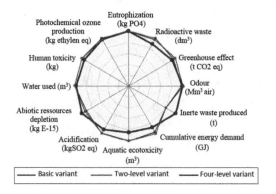

Figure 7. LCA of the basic variant with different densities.

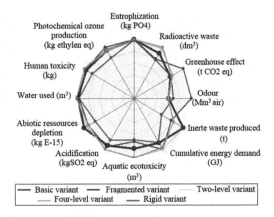

Figure 8.   LCA comparing all variants.

The results show that the generation of impacts between the two densified variants is almost identical when this difference is more important compared to the basic shape. Finally, this one generates seven impacts less than the other two variants. Consequently, the most optimal variant is the basic form. Despite of its important heat requirements, this last one is still the lowest energy-consuming variant. Therefore, the generation of environmental impacts is closely correlated to the energy balance. Finally, a high- densified form could reduce better the heating needs, whereas a less densified one reduces cooling needs. At this stage research, it is important to determine the optimal configuration which could reduce better the environmental impacts (Figure 8).

Finally, the most optimal form for reducing environmental impacts is the rigid variant, which is recommended such as compact and rigid form for better environmental optimization. However, this optimization is only reliable through the identification of the sources of impacts and their contributions to the overall balance sheet.

3.4   *Identification and contribution of sources of impacts throughout the whole life cycle*

In order to identify the sources of impacts and their contribution to the overall balance of the different variants, we have carried out several series of simulations for three main variants.

Table 2 summarizes the results obtained during the use phase. it is the one that generates the most impact, except for the impact of inert waste, which is generated during the demolition phase. Beside the transport of materials is the most important source all other impacts.

We can summarize the sources of impacts of the main variants as follows:

- Specific electricity: is the main contributor to seven impacts generated by the basic and rigid variants and four impacts generated by the fragmented variant. It is also the only contributor of radioactive waste for all three variants.
- Heating: is the main contributor to most of the impacts generated by the fragmented variant, but it is only responsible for one impact for the basic variant (odours) and two impacts for the rigid variant (greenhouse effect and odours).
- DHW (Domestic Hot Water): is responsible for generating the first five impacts engendered by the basic and rigid variants with different contributions, second three impacts engendered by the fragmented variant.

Based on this comparison, the results show that it is important to reduce water and energy consumption in order to limit the generation of almost all the impacts for the different variants, particularly during the use phase.

Table 2. Comparison of the contribution of the main sources of impacts of different variants.

Impacts	Contribution of main sources of impacts by variants and impacts (%)					
	Basic variant		Rigid variant		Fragmented variant	
Greenhouse effect	DHW* 38.16	Electricity 21.34	DHW* 35.23	Heating 31.56	Heating 51.23	DHW* 26.58
Acidification	Electricity 34	cooling 20.44	Electricity 40.58	Water 20.73	Heating 31.51	Electricity 29.28
Cumulative energy demand	Electricity 48.31	DHW 20.95	Electricity 50.46	DHW 20.72	Electricity 37.27	Heating 34.63
Water used		Water 91.21		Water 92.42		Water 93.21
Inert waste produced	Façades 33.29	Low floor 32.51	Low floor 37.93	Walls 34.06	Façades 47.95	Low floor 27.24
Abiotic resources depletion	Electricity 57.11	DHW* 17.01	Electricity 57.69	DHW* 17.53	Electricity 46.58	Heating 29.71
Eutrophisation		Water 96.79		Water 96.82		Water 96.25
Photochemical ozone production	Electricity 29.46	DHW* 27.96	Electricity 33.2	DHW* 27.43	Heating 43.61	DHW* 22.63
Aquatic ecotoxicity	Electricity 43.57	Cooling 24.86	Electricity 54.12	Water 15.77	Electricity 42.93	Heating 24.12
Radioactive waste		Electricity 67.14		Electricity 71.54		Electricity 66.15
Human toxicity	Electricity 32.67	Water 20.82	Electricity 38.63	Water 21.47	Heating 32.47	Electricity 27.53
Odour	DHW* 52.69	Heating 18.21	DHW* 45.04	Heating 40.33	Heating 59.36	DHW* 30.8

*DHW: Domestic Hot Water

3.5 *Limitation of impacts by control of the main identified impacts sources*

At this stage of analysis, we have identified the sources of impacts and their contribution to the overall balance sheet, which allows to limit the impacts generated by the variants. In this context, we occur at the identified sources. The results obtained are summarized through an LCA (Figure 9) and Table 3 summarizes their percentages of improvement.

The results obtained show a significant reduction in all environmental impacts through the treatment of certain sources previously identified as responsible for generating impacts, in particular the transport of materials, energy and water consumption. The improvement of the rigid form is finally presented as the best improved variant compared to the other alternatives.

4 CONCLUSION

The present study allowed to know the influence of the morphology of a neighbourhood on the generation of impacts in one hand, and to determine the most optimal shape for reducing them on the other hand. As a conclusion, the rigid variant shape is generating the lowest environmental impacts compared to all the variants studied.

According to the previous results, we conclude that we must avoid fragmenting neighbourhoods in order to reduce greenhouse gas emission and energy consumption. Whereas, when the form is rigid, it should be densified.

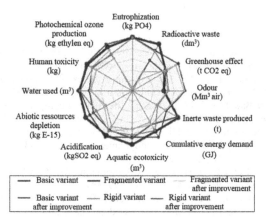

Figure 9. LCA before and after improvement of variants.

Table 3. Percentages of improvement of the different variants.

Impacts	Basic variant %	Rigid variant %	Fragmented variant %
Greenhouse effect	5.45	4.91	4.85
Acidification	12.22	10.76	12.41
Cumulative energy demand	17.13	21.95	27.16
Water used	19.60	20.43	21.63
Inert waste produced	1.83	2.73	2.65
Abiotic resources depletion	15.62	26.53	33.90
Eutrophisation	95.52	96.05	95.33
Photochemical ozone production	16.03	12.32	13.43
Aquatic ecotoxicity	11.78	13.55	17.60
Radioactive waste	33.99	50.77	73.68
Human toxicity	10.94	9.95	11.31
Odours	2.27	2.19	1.10

The different sources responsible for generating the impacts of the district studied were identified; in this context, during the exploitation phase, it is necessary to give a particular attention to specific electricity, domestic hot water, air conditioning and water consumption in order to reduce particularly radioactive waste, photochemical ozone, acidification and eutrophication.

During the demolition phase, it is important to reduce the materials' transport distance. This is the main source which is generating 93% of all the environmental impacts.

The identification of sources allows to improve the environmental balance according to new scenarios. These are: 10 km distance between the site and the inert landfill at the end of its life, 10 km distance to the incinerator, 20 km distance to the recycling centre, 0.12 of electricity efficiency of cogeneration 30 and 60 liter/day/person respectively of hot and cold water and 70% waste incineration recovery. We got a 10% improvement for most impacts. However, we got 15.62%, 26.53% and 33.90% reduction of depletion resources for the basic form, rigid form and fluid form respectively., 33.99%, 50.77% and 73.68% reduction in the impact of radioactive waste were obtained for the basic form, rigid form and fluid form respectively and more than 95% eutrophication for all variants.

According to the various results, elements which require particular attention during the impact reduction process were identified. This contributes to improve the environmental balance of all variants through a significant limitation of all impacts generated during the life cycle.

REFERENCES

Anderson, J.E., Wulfhorst, G. & Lang, W. 2015. Energy analysis of the built environment– A review and outlook. *Renewable and Sustainable Energy Reviews* 44(C): 149–158.

Baynes, T.M. & Wiedmann, T. 2012. General approaches for assessing urban environmental sustainability. *Current Opinion in Environmental Sustainability* 4(4): 458–464.

Kang, J.E., Yoon, D.K. & Bae, H.J. 2017. Evaluating the effect of compact urban form on air quality in Korea. *Environment and Planning B: Urban Analytics and City Science* 46(1): 179–200.

Kent, C.W., Grimmond, S., Gatey, D. & Hirano, K. 2019. Urban morphology parameters from global digital elevation models: implications for aerodynamic roughness and for wind speed estimation. *Remote sensing of environment* 221: 316–339.

Lausselet, C., Borgnes, V. & Brattebø, H. 2019. LCA modelling for Zero Emission Neighbourhoods in early stage planning. *Building and Environment* 149: 379–389.

Long, N., Gardes, T., Hidalgo, J., Masson, V. & Schoetter, R. 2018. Influence of the urban morphology on the urban heat island intensity: an approach based on the local climate zone classification. *PeerJ Preprints* 6: e27208v1.

Loiseau, L., Junqua, G., Roux, P. & Bellon-Maurel, V. 2012. Environmental assessment of a territory: An overview of existing tools and methods. *Journal of Environmental Management* 112: 213–225.

Mouzourides, P., Kyprianou, A. & Ching, J. 2019. Linking the urban-scale building energy demands with city breathability and urban form characteristics. *Sustainable cities and society* 10.1016/j.scs.2019.

Muñiz, I. & Garcia-López, M.A. 2019. Urban form and spatial structure as determinants of the transportation. *Research Part D: Transport and Environment* 67: 334–350.

Murshed, S., Duval, A., Koch, A. & Rode, P. 2019. Impact of urban morphology on energy consumption of vertical mobility in Asian cities-a comparative analysis with 3D city models. *Urban Science* 3(1), 4.

Solutions for Sustainable Development – Szita, Jármai & Voith (eds.)
© 2020 Taylor & Francis Group, London, ISBN 978-0-367-42425-1

Life cycle extension of damaged pipelines using fiber reinforced polymer matrix composite wraps

J. Lukács & Zs. Koncsik
Faculty of Mechanical Engineering and Informatics, Institute of Materials Science and Technology, University of Miskolc, Miskolc, Hungary

P. Chován
FGSZ Ltd., Siófok, Hungary

ABSTRACT: Life time management and life cycle extension of different structures is one of the most important technical-economic problems nowadays. High pressure, hydro-carbon transporting pipelines have strategic importance, accordingly, their safe and practically fail-safe operation is elemental economic interest. For the developing both an Integrity Management Plan or a Pipeline Integrity Management System, different data, frequently experimental data should be collected and evaluated. The purpose of the paper is to present the role of external reinforcing technics on the life cycle extension of damaged transporting steel pipelines, based on full scale tests. Externally reinforced seamless and seam welded steel pipeline sections were investigated; fatigue and burst tests were executed. Different fiber reinforced polymer matrix composite wrap systems were used; both separate and interacting metal loss defects were investigated. Safety factor was calculated after the investigations, which is applicable for the assessment of the effectiveness of the different reinforcing systems, furthermore for the ranking of these systems.

1 INTRODUCTION

Natural gas nowadays is one of the most important energy sources for different consumers. Natural gas has advantageous and favorable environmental and physical properties and due to these properties, it is more and more dominant among the applicable energy sources. The share of natural gas in the applied energy sources depends on the geographical and geopolitical conditions as well as the resource supply of the different countries.

2017 was a bumper year for natural gas, with consumption (*3.0%, 96 bcm*) and production (*4.0%, 131 bcm*) both. The growth in consumption was led by Asia, with particularly strong growth in China (*15.1%, 31 bcm*), supported by increases in the Middle East (Iran *6.8%, 13 bcm*) and Europe. The growth in consumption was more than matched by increasing production, particularly in Russia (*8.2%, 46 bcm*), supported by Iran (*10.5%, 21 bcm*), Australia (*18.0%, 17 bcm*) and China (*8.5%, 11 bcm*) (BP 2018).

In Hungary, the total primary energy consumption in 2017 is *23.2 million tons* oil equivalent, which is *1.18%* of the total Europe demand. Near *40%* of Hungarian need is covered by natural gas supply and only *3%* by renewable energy sources (BP 2018).

In Hungary, the FGSZ Natural Gas Transmission Closed Company Limited (FGSZ Ltd.) is responsible for transporting the natural gas from the import, gas storage, and domestic production entry points to the customers. FGSZ Ltd. is the operator of a fully integrated natural gas transmission system which contains inlet points, compressor stations, pipeline nods, high-pressure pipelines and gas delivery stations. This system is connected to almost all the natural gas transmission systems of the neighboring countries and delivers gas to the gas distribution companies, power plants and some huge industrial consumers. The gas is fed into FGSZ Ltd.'s high-pressure pipeline system through the gas inlet points, supplied from import sources, domestic gas fields and domestic gas storage facilities. The whole system consists of only steel pipelines, and the typical operating pressure is *63 bar* (*6.3 MPa*). The major data of the system are as follows (Chován, 2011, FGSZ 2016):

– length of the system: *5782 km*;
– typical diameter of the system: *100-1400 mm*;
– operating pressure range: *40-75 bar* (*4-7.5 MPa*);
– average age of the pipeline sections: *25 years*;
– number of transfer stations: *nearly 400 items*;
– number of dispatcher center: *1 item*;
– number of gas transmission plants: *6 items*;
– number of compressor stations: *6 items*.

Generally, steel pipelines are commonly used for transportation over long distance due to their applicable mechanical properties. One of the disadvantages of the steel material is its low corrosion resistance (de Barros et al. 2018). Gas transporting pipelines operate often in a harsh environment, which can lead, especially after a long-time application, to through-wall corrosion defects, causing leakage (Kumar, 2016), or cracks (Medjdoub, 2018). Traditionally, the pipelines with corrosion defects have to either been repaired using welding or reinforced techniques or replaced by new ones.

The present study contains information about our full scale investigations of composite pipeline reinforcing methods (wrap systems), used for repairing of pipeline sections with artificial surface and through-wall, separate and interacting defects. The main objective of the paper is data retrieval for integrity management tasks (in other words for Pipeline Integrity Management System), which can help in a decision situation, if the composite repairing system can operate for a long time, or it is only appropriate for a short-term maintenance period.

2 INVESTIGATIONS

Seam welded (HFW) and seamless steel pipeline sections without girth welds were examined. The main characteristics of the investigated pipeline sections and the internal pressure values are summarized in Table 1. Different types of fiber reinforced polymer matrix composite systems (wrap systems, *W1-W4*) were used for external reinforcing of artificial flaws, one unreinforced and seven reinforced pipeline section were tested. In Table 1 d_k is the external/outside diameter of the steel pipe (*OD*), t_a is the wall thickness of the steel pipe, p_{min} and p_{max} are the minimum and maximum internal pressure values during the fatigue tests.

Table 1. The main characteristics of the investigated pipeline sections and the internal pressure values during fatigue tests.

Pipe section	Pipe material	Pipe type	d_k mm	t_s mm	Wrap system –	p_{min} bar (MPa)	p_{max} bar (MPa)
S2-A1	L360MB	seam welded	323.9	7.1	no wrap (0W)	31.5 (3.15)	63.0 (6.30)
S2-A2			323.9	7.1	type1 (W1)	31.5 (3.15)	63.0 (6.30)
S2-B11	X52	seamless	324.7	10.65	type2 (W2)	51.0 (5.10)	63.0 (6.30)
S2-B12			324.7	8.4	type2 (W2	39.0 (3.90)	63.0 (6.30)
S2-B21	X52	seamless	327.2	10.2	type3 (W3)	51.0 (5.10)	63.0 (6.30)
S2-B22			327.2	8.45	type3 (W3)	39.0 (3.90)	63.0 (6.30)
S2-B31	X52	seamless	327.9	10.4	type4 (W4)	51.0 (5.10)	63.0 (6.30)
S2-B32			327.9	8.5	type4 (W4	39.0 (3.90)	63.0 (6.30)

Table 2. Details of artificial flaw characteristics.

Pipe section	Artificial flaw characteristics	Wrap system
S2-A1-No1	longitudinal gouge: h_k = 100 mm, m_k = 3 mm circumferential gouge: h_k = 150 mm, m_k = 4.7 mm interacting circumferential gouges: h_k = 2 * 130 mm, m_k = 4 mm	no wrap (0W)
S2-A1-No2	longitudinal gouge: h_k = 100 mm, m_k = 3 mm	no wrap (0W)
S2-A1-No3	new longitudinal gouge: h_k = 100 mm, m_k = 3.1 mm	no wrap (0W)
S2-A2-No1	longitudinal gouge: h_k = 100 mm, m_k = 3 mm circumferential gouge: h_k = 150 mm, m_k = 4.7 mm interacting circumferential gouges: h_k = 2 * 130 mm, m_k = 4 mm	type1 (W1)
S2-A2-No2	new longitudinal gouge: h_k = 100 mm, m_k = 3.1 mm	type1 (W1)
S2-B11	through hole d = 3 mm	type2 (W2)
S2-B12	through hole d = 3 mm	type2 (W2)
S2-B21	through hole d = 3 mm	type3 (W3)
S2-B22	through hole d = 3 mm	type3 (W3)
S2-B31	through hole d = 3 mm	type4 (W4)
S2-B32	through hole d = 3 mm	type4 (W4)

For executing fatigue tests MTS type electro-hydraulic equipment was applied. The planned load was a sinusoidal type, for *100,000 cycle*. For the burst tests, an own developed testing apparatus was used, with a maximal applicable pressure of *700 bar (70 MPa)*.

Different types of artificial flaws, as characteristic metal losses were investigated. Table 2 summarizes the details of the artificial flaws, Figure 1 and Figure 2 show the geometry and

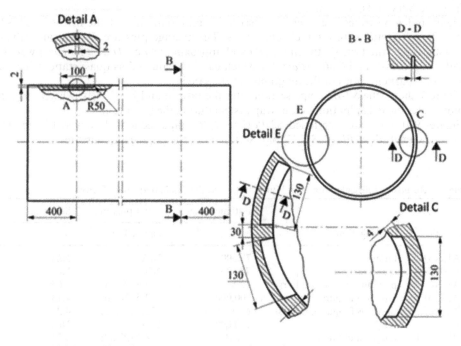

Figure 1. Artificial metal loss flaws on *S2-Ai (i = 1, 2)* pipe sections.

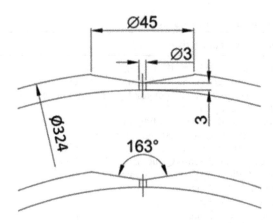

Figure 2.   Artificial metal loss flaws on *S2-Bi* (*i* = *11, 12, 21, 22, 31, 32*) pipe sections.

the location of the flaws on the *S2-Ai* (*i* = *1, 2*) and the *S2-Bi* (*i* = *11, 12, 21, 22, 31, 32*) pipe sections, respectively.

3   TEST RESULTS

The results of the full scale tests executed on unreinforced and reinforced pipeline sections are summarized in Table 3, where safety factor means burst pressure divided by Maximum Allowable Operating Pressure (*MAOP*).

Figures 3-5 demonstrate *S2-A1-No3, S2-A2-No1* and *S2-A2-No2* pipe sections after the fatigue and burst tests, respectively.

Figure 6 demonstrates the internal pressure vs. time diagrams of the investigated *S2-Ai* (*i* = *1, 2*) pipe sections during their burst tests. The average pressure growth rate values in the first stage of the tests (between *0 MPa* and approximately *6 MPa (0-60 bar)*) were *0.36 MPa/s, 0.35 MPa/s* and *0.30 MPa/s*, for which can be evaluated as quasi-static values, and the another particularities of the diagrams are the same.

Figure 7 shows the *S2-B12* pipe section after the fatigue and burst tests, whereas Figure 8 demonstrates the delamination of the wrap system during the burst test.

Figures 9-10 illustrates the failed *S2-Bi* (*i* = *21, 22*) pipe sections, however, Figures 11-12 shows the damaged *S2-Bi* (*i* = *31, 32*) pipe section after the fatigue and burst tests.

Table 3.   Results of full scale fatigue and burst tests, and the calculated safety factors.

Pipe section	Investigation	Fatigue cycles cycle	Burst pressure bar (MPa)	Safety factor –
S2-A1-No1	fatigue	79,500	N/A	N/A
S2-A1-No2	fatigue	93,600	N/A	N/A
S2-A1-No3	previous fatigue and burst	N/A	233.5 (23.35)	3.65
S2-A2-No1	fatigue and burst	100,000	264.6 (26.46)	4.13
S2-A2-No2	previous fatigue and burst	N/A	273.2 (27.32)	4.27
S2-B11	fatigue	100,000	N/A	N/A
S2-B12	fatigue and burst	100,000	278.0 (27.80)	4.34
S2-B21	fatigue and burst	100,000	380.0 (38.80)	5.94
S2-B22	fatigue and burst	100,000	330.0 (33.00)	5.16
S2-B31	fatigue and burst	100,000	315.1 (31.51)	4.92
S2-B32	fatigue and burst	100,000	326.0 (32.60)	5.09

Figure 3.   Unreinforced (*0W*) pipe section *S2-A1-No3* after the burst test.

Figure 4.   Reinforced (*W1*) pipe section *S2-A2-No1* after the first burst test.

Figure 5.   Reinforced (*W1*) pipe section *S2-A2-No1* after the second burst test.

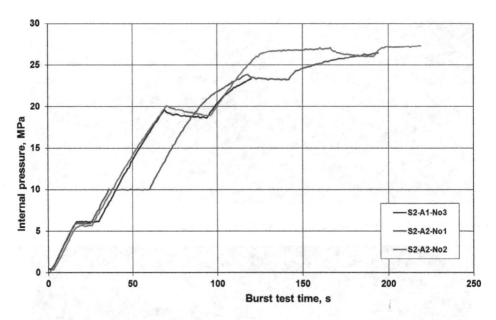

Figure 6. Internal pressure vs. time diagrams during the burst tests of the *S2-Ai* (*i = 1, 2*) pipe sections.

Figure 7. Reinforced (*W2*) *S2-B12* pipe section after the fatigue and burst tests.

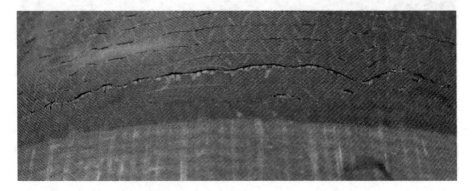

Figure 8. Reinforcing wrap system (*W2*) delamination during the burst test on the *S2-B12* pipe section.

Figure 9.   Reinforced (*W3*) *S2-B21* pipe section after the fatigue and burst tests.

Figure 10.   Reinforced (*W3*) *S2-B22* pipe section after the fatigue and burst tests.

Figure 13 summarizes the internal pressure vs. time diagrams of the investigated *S2-Bi* (*i = 12, 21, 22, 31, 32*) pipe sections during their burst tests. The average pressure growth rate values in the first stage of the *S2-Bi2* (*i = 1, 2, 3*) pipe sections were *0.29-0.36 MPa/s*, nevertheless of the *S2-B21* pipe section was lower (*0.15 MPa/s*) and of the

Figure 11.   Reinforced (*W4*) *S2-B31* pipe section after the fatigue and burst tests.

Figure 12.   Reinforced (*W4*) *S2-B32* pipe section after the fatigue and burst tests.

S2-B31 pipe section was higher (*0.51 MPa/s*). These differences do not cause significant effects on the damage process of the pipe sections. The behavior of the pipe sections during the burst tests was essentially analogical, unfortunately, we have non-return valve closing difficulty in the middle part of the burst test of *S2-B22* pipe section.

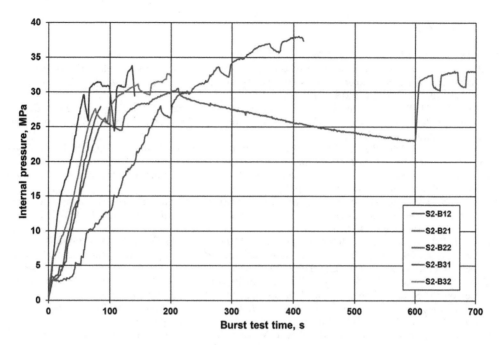

Figure 13.   Internal pressure vs. time diagrams during the burst tests of the *S2-Bi* (*i = 12, 21, 22, 31, 32*) pipe sections.

4   SUMMARY AND CONCLUSIONS

Based on our full scale investigations and their results, the following conclusions can be drawn.

The types of metal loss flaws (surface – through, longitudinal – circumferential, separate – interacting) have significant effects on the behavior and the effectiveness of the reinforcing wrap systems.

The investigated reinforcing wrap systems can be used basically for transporting and industrial steel pipelines, for a wide variety of pipe diameters and lengths, for both quasi-static and cyclic loaded pipeline sections, for both workshop-work and field-work.

The defined safety factor and their calculated values demonstrate both the reserves of the steel pipes and the usefulness of the reinforcing materials and technologies. Furthermore, the safety factor is applicable for the ranking of these reinforcing systems.

Results of full scale tests correspond with results of numerical investigations (Égert & Pere 2009, Pere, Égert & Szabó 2009, Lukács et al. 2010) in case of externally reinforced damaged pipelines.

Databases and especially experimental data have a determinant role in the integrity assurance of different structures, like pipeline systems (Lukács, Nagy & Török 2008). With the help of these databases and frequently with the use of the experimental data, integrity management tasks can be solved (Lukács 2005).

ACKNOWLEDGEMENTS

The presented work was carried out as part of the EFOP-3.6.1-16-2016-00011 "*Younger and Renewing University – Innovative Knowledge City – institutional development of the University of Miskolc aiming at intelligent specialization*" project implemented in the framework of the Széchenyi 2020 program. The realization of this project is supported by the European Union, co-financed by the European Social Fund.

REFERENCES

BP 2018. *Statistical Review of World Energy.* 67th edition.

de Barros, S. et al. 2018. An assessment of composite repair system in offshore platform for corroded circumferential welds in super duplex steel pipe. Technical note. *Frattura ed Integrità Strutturale* 44: 151–160. doi: 10.3221/IGF-ESIS.44.12.

Chován, P. 2011. Risk based reconstruction planning. In *Proceedings of The First Central and Eastern European International Oil and Gas Conference and Exhibition.* Available: http://www .oilgasconf.montanpress.hu

Égert, J. & Pere, B. 2009. Repair of Internal and External Longitudinal Failures of Pipes by Fabric Composite Reinforcement. *Acta Technica Jaurinensis* 2(1): 3–17.

FGSZ 2016. *FGSZ Ltd.'s Annual Report 2016.* Available online, at 24. June 2019: https://fgsz.hu/file/docu ments/0/0917/fgsz_annual_report_2016.pdf

Kumar, P.N. et al. 2016. Repair of trough thickness corrosion/leaking defects in corroded pipelines using Fiber Reinforced Polymer overwrap. *Materials Science and Engineering* 346, IMMT 2017, doi: 10.1088/1757-899X/346/1/012016.

Lukács, J. 2005. Dimensions of lifetime management. *Materials Science Forum* 473–474: 361–368.

Lukács, J. et al. 2010. Experimental and Numerical Investigations of External Reinforced Damaged Pipelines. *Procedia Engineering* 2(1): 1191–1200.

Lukács, J., Nagy, Gy. & Török, I. 2008. The Role of Process Models, Flow Charts and Material Databases on the Structural Integrity of High Strength Steel Pipeline Systems. In *Proceedings of International Conference on New Developments on Metallurgy and Applications of High Strength Steels, Buenos Aires,* 26–28 May 2008. Minerals, Wiley, John & Sons, Incorporated.

Medjdoub, S.M., Bouadjra, B.B. & Abdelkader, M. 2018. Optimization of the geometrical parameters of bonded composite wrap for repairing cracked pipelines. *Frattura ed Integrità Strutturale* 46: 102–112. doi: 10.3221/IGF-ESIS.46.11.

Pere, B., Égert, J. & Szabó, T. 2009. Reinforcement of inner and outer circular failures of pipes by textile composite layers. *Journal of Computational and Applied Mechanics* 10(1): 1–15.

Solutions for Sustainable Development – Szita, Jármai & Voith (eds.)
© 2020 Taylor & Francis Group, London, ISBN 978-0-367-42425-1

Innovative solutions for the building industry to improve sustainability performance with Life Cycle Assessment modelling

V. Mannheim & Z.S. Fehér
Higher Education Industrial Cooperation Centre, University of Miskolc, Hungary

Z. Siménfalvi
Faculty of Mechanical Engineering and Informatics, University of Miskolc, Hungary

ABSTRACT: The Life Cycle Assessment for cement as binder and concrete base material is an important part of the total LCA of buildings. In the LCA Research Team at University of Miskolc the total life cycle model of buildings has been developed in a complex way from the production phase of cement and concrete through the phase of usage to the end-of-life stage with consideration of the information models of the Environmental Product Declaration. Within the total lifecycle, the cement production phase and the end-of-life phase play an important role, because cement production is responsible for 8 % of global carbon-dioxide emissions. With the application of LCA software GaBi 8 thinkstep for the production technologies, their environmental efficiency can be determined. After the development of the LCA model for the production of cement and concrete elements, the possible innovative solutions for improving sustainability performance are presented by quantifying environmental impacts and energy optimization. Despite of the fact that building industry and environmental management are closely interlocked, there is fairly poor national and international professional literature available about the two connected professions.

1 INTRODUCTION

In order to analyse completely a building from environmental effects point of view (e.g. greenhouse gases per functional unit - kg equivalent/functional unit), the life cycle evaluation data of the prefabricated concrete formulas are needed. The environmental effects can be defined numerically due to environmental effect points (ecological absence method), primary energy, renewable and non-renewable energy sources and due to emission of greenhouse gases (IPCC 2013). During the development based on life cycle assessment of an innovative and environmentally friendly concrete structural element, the total life cycle of each product has to be considered so we can focus on the stage of becoming waste within the stages of the total life cycle (Tóth Szita, K. 2017). The Environmental Product Declaration (EPD) concerning the construction sector sets the stage of bases of building analyses furthermore specializes the information modules belonging to different life cycle stages. In this research, the total life cycle stages of concrete were elaborated considering the EPD modules. After the development of the steps following the LCA of the production of concrete elements and its methodological approaches, the results of the LCA of prefabricated concrete elements from normal and high strength concrete are summarized. By quantifying environmental impacts and comparing total primary energy need and greenhouse gas emissions (Tomporowski, A. et al. 2018). The life cycle analysis of the investigated technologies was carried out by applying GaBi 8 thinkstep software and an additional database from the building industry. The LCA software with the latest database assures the life cycle assessment as a tool of basic importance aiming more sustainable production and process development. Among the goals of the LCA research group established in Higher Education Industrial Cooperation Centre (HEICC) are elaborations of such innovative technologies that can provide theoretical and practical implementation of environmental friendly and optimal engineering solutions. This research work is strongly connected to LCA-based researches of the University of Miskolc and it is extremely relevant because the prevention and economization of limited resources become conspicuous with the application of life cycle analysis.

2 SCOPE AND GOAL OF LIFE CYCLE ASSESSMENT

The Life Cycle Assessment addresses the environmental aspects and potential environmental impacts throughout a product's life cycle from raw material acquisition through production, use, end-of-life treatment, recycling and final disposal. LCA is one of several environmental management techniques (e.g. environmental performance evaluation and environmental impact assessment). The life cycle approaches and methodologies can be applied to several aspects (Guinée, J. 2002). The scope of LCA depends on the subject and the intended use of the study (including the system boundary and level of detail). The depth and the breadth of LCA can differ considerably depending on the goal of a particular LCA. Based on LCA the investigation of environmental impacts of different technologies becomes possible, therefore its application is most adequate in case of substituting products and processes (Mannheim 2014). With its support, it can be quantified, which environmental load is caused during the total lifetime of a given technology or product. For practitioners of LCA, ISO 14044 details the requirements for conducting an LCA. The life cycle inventory (LCI phase) is an inventory of input/output data with regard to the system being studied. It involves the collection of the data necessary to meet the goals of the research study. The purpose of the life cycle impact assessment (LCIA phase) is to provide additional information to help assess a product system's LCI results. The life cycle interpretation (the final phase) summaries the conclusion in accordance with the goals. The LCA can assist in the building industry:

- Identifying opportunities to improve the environmental performance of products or technologies at various points in their life cycle.
- Informing decision-makers in building industry (strategic planning, priority setting, product or process design or redesign).
- The selection of relevant indicators of environmental performance, including energy efficiency and costs.
- Marketing goals (e.g. making an environmental claim or producing an environmental product declaration).

LCA models the life cycle of a product as its product system, which performs one or more defined functions. The essential property of a product system is characterized by its function and cannot be defined solely in terms of the final products (Cooper, J.S. et al. 2008). Figure 1 shows an example of a product system. Product systems are subdivided into a set of unit processes. Unit processes are linked to one another by flows of intermediate products and/or waste for treatment, to other product systems by product flows, and to the environment by elementary flows (Mannheim & Siménfalvi (2012)).

3 TOTAL LIFE CYCLE OF CEMENT AND CONCRETE

The compressive strength of concrete by cylinder is investigated on a sample stored under water with a height of 300 mm and a diameter of 150 mm. According to the European Norm the compressive strength behavior of concrete has to be tagged by cubic where the cubic strength is determined by a specimen with a length of 150 mm. The strength classes are marked in the notation with numbers behind the concrete identification sign (compressive strength of concrete cylinder/compressive strength of concrete cubes). In case of notation according to the valid standard, the strength properties of cylinder and cube samples with the largest grain size of the additive and the consistency have to be indicated with the slump flow classes. The consistency of concrete

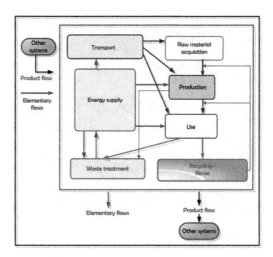

Figure 1.   Example for a product system.

influences the miscibility, transportability, workability, compatibility, stability of fresh concrete. The valid standard MSZ EN 206-1:2002 indicates the consistency classes with a combination of letters and numbers so measures according to aquitard, cold-resistant, wear resistant properties of concrete can also appear behind the sign of consistency class regarding the classification of different environmental effects. In case of lightweight concrete, the density also has to be indicated (e.g. LC1600-10-16/K-LC, lightweight concrete with high laudability). Based on density the concrete can be classified as presented in Table 1. During operating lifetime (by normal operating circumstances) the concrete will be durable if it can withstand environmental effects without significant damaging. When more different environmental effects occur the combination of environmental classes has to be applied (Buday-Malik et al. 2018).

Figures 2 and 3 illustrate the total life cycle of cement and the total life cycle of concrete based on the applied EPD modules. Essential to concrete production the product unit, cement, its total life cycle representing flow chart is already published in a previous research.

The total life cycle of concrete can be divided into three stages, and numerous factors and loads must be considered. The applied concrete formwork for the production of concrete elements can be removed easily after solidification applying a release agent. Formwork made of wood can be recycled 20 times while formwork made of steel approximately 700 times. As for high strength concrete elements the rate of wooden formwork is low, at the life cycle analysis of high strength concrete elements this input unit can be neglected. At the same time nor the steel formwork will be considered. Even in case of life cycle assessment of normal strength concrete elements, their application rate is infinitesimally small. Additive materials are given in higher amount aiming to reach or improve certain properties: mineral fine matters (filler materials), for example, rock flour, pozzolans, pigments and organic materials. Additive agents are given in

Table 1.   Classification of concrete based on their density property.

Designation	Sign	Density	Application field
Lightweight concrete	LC	800–2000	By building of thermal insulator structures, adding appropriate additive.
Normal concrete	C	2001–2600	Most important type of concrete nowadays, the structural concrete.
Heavyweight concrete	HC	>2600	Especially for radiation protection function.

* The unit of density is in kg/m^3.

247

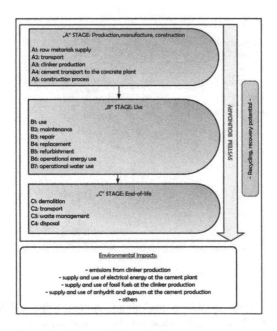

Figure 2.　The total life cycle of cement.

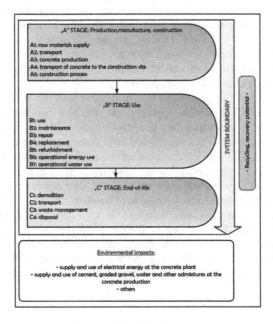

Figure 3.　The total life cycle of concrete.

a small amount, which influences certain properties of the concrete by the physical and/or chemical effect. Additive agents are plasticizers, air entrainers, accelerators, retarders, etc. Plasticizers in concrete factory as polycarboxylates are often applied. It is assumed that for cleaning up the production hall together with the concrete production, a small amount of water will be consumed. Sewage water arises only in small amount, as the washing water will be reused for concrete production furthermore it will be partially vaporized. The steel for concrete and the concrete can be recycled. Their recycling costs do not count to the production of concrete elements; therefore they do not appear in the life cycle assembly. The surface of the concrete element will be reconstructed and stored before transporting it to the construction site. At transportation, wood is necessary for fixation of the concrete element, which will be treated as construction waste.

By the consideration of environmental effects, the method of ecological absence (the value from the difference between actual state and the appointed goal) will be applied, by which the dissipative resource consumption will be assessed (during moving and temporal changes the total energy of the system will be decreased by processes with some losses). As for the financial consumption of the resources, it is not the extraction of the resources that counts, but the fact, that extent of losses of the extracted and processed resources come to pass, and so it will not be available in future consumption. The dissipative consumption will be modelled in the life cycle analysis applying resources-corrections. If the metals are 100 % recyclable, the resources-correction practically has to be given according to the metal portion. If the concrete is 90% recyclable, the resources-correction is applied for the 90% portion of gravel and sand content. The indices related to primer energy consumption and emission of greenhouse gases are not influenced by this resources-correction (Reiners, J. 2011). The determination of critical loads is based on valid environmental protection laws and international environmental protective agreements. The costs of concrete recycling do not charge the production of concrete elements, therefore it is not necessary to consider in the life cycle assembly (Grieder, A. et al. 2016). During the life cycle analysis of a concrete element, the following regularization and standards have to be considered:

- EPD and information modules regarding to construction sector
- ISO 14025:2006
- Basic regularization of product categories (PCR)
- ISO 14040:2006
- ISO 14044:2006
- ISO 14045:2012
- EN 15804
- MSZ EN15978:2012+A1:2014
- Environmental criteria in accordance with environmental protective law
- MSZ EN ISO 14020:2002 (environmental labels and declaration)

Within the total lifecycle, the end-of-life phase plays an important role, where the dissipative use is evaluated. By the development of the life cycle steps of concretes and its methodological approaches, the results of the LCA of concrete formulations from CEM II. B-M, 32.5 R (V-LL) are compared by quantifying environmental impacts and greenhouse gas emissions.

4   LIFE CYCLE ANALYSIS RESULTS OF CONCRETE FORMULATIONS

For a specific research task, LCI and LCA analyses of two cement-based formulations were made with the available GaBi 8 thinkstep LCA software at the University of Miskolc. The GaBi 8 ts software came into the market in 2018 and has several advantages for its former versions (Kupfer, T. et al. 2018). The Software with databases 2017 establishes Life Cycle Assessment as an essential tool to develop more sustainable products and processes while increasing resource efficiency, reducing material, energy and cost. With the 2017 release of GaBi databases, the assessment of land use has made a big step forward. The new professional database incorporates world-leading product sustainability intelligence (Mannheim & Siménfalvi (2018). The professional database concept is an important aspect, which ensures the accuracy and relevancy of results. The value of the functional unit in LCA is 1 kg of concrete and the input energy value was 0.036 MJ. The concrete formulas examined CEM II. B-M, 32.5 R (V-LL) type cement. The applied impact assessment method was CML 2001 (January 2016). CML uses the GWP factors published by IPCC. According to measurements the Global Warming Potential (GWP in CO_2-equivalent), the Acidification Potential (AP in SO_2-equivalent), the Eutrophication Potential (EP in Phosphate-equivalent) and the

Ozone Layer Depletion Potential (ODP in R11-equivalent) were determined. The main environmental impact category diagrams show the GWP, AP, EP and ODP values in order to compare three concrete formulas where a certain proportion of cement with a renewable material (biomass, untreated renewable primary product), with a mineral product (processed mineral) as well as natural and artificial materials (mineral product + product from waste + polymer). Global Warming Potential is the most important category, and its weighting factor is 10. Several time perspectives are available with the GWPs for 100 years recommended as the baseline characterisation method for climate change. In the implementation of the CML version in January 2016, the GWP factors are upgraded to AR5; earlier methods are based on Assessment Report (Tomporowski, A. et al. 2017). By default, CML includes biogenic carbon at the same level as fossil carbon; hence CO_2 uptake has a GWP of 1 kg CO_2-equivalent. We would like to highlight the Global Warming Potential (GWP) values for three concrete formulas. In the second and third cases, the GWP values (incl. biogenic CO_2, GWPs: 0,45 and 0,36 kg CO_2-equivalent) are obtained, compared to the first GWP values (total GWP: 0,35 kg CO_2-equivalent). The biogenic carbon dioxide emissions are tracked separately from the fossil carbon dioxide emissions. Concerning biotic CO_2, the uptake and emission have to be considered. Generally, in GaBi the carbon uptake from the atmosphere and the biotic emissions are modelled. Figures 4–7 show the different environmental impact category diagrams of the LCA software for GWP, AP, EP and ODP values. The investigations show that by concrete formula from natural material the value of GWP is the best. Investigating the emission of greenhouse gases, in relation to concrete formulas in case of formula I. the SO_2-equivalent is $6{,}29.10^{-4}$ kg, while in case of formula III. this value is $5{,}81.10^{-4}$ kg. The Eutrophication Potential in relation to concrete formulas in case of formula I. the phosphate-equivalent is $9{,}1410^{-5}$ kg, while in case of formula III. this equivalent is $9{,}26.10^{-5}$ kg. Referring to Ozone Layer Depletion Potential in relation to concrete formulas in case of formula I. the R11-equivalent is $1{,}62.10^{-13}$ kg, while in case of formula II. it is $1{,}87.10^{-13}$ kg.

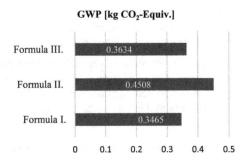

Figure 4.   GWP values of concrete formulas.

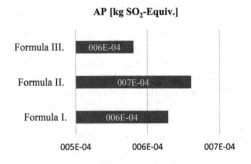

Figure 5.   AP values of concrete formulas.

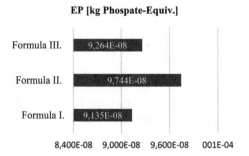

Figure 6.   EP values of concrete formulas.

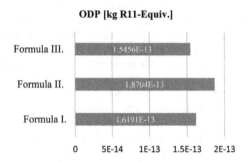

Figure 7.   ODP values of concrete formulas.

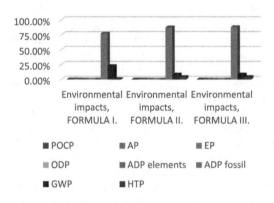

Figure 8.   Percentage distribution with application of weighting factors.

Figure 8 shows the percentage distribution of the environmental impacts with the application of weighting factors (final weighting factors for all midpoint impact categories). The weighting sets are used the EF 3.0 that includes also aspects of the robustness of the results and presented including toxicity related impact categories (JRC Technical Reports, 2018). In relation to all investigated environmental impact category, the concrete formula with mineral product + product from waste + polymer presents substantially higher environmental intensity than the concrete formula with renewable primary product. In case of Global Warming Potential, the formula II. indicates approximately 23% higher environmental intensity than the formula I. with biomass. Referring to Ozone Layer Depletion Potential, the formula II. denotes approximately 13% higher environmental intensity than the Formula I. with a renewable material.

5 CONCLUSIONS

The decision about the environmentally friendly nature of a technology is a complex task and the fact that a given technology is environmentally friendly can be only declared when circumspect analyses are already available. In a long term, those environmentally friendly and energy efficient technologies are the most economical which implementation and development infer the economic efficiency and innovation index of a company. Regarding that among the initiated products and the applicable technologies priorities must be formulated, the implementation of life cycle analysis is essential according to certain product and technologies. The Life Cycle Assessment for cement as a binder and concrete as a building material is undoubtedly an important part of the LCA of buildings. Based on LCA analyses the investigation of the environmental impact of different technologies becomes possible therefore its application is the most adequate in case of substituting products and processes. With its support, it can be quantified that during the total lifetime of a given technology or a given product the environmental loads are caused, furthermore what kind of and how many natural resources are consumed, included energy expenses. In the framework of LCA researches at the University of Miskolc, the total life cycle of concrete has been developed in a complex way from the production phase through the usage to the end-of-life stage with the consideration of information modules of the Environmental Product Declaration (EPD). Within the total lifecycle, the end-of-life phase plays an important role where the dissipative use is evaluated. After the development of the steps following the Life Cycle Assessment of the production of cement and concrete, the results of the LCA of concrete formulations from CEM II. are compared by quantifying environmental impacts and greenhouse gas emissions. It was determined the percentage distribution of the largest environmental impacts (Global Warming Potential, Abiotic Depletion Potential for fossil fuels and Human Toxicity Potential) with the application of weighting factors. The applied weighting set, robustness factors and final weighting coefficients are reported for all impact categories including toxicity-related impact categories. With application of weighting factors, the research results show that the percentage distribution values for Abiotic Depletion Potential are between 77% and 88% and the values of Global Warming Potential are between 8% and 22% for concrete formulas. One of the main research fields of LCA research group of the University of Miskolc is the LCA-based comparative investigation tends to the development of construction structural elements, products and production technologies. During the LCA analysis related to the production of prefabricated concrete elements, the environmental impact categories are determined. By presentation of total life cycle analyses and documentation of modelling requirements. The assessed dataset and the environmental impacts arising from that can be discussed. The results of the research group operating in the field of life cycle analysis can expectedly widely contribute to the environmental protection innovation of construction structural elements, products and production technologies.

ACKNOWLEDGEMENTS

This research was supported by the European Union and the Hungarian State, co-financed by the European Regional Development Fund in the framework of the GINOP-2.3.4-15-2016-00004 project, aimed to promote the cooperation between the higher education and the industry.

REFERENCES

Buday-Malik, A. Tóth Szita, K., Velősy, A., Zajáros, A. & Terjék, A. 2018. Role of the construction and building sector in climate change - What should we do? In Tünde Alapi és István Ilisz (2018), *24th International Symposium on Analytical and Environmental Problems*: 331–335. Szeged. ISBN 978-963-306-623-2. http://acta.bibl.u-szeged.hu/view/year/2018.type.html.

Cooper, J.S., Godwin, C. & Hall, E.S., 2008. Modeling process and material alternatives in life cycle assessments. *International Journal of Life Cycle Assessment* 13 (2): 115–123. DOI: 10.1065/lca2007.06.341.

Grieder, A., Hubler, P. & Pöll, M. 2016. Ökobilanz Betonfertigteile. *Schlussbericht. Stadt Zürich, Amt für Hochbauten. Zürich.* https://www.stadt-zuerich.ch/hbd/de/index/hochbau/bauen-fuer-2000-watt/grundlagen-studienergebnisse/2016-11-nb-betonrechner.html.

Guinee, J. B. 2002. Handbook on life cycle assessment - operational guide to the ISO standards. *The international journal of life cycle assessment.* DOI: https://doi.org/10.1007/BF02978897.

Kupfer, T, Baitz, M, Makishi Colodel, C., Kokborg, M., Schöll, S., Rudolf, M., Thellier, L., Bos, U., Bosch, F., Gonzalez, M., Schuller, O., Hengstler, J., Stoffregen, A. & Thylmann, D. 2018: GaBi Database & Modeling Principles. GaBi thinkstep AG. http://www.gabi-software.com/fileadmin/gabi/Modeling_Principles_2019.pdf.

Mannheim, V. 2014. Examination of thermic treatment and biogas process by LCA. *ANNALS of Faculty of Engineering Hunedoara - International Journal of Engineering.* XII (2): 225–234. ISSN: 1584-2665. http://annals.fih.upt.ro/pdf-full/2014/ANNALS-2014-2-37.pdf.

Mannheim, V. & Siménfalvi, Z. 2012. Determining a priority order between thermic utilization processes for organic industrial waste with LCA. *Waste Management and the Environment VI.* 163: 153–166. ISBN: 978-1-84564-606-6. ISSN: 1743-3541. DOI: 10.2495/WM120151.

Mannheim, V. & Siménfalvi, Z. 2018. Life Cycle Assessment for Waste Management Processes. *The 9th annual international experts conference Enviromanagement, Strebsko Pleso, 10–11 October 2018*: 1–17. ISBN 978-80-85655-36-0. http://nmc.sk/wp-content/uploads/2018/09/ANOTACIA_MANNHEIM.pdf.

Reiners, J. 2011. Stellung von Zement und Beton in der Nachhaltigkeitsdiskussion. *Technisch-Wissenschaftliche Zementtagung, Düsseldorf, 28 September 2011.* https://www.vdz-online.de.

Tomporowski, A., Flizikowski, J, Opielak, M., Kasner, R., Kruszelnicka, W. 2017. Assessment of energy use and elimination of CO_2 emissions in the life cycle of an offshore wind power plant farm. *Polish Maritime Research*, 24(4): 93–101. DOI: 10.1515/pomr-2017-0140.

Tomporowski, A.; Flizikowski, J.; Kruszelnicka, W.; Piasecka, I.; Kasner, R.; Mroziński, A.; Kovalyshyn, S. 2018. Destructiveness of Profits and Outlays Associated with Operation of Offshore Wind Electric Power Plant. Part 1: Identification of a Model and its Components. *Polish Maritime Research* 25: 132–139. DOI: 10.1515/pomr-2015-0021.

Tóth Szita, K. 2017. The Application of Life Cycle Assessment in Circular Economy. *Hungarian Agricultural Engineering* 31: 5–9. DOI: 10.17676/HAE.2017.31.5.

Solutions for Sustainable Development – Szita, Jármai & Voith (eds.)
© *2020 Taylor & Francis Group, London, ISBN 978-0-367-42425-1*

Investigation of different foam glasses with Life Cycle Assessment method

A. Simon
Faculty of Materials Science and Engineering, University of Miskolc, Hungary

K. Voith
Faculty of Mechanical Engineering and Informatics, University of Miskolc, Hungary

V. Mannheim
Higher Education Industrial Cooperation Centre, University of Miskolc, Hungary

ABSTRACT: Foam glass is an increasingly popular material that is applied mainly in the construction and building industry. The aim of the research is to investigate the production of glass foams from different glass wastes according to the Life Cycle Assessment method. This method was chosen because the lifecycle approach to waste management and the circular economy are in the focus of the EU. Their goal is to reduce emissions and move towards a resource-efficient society. The new circular economy package of the EU sets out ambitious recycling targets for the member countries. The package envisaged is a common EU target for recycling 65% of packaging waste by 2025 and 70% by 2030. The targets for glass waste are 70% and 75%, respectively. Since the glass is recyclable in 100%, the glass waste is the target material of this study. After the separation phase - from metal and paper – the shredded and milled glass powder is mixed with SiC as a foaming agent and fired for 10 minutes in a laboratory furnace between 835 – 955°C. The following waste glasses were used for the laboratory experiments:

1. packaging (container) glass
2. float (flat) glass
3. ampoule glass

Both the properties of the glass (density, thermal conductivity, strength, and microstructure) and the total life cycle of glass waste are examined. For Life Cycle Analyses, Gabi 8.0 thinkstep software is used. As a result of this study, the waste preparation process can be optimized and the best foam glass type can be determined.

1   INTRODUCTION

In this research work, we focus on the preparation and production of glass foam from domestic glass wastes. There are several reasons for our choice. The main reason is the fact that the packaging waste glass of households is 100% recyclable. This is an important aspect since according to the Directive 2018/852 of the European Parliament and of the Council of 30 May 2018 (amending Directive 94/62/EC on packaging and packaging waste), each member state should strive to prevent generating packaging waste and, in addition, the focus must be on the recycling and reutilization of packaging waste, preventing the final disposal of packaging waste. In this way, secondary raw materials from waste glass contribute to the circular economy. Packaging glass waste is basically one of the best waste types for recycling. Hungary, as a member country of the EU, has an obligation to recycle at least 55% of packaging waste by 2020. Directive 2010/75/EU comprises eight sectors - referred to as 'the EU glass industry' - based on the manufactured products. These sectors are container glass, flat glass, continuous filament glass fibre, domestic glass, special glass (without water glass), mineral wool (with two divisions, glass wool and stone wool), high temperature insulation wools (excluding polycrystalline wool) and frits (Scalet B. M. et al. 2013). Container glass is the largest sector of the EU glass industry, representing almost the 60% of the total glass production, depending on the reference year. The most important products of the container glass sector with approximately 75% of the total production are bottles for beverages (e.g. wines, beers, spirits, soft drinks, mineral water) and wide neck jars for food industry (e.g. jams, oil, vinegar, dressings, baby food) with 20%. The sector covers the production of glass packaging, i.e. bottles and jars used for packaging food, drink, cosmetics and perfumes, pharmaceuticals and technical products. The flat glass sector is the second largest sector of the European glass industries, which represented about 25–30% in both tonnage and value. Flat glass is mainly used in buildings and automotive industries, but later also solar-energy application and urban and domestic furniture and home decoration applications have been developed. Flat glass is manufactured in the EU by 7 companies operating some 50 float plants. Float flat glass can be used for various applications e.g. window glass or security glass. Float glass represents the main product; while rolled glass is only about 3.5% of the total and is declining, while the production of float glass has increased over the years (Scalet B. M. et al. 2013). Borosilicate glasses can be considered to incorporate boron and silicon oxides. A typical composition is 70–80% SiO_2, 7–15% B_2O_3, 4–8% Na_2O or K_2O, and 2–7 % Al_2O_3. Glasses with this composition show high resistance to chemical corrosion and temperature change. Applications include chemical process components, laboratory equipment, pharmaceutical containers, lighting, cookware, and oven doors and hobs. Many of the borosilicate formulations are for low volume technical applications and are considered to fall into the special glass category (Scalet B. M. et al. 2013). The EU association for flat glass production is Glass for Europe. Production of glass fibers (continuous filament glass fibre - CFCG) is one of the smallest sectors of the glass industry in terms of tonnage (2%) although the products have a relatively high value to mass ratio (HORIZON 2020 Project Nr. 693845).

During the foaming process, nucleation and coalescence of the bubbles generated by the released foaming gas strongly depend on the viscosity of the melted glass. Alkali oxides significantly decrease the viscosity by forming non-bridging oxygens thus decreasing the connectivity of the glass network. Alkali and alkaline earth oxides are modifying oxides which means their ions settle into the interstices of the network, reducing the unoccupied free volume of the structure. (Shelby 2005, da Silva et al. 2019). During the foaming, in the glass melt having low viscosity the nucleation of interstitial pores is facilitated. However, the higher is the viscosity, the more hindered is the nucleation and growth of the bubbles. As the generated structure is not homogenous, the amount and the pressure of the released gas is diverse leading to the rupture of the formed bubbles and increment of the open porosity (da Silva et al. 2019). According to Wang et al (Wang et al, 2018), at high sintering temperature, the SiO_2 layer enveloping on the surface of the SiC particles are removed by the formed liquid phases leading to a more intense foaming process and gas release. Being a diffusion controlled process, the nucleation and coalescence of the bubbles considerably depend on the temperature.

The life cycle approach plays an outstanding role in the no. Act CLXXXV of 2012 on waste management ("Waste Management Act"), thus it has become an important part of environmental protection. The Life Cycle Assessment is the most promising and most useful method in case of substitute products and processes. Unit processes in process engineering are linked to one another by flows of intermediate products and waste for treatment by product flows (Mannheim & Siménfalvi, 2012). Using life cycle analysis we have the opportunity to determine the priority of not only the individual waste management procedure but the priority order among the recycling methods (Mannheim, V. 2014). After developing LCA of the process engineering of foam glass, the results are summarized by quantifying environmental impacts. The life cycle analysis of the investigated technologies was carried out by applying GaBi 8 software at the University of Miskolc. This software assures the life cycle assessment as a tool of basic importance aiming sustainable development.

2 GOAL OF LIFE CYCLE ASSESSMENT

LCA is one of several environmental management techniques (e.g. environmental perform-ance evaluation and environmental impact assessment). The essence of LCA is to determine the environmental load, the usage of natural resources - including energy costs - a certain product cause during its entire life cycle, so from its production to the disposal of its waste. The values of the integrated environmental efficiency of offshore wind power installation with CO_2-eq emissions in comparison with a conventional power plant utilizing oil or natural gas attain slightly lower values (Tomporowski, A. et al. 2017). In respect of non-functionality impact, higher levels of adverse impact on the environment and depletion of resources were found for a land-based wind farm, whereas an offshore wind power plant was found to pose more threat to human health (Piasecka, I. et al. 2019).

Life Cycle Assessment can be the basis to:
- identify the opportunities that increase the environmental impact of a product at different points of its life cycle
- inform policymakers in industry, government, and NGOs (strategic planning, prioritiza-tion, product or process planning, or re-planning)
- select relevant indicators of environmental impacts, including measurement technologies
- and marketing (elaboration of an ecological classification scheme, environmental stress or the determination of environmental product definition).

During the life cycle analysis of glass elements, the following regularization and standards have to be considered:

- ISO 14025:2006 (Environmental labels and declarations – Type III environmental declar-ations – Principles and procedures)
- ISO 14040:2006 (Environmental management – Life cycle assessment – Principles and framework)
- ISO 14044:2006 (Environmental management – Life cycle assessment – Requirements and guidelines)
- ISO 14045:2012 (Environmental management – Eco-efficiency assessment of product sys-tems – Principles, requirements and guidelines)
- MSZ EN ISO 14020:2002 (Enviromental labels and declarations. General principles)
- Industrial Emission Directive 2010/75/EU (Integrated Pollution Prevention and Control), JRC reference report 2013
- Best Available Techniques (BAT) Reference Document for the Manufacturer of Glass (2013)

The Life Cycle Assessment was carried out according to ISO 14040 and ISO 14044. During the development based on life cycle assessment of an innovative and environmental friendly element, the total life cycle of each product has to be considered (Tóth Szita, K. 2017). Back-ground data were used from the GaBi 8 database (2018). European (EU-28) data sets were used for raw materials, auxiliary materials, energy, water and transport. Data choices focussed on including the best fitting alternatives for all processes in the LCA-model. The life cycle inventory (LCI) results were modelled and calculated using the GaBi software tool and a life cycle impact assessment (LCIA) were performed (Eyerer & Reinhardt, 2000).

3 PROCESS TECHNOLOGY AND LABORATORY EXPERIMENTS

3.1 *Glass flour from glass waste*

Currently, in Hungary, there are some glass waste processing plants where only white glass waste is processed. Most of the manufacturers colour glass waste processing is not yet possible in any Hungarian company, so coloured glass waste is sold abroad (Adamovics et al. 2012, Tóth Kiss K. 2007). However, glass waste has another way to be utilized: producing foam glass, which is increasingly accepted by the construction industry due to its relatively high strength and low density. Glass foam was produced for the first time in the United States from glass tiles, foaming agents and other additives. In our research, we have chosen an engineering technology where the first step is the preparation of the glass waste to produce glass flour for the foaming process. Firstly, the received glass waste must be crushed to suitable particle size and the foreign materials must be separated. It is assumed that there is no incoming glass waste piece larger than 300 mm. It is also assumed that 85% of the glass waste arrives in a broken state in the preparatory plant. According to these assumptions, the incoming glass waste is comminuted in two steps. It is practical to use vibration sieve between the shredding stages to provide a uniform particle size. From the glass waste (< 10 mm) the metallic waste is separated by a magnetic separator. The next important step of this enviro-friendly process engineering technology is the colour sorting by the optical separator. The white and coloured glass waste separated is being further ground in a ball mill. The properties of product produced by fine grinding depend on the speed and the degree of the primary and secondary mechanochemical processes. The grinding process can be directed by helping of primary processes and reducing of secondary processes (Mannheim, V. 2011). The particle size of the glass flour as an end product of this procedure is less than 1 mm. An important step in production is the fine grinding of the materials into the required fraction. The particle size distribution can be influenced by design and operating parameters, by grinding media (especially diameter of the grinding balls, density, hardness and filling ratio) and by the feed material itself (Mannheim, V. 2007). This glass flour will be one of the basic materials of the glass foam as the end product for recycling. Glass viscosity, foaming temperature, and residence time are strongly related. The optimum foaming temperature must be selected by considering the maximum foam stability, which is controlled by viscosity (Mucsi, G. et al. 2013). Figure 1 shows the process of engineering technology for glass waste.

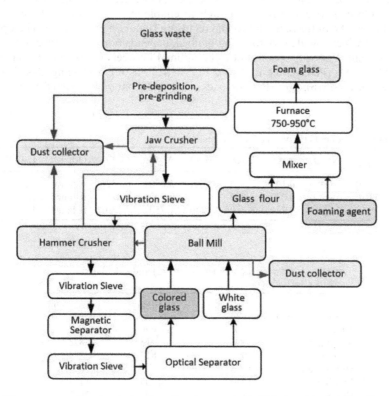

Figure 1.   The process engineering technology for glass waste. (self-designed technology flow chart).

3.2 Experiments

In the first step, the composition of the waste glasses (Table 1) was determined with ICP analysis. From the components, silica is the primary glassforming oxide. Pure silica glass offers excellent properties but has a high melting temperature (<2000°C) which can be decreased by adding fluxes. Alkali oxides (Na_2O, K_2O) are one of the most common fluxes. While the addition of fluxes decreases the melting temperature thus the costs of production, the chemical durability of the glass is impaired at the same time. Weakened chemical properties and the structure of the glass can be modified and stabilized by the addition of property modifiers that are mainly alumina, alkaline earth or transition metal oxides (Shelby 2005).

In our experiments, every foam mixture consisted of 98 wt% glass powder and 2 wt% SiC as a foaming agent. First, a cylindrical sample from every mixture was placed into a heating microscope (Camar Elettronica) to determine its characteristic temperatures (Table 2). Softening temperature indicates when liquid phases appear. When almost the entire volume is filled with liquid phase, the sample reaches the sphere state. Melting stage shows that the sample has reached its melting point. The foaming temperature has to exceed the softening point to ensure the formability of the glass but does not allow to reach the melting point to be able to preserve the foamed structure. Float glass has the lowest and ampoule glass has the highest softening and melting temperatures.

As for the mixtures, foaming behaviour was characterized by the changes in the sample's height in the function of the temperature (Figure 2). Float glass mixture presented the most intense foaming. The lowest (835°C) and the highest (955°C) foaming temperature belongs to the float and to the packaging glass mixtures, respectively. Ampoule glass mixtures were foamed at 905°C. When heating up to the foaming temperature, SiC decomposes and reacts with glass. As a result, SiO_2 and CO_2 or CO are formed. The higher is the temperature, the more gas is released (Yang et al, 2014). The foaming temperature affects the viscosity of the glass melt as well. When the viscosity is high (10^{10}– 10^8 Pas) the foaming process is slow. At low viscosity, the nucleation and expansion of the bubbles are very fast, thus the coalescence rate is high. Optimal viscosity is low enough to enable expansion of the bubbles and high enough to hinder the outgassing and foam collapsing (Petersen et al, 2017). The optimal viscosity range, at which low-density foam is obtained, depends both on the type of the glass and the foaming agent. Viscosity was found to be optimal between 10^4-10^6 Pas for labware glass foamed with $CaCO_3$ (Petersen et al, 2017), $10^{3.3}$– $10^{4.5}$ Pas for packaging glass foamed with SiC (Petersen et al, 2016), $10^{3.3}$ Pas for packaging and float glass foamed with SiC (Mear,

Table 1. Main components (in wt%) of the glasses.

	SiO_2	Na_2O	K_2O	CaO	MgO	Al_2O_3	Fe_2O_3	Other	R_2O/RO ratio
Ampoule	81.70	7.050	0.145	2.160	0.124	3.070	0.032	balance	3.2
Float	73.40	13.200	0.430	8.550	1.890	0.900	0.025	balance	1.3
Packaging	71.50	12.500	0.720	8.750	2.440	1.750	1.150	balance	1.2

Table 2. Characteristic temperatures (°C) and calculated viscosity (Pas) of the glasses.

	Ampoule	Float	Packaging
Sintering	736	696	702
Softening	849	715	828
Sphere	1099	817	884
Half-sphere	1196	983	1014
Melting	1240	1009	1112
Viscosity	~$10^{5.5}$	10^5	~10^4

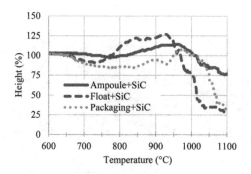

Figure 2.   Foaming behaviour of the glass+SiC mixtures.

2004) and $10^{3.4}$–$10^{3.9}$ Pas for soda-lime-silica glass foamed with SiC (Fernandes et al, 2009; Bernardo et al 2010). The viscosity of the glasses investigated in this research was calculated for the compositions of Table 1 by using a free calculator (Viscosity calculator, 2019) based on the extended experimental database of Fluegel (Fluegel, 2007). The calculated viscosity at the foaming temperature is 10^5 Pas for the float glass, and less than 10^6 Pas and 10^4 Pas for the ampoule and packaging glass, respectively.

After determining the foaming temperatures, three samples were pressed at 30 MPa pressure from each mixture. Diameter, height and weight were measured with a calliper and a balance, respectively for each sample both after pressing and firing to calculate the change in volume and the density. Density affects the thermal conductivity and the compressive strength of the foams. The main requirements for foams are contradictory, as low density ($\leq$0.2 g/cm^3) and good strength (>1–2 MPa) are expected at the same time. Amongst the properties, density is the most important as it is strongly correlated to the closed porosity, thermal conductivity and mechanical strength of the glass foam (Petersen et al, 2016) Pressed samples have almost the same volume and density (1.6–1.8 cm^3 and 1.7–1.8 g/cm^3, Figure 3). After foaming, the most and the less intense increase in volume belongs to the packaging and ampoule glass foam, respectively. The change in density is contrary to that, ampoule glass foam has the highest (0.5 g/cm^3) while packaging glass has the lowest (0.2 g/cm^3) density.

Macrographs of the foamed samples were taken with a Canon EOS 700D camera. Macrostructure of some representative samples is shown in Figure 4. Float glass foam has the most homogenous structure with small, uniformly distributed cells. Ampoule and packaging glass foams show bigger sized cells. Packaging glass foam has even mm sized cells opened to both surfaces of the pressed sample. Despite the highest foaming temperature, the viscosity of melted packaging glass was still high (10^6 Pas) although the viscosity of glass melts exponentially decrease with temperature. Due to the high temperature and viscosity, more gas was released and more bubble nucleated resulting bigger foam cells with open pores.

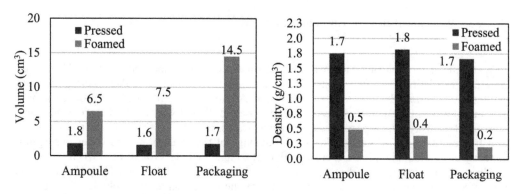

Figure 3.   Change in volume and density of the glasses foams.

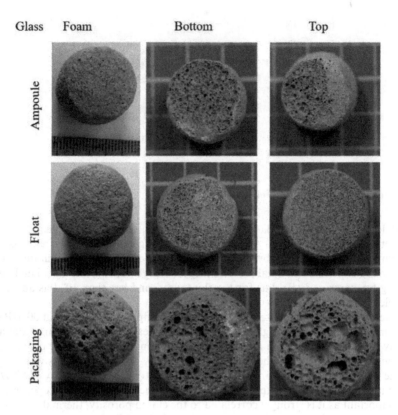

Glass	Foam	Bottom	Top
Ampoule			
Float			
Packaging			

Figure 4.   Macrostructure of the glass foams.

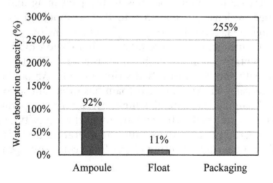

Figure 5.   Water absorbance capacity of the glass foams.

Water absorption capacity test was carried out according to method A of the Hungarian Standard MSZ EN 1217 2000. Test results are represented in Figure 5. Packaging glass foam presented exceptionally high water absorption. Ampoule glass foam also absorbed a high amount of water. Based on these result, we suppose that these foams contain more open, inter-connected porosity. Open porosity is adverse as water can set into the inner part of the foam and freeze during the winter leading to the damage of the insulation. The water absorption capacity of the float glass foam is close to that of commercial glass foams (≤10 wt%).

Thermal conductivity (k) of the foamed glasses, measured at 24°C and 50°C (Figure 6), is not affected by the ambient temperature. From the components of Table 1, alkali oxides (ions) are the most mobile. Less mobile alkaline earth ions (Ca^{2+}, Mg^{2+}) act as stabilizers in

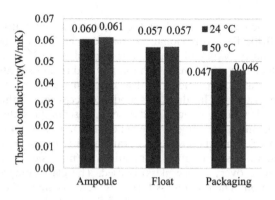

Figure 6.   Thermal conductivity of the glass foams at 24°C and at 50°C.

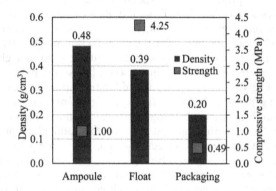

Figure 7.   Density versus strength of the glass foams.

the glass network and reduce the mobility of modifier ions as well (Shelby 2005). Ampoule and float glass foam, presenting similar microstructure and comparable density, have close results. Ampoule glass foam had the highest density, with a homogenous cell structure and the highest R_2O/RO (alkali/alkaline earth oxide) ratio (3.2). Float and packaging glass have similar R_2O/RO ratio (1.3 and 1.2, respectively) but a very different microstructure. Float glass foam has a more compact structure, less open porosity, higher alkali oxide and lower alkaline earth content than packaging glass which leads to slightly more effective thermal conductivity.

Despite its rather high density, the compressive strength of the ampoule glass foam is slightly modest (Figure 7). Packaging glass foam presented the lowest density and compressive strength. Optimal combination belongs to float glass foam – moderate density and enough high strength. The homogeneity of the microstructure – with special regard to the ratio of open/closed porosity – affects the mechanical properties of the foams more than the composition of the glass.

3.3   *Results of the LCA measurements*

The GaBi LCA software models every element of a product or system from a life cycle perspective, equipping businesses to make the best-informed decisions on the manufacture and lifecycle of any manufactured product. In addition, it examines the environmental impacts and presents alternative options for manufacturing, distribution, recycling and sustainability. In order to analyse completely a technological process from an environmental effects point of view, the value of a functional unit is needed. The GaBi 8 ts software came into the market in 2018 and has several advantages for its former versions (Kupfer, T. et al. 2018). The new professional database incorporates world-leading product sustainability intelligence (Mann-heim & Siménfalvi, 2018). The professional database concept is an important aspect, which ensures the accuracy and relevancy of results. The applied data set represents the European situation (in EU 28), focusing on the main technologies and region-specific characteristics. All relevant background data such as energy and auxiliary material are taken from the GaBi Databases, keeping consistency. This dataset is modelled according to the European Standard EN 15804 for Sustainable Construction. The dataset represents the state-of-the-art in view of a referenced functional unit. The value of the functional unit was 1 kg of glass foam from waste glass. The LCA of foam glass (4–8mm, granulation) covers the cradle-to-gate phase, that is, particularly the processes grinding, granulation, burning, sieving and shipping. For the LCA process, glass for recovery (waste glass) is used. The glass composition can vary as it depends on the application. The foam glass formulas were examined from three type glasses: EU 28 expanded glass granulates (ampoules), EU 28 float flat glass (secondary material) and packaging (container) glass. The system boundaries are represented by the finished product at the factory gate. For the production of expanded glass, allocation according to mass is applied, due to the co-product filter dust. Float flat glass can be used for various applications e.g. window glass or security glass. The float flat glass process is mainly based on the best available techniques reference document for the manufacture of glass by the industrial emis-sion directive 2010/75/EU. In the float glass process quartz sand and glass for recovery are melted with additional materials e.g. dolomite, kaolin, limestone, potassium carbonate, soda and sodium sulphate. The various materials are added to help to improve the performance of the glass manufacturing process and to optimize the glass properties. The most important application field of container glass is beverage bottles. Container glass is the largest sector of the EU glass industry, representing almost 60% of the total glass production. The most important products of the container glass sector with approximately 75% of the total produc-tion are bottles for beverages (e.g. wines, beers, spirits, soft drinks, mineral water) and wide neck jars for food industry (e.g. jams, oil, vinegar, dressings, baby food) with 20%. Glass making is a very energy intensive activity and the choice of energy source, heating technique and heat recovery method are central to the design of the furnaces. The same choices are also some of the most important factors affecting the environmental performance and energy effi-ciency of the melting operation. The main energy sources for glass making are natural gas, fuel oil and electricity. The input energy value was 2,52 MJ by glass foam formulas. Electricity is modelled according to the individual country-specific situations. The individual energy car-rier specific power plants are modelled according to the current EU 28 electricity grid mix. Modelling the electricity consumption mix includes transmission/distribution losses and its own use by energy producers, as well as imported electricity. By LCA all relevant and known transport processes are included. Ocean-going and inland ship transport, as well as rail, truck and pipeline transport of bulk commodities, are considered. The energy carriers (diesel fuel, gasoline, technical gases and fuel oils) are modelled according to the specific supply situation (Kupfer, T. et al. 2018). The LCI method applied is in compliance with ISO 14040 and 14044. The applied impact assessment method was CML 2001 (January 2016). According to measure-ments the Global Warming Potential (GWP in CO_2-equivalent), the Acidification Potential (AP in SO_2-equivalent), the Eutrophication Potential (EP in phosphate-equivalent) and the Ozone Layer Depletion Potential (ODP in R11-equivalent) were determined. Several time per-spectives are available with the GWPs for 100 years recommended as the baseline character-isation method for climate change. The global warming potential is calculated in carbon dioxide equivalents. This means that the greenhouse potential of emission is given in relation to CO_2. Since the residence time of the gases in the atmosphere is incorporated into the calcu-lation, a time range for the assessment must also be specified (Kreißig & Kümmel, 1999). Fig-ures 8–11 show the different environmental impact category diagrams for three glass foam formulas by GWP, AP, EP and ODP values. The investigations show that by glass foam from ampoule the value of GWP is the best. In case of Global Warming Potential, the environmen-tal load of packaging glass foam indicates almost three times more than the ampoule glass foam. Investigating the emission of greenhouse gases, in relation to foam glass formulas in case of ampoule glass foam the SO_2-equivalent is $1,05.10^{-3}$ kg, while in case of float glass foam this value is $9,3.10^{-3}$ kg. The Eutrophication Potential in relation to glass foam formulas in case of ampoule glass foam the phosphate-equivalent is $2,61.10^{-4}$ kg, while in case of packaging glass foam this equivalent is $7,95.10^{-4}$ kg. Referring to Ozone Layer Depletion Potential in relation to glass foam formulas in case of ampoule the R11-equivalent is $5,94.10^{-11}$ kg, while in case of packaging glass foam is $6,14.10^{-11}$ kg. Figure 12 shows the percentage distribution of the largest environmental impacts (Global Warming Potential, Abiotic Depletion Potential for fossil fuels and Human Toxicity Potential) with the application of weighting factors (final weighting factors for all midpoint impact categories). The weighting sets are used the EF 3.0 that includes also aspects of the robustness of the results and presented including toxicity related impact categories (JRC Technical Reports, 2018).

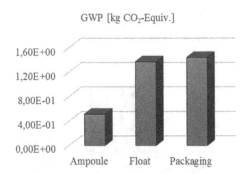

Figure 8.   Global Warming Potential (GWP) values.

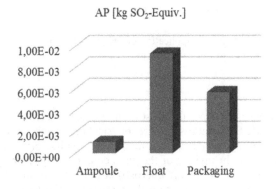

Figure 9.   Acidification Potential (AP) values.

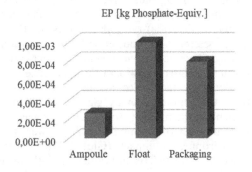

Figure 10.   Eutrophication Potential (EP) values.

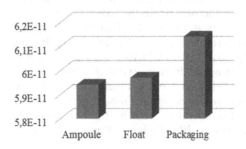

Figure 11.   Ozone Layer Depletion Potential (ODP) values.

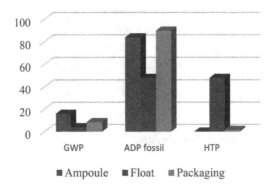

Figure 12.   Percentage distribution with application of weighting factors.

4   CONCLUSIONS

In this research work, different types of glasses (namely ampoule, float and packaging glass) were foamed to investigate their microstructure, physical and mechanical properties. Foaming temperatures were determined via heating microscopy results. Flux content of the glasses affects strongly the softening behaviour of the glass consequently their optimal foaming temperature as well. The homogeneity of the microstructure, the amount of small-sized cells and closed pores have an advantageous effect on the strength of the foams. Homogenous microstructure - even in case of low density - results higher strength and better thermal insulation. The viscosity of glass foams containing more alkali earth oxides instead of alkali oxides is higher even at elevated foaming temperatures resulting more open pores in the foamed structure. R_2O/RO (alkali/alkaline earth oxide) ratio also affects the thermal conductivity of the foams. For the Life Cycle Assessment method, glass foams from ampoule, float and packaging glass in EU 28 was used. The main environmental impact category diagrams show the GWP, AP, EP and ODP values of three glass foam formulas. The investigations show that in case of Global Warming Potential, the environmental load of glass foam from packaging glass indicates almost three times more than by glass foam from ampoules. Referring to Ozone Layer Depletion Potential, the product from packaging glass denotes 3% higher environmental intensity than the product from ampoule glass. It was determined the percentage distribution of the largest environmental impacts (Global Warming Potential, Abiotic Depletion Potential for fossil fuels and Human Toxicity Potential) with the application of weighting factors. The applied weighting set, robustness factors and final weighting coefficients are reported for all impact categories including toxicity-related impact categories. With application of weighting factors, the analytical results show that the percentage distribution values of Abiotic Depletion Potential for fossil fuels are between 48% and 90% for glass foams.

ACKNOWLEDGEMENTS

This research was supported by the European Union and the Hungarian State, co-financed by the European Regional Development Fund in the framework of the GINOP-2.3.4-15-2016-00004 project, aimed to promote the cooperation between the higher education and the industry.

The described article was carried out as part of EFOP-3.6.1-16-2016-00011 "Younger and Renewing University-Innovative Knowledge City-institutional development of the University of Miskolc aiming at intelligent specialisation" project implemented in the framework of the Szechenyi 2020 program. The realization of this project is supported by the European Union, co-financed by the European Social Fund.

REFERENCES

A. Fluegel: Glass Viscosity Calculation Based on a Global Statistical Modeling Approach; (PDF, 3 MB), Glass Technol.: Europ. J. Glass Sci. Technol. A, vol. 48, 2007, no. 1, p 13–30.

Adamovics, O., Bejenaru-Sramkó, Gy., Tóthné, Kiss K., Tóth, I.: Útmutató az elérhető legjobb technika meghatározásához az üveggyártás engedélyeztetése során. Vidékfejlesztési Minisztérium. 2012 Budapest, Hungary. (Accessed on 22/ 05/2019) https://ippc.kormany.hu/.

E. Bernardo, G. Scarinci, P. Bertuzzi, P. Ercole, L. Ramon: Recycling of waste glasses into partially crystallized glass foams. Journal of Porous Materials June 2010, Volume 17, Issue 3, pp 359–365.

Eyerer P., Reinhardt, H.-W. 2000. Ökologische Bilanzierung von Baustoffen und Gebäuden - Wege zu einer ganzheitlichen Bilanzierung, Bau Praxis, Birkhäuser-Verlag, Basel, CH.

F. Mear, Etude de mousses de verres issus de Tubes à Rayons Cathodiques (TRC) en fin de vie contenant de l'oxyde de plomb: Elaboration, caractérisations physicochimiques et applications, (PhD Thesis), Université de Montpellier II, 2004.

Hao Wang, Ziwei Chen, Lili Liu, Ru Ji, Xidong Wang: Synthesis of a foam ceramic based on ceramic tile polishing waste using SiC as foaming agent. Ceramics International 44 (2018) 10078–10086.

H.R. Fernandes, D.U. Tulyaganov, J.M.F. Ferreira: Production and characterisation of glass ceramic foams from recycled raw materials. Advances in Applied Ceramics 108 (2009) 9.

Jianguang Bai, Xinghua Yang, Shaochun Xu, Wenjia Jing, JianfengYang: Preparation of foam glass from waste glass and fly ash. Materials Letters 136 (2014) 52–54.

Kreißig, J. & Kümmel, J. 1999. Baustoff-Ökobilanz. Wirkungsabschätzung und Auswertung in der Steine-Erden-Industrie. Bundesverband Baustoffe Steine + Erden e.V.

Kupfer, T, Baitz, M, Makishi Colodel, C., Kokborg, M., Schöll, S., Rudolf, M., Thellier, L., Bos, U., Bosch, F., Gonzalez, M., Schuller, O., Hengstler, J., Stoffregen, A. & Thylmann, D. 2018: GaBi Database & Modeling Principles. GaBi thinkstep AG. http://www.gabi-software.com/fileadmin/gabi/Modeling_Principles_2019.pdf.

Mannheim, V. 2007. Empirical modeling and determination of the grindability in stirred ball mills. Építőanyag, 59(2): 36-40. doi: 10.14382/epitoanyag-jsbcm.2007.6.

Mannheim, V. 2011. Empirical and scale-up modeling in stirred ball mills. Chemical Engineering Research and Design, 89(4): 405-409. DOI: 10.1016/j.cherd.2010.08.002.

Mannheim, V. 2014. Examination of thermic treatment and biogas process by LCA. ANNALS of Faculty of Engineering Hunedoara – International Journal of Engineering. XII (2): 225–234. ISSN: 1584-2665. http://annals.fih.upt.ro/pdf-full/2014/ANNALS-2014-2-37.pdf.

Mannheim, V. & Siménfalvi, Z. 2012. Determining a priority order between thermic utilization processes for organic industrial waste with LCA. Waste Management and the Environment VI. 163: 153–166. ISBN: 978-1-84564-606-6. ISSN: 1743-3541. DOI: 10.2495/WM120151.

Mannheim, V. & Siménfalvi, Z. 2018. Life Cycle Assessment for Waste Management Processes. The 9th annual international experts conference Enviromanagement, Strebsko Pleso, 10-11 October 2018: 1-17. ISBN 978-80-85655-36-0.

Mucsi, G., Csőke, B., Kertész, M., Hoffmann, L. 2013. Physical Characteristics and Technology of Glass Foam from Waste Cathode Ray Tube Glass. Journal of Materials vol. 2013. Article ID 696428, 11 pages. DOI: 10.1155/2013/696428.

Piasecka, I.; Tomporowski, A.; Flizikowski, J.; Kruszelnicka, W.; Kasner, R.; Mroziński, A. 2019. Life Cycle Analysis of Ecological Impacts of an Offshore and a Land-Based Wind Power Plant. Applied Sciences 9, 231. doi: 10.3390/app9020231.

Rasmus R. Petersen, Jakob König, Yuanzheng Yue: The viscosity window of the silicate glass foam production. Journal of Non-Crystalline Solids Volume 456, 15 January 2017, Pages 49–54.

Rasmus R. Petersen, Jakob König, Yuanzheng Yue: Evaluation of Foaming Behavior of Glass Melts by High-Temperature Microscopy. International Journal of Applied Glass Science, 1–8 (2016).

Robson C. da Silva, Evaldo T. Kubaski, Sergio M. Tebcherani: Glass foams produced by glass waste, sodium hydroxide, and borax with several pore structures using factorial designs. Int J Appl Ceram Technol. 2019; 1–9.

Scalet B.M., Garcia Munoz, M., Sissa, A.Q., Roudier, S., Delgado Sancho L. 2013. JRC reference report. Best available Techniques (BAT) Reference Document for the Manufacturer of Glass (2013). Industrial Emissions Directive 2010/75/EU Integrated Pollution Prevention and Control.

Shelby, J.E. 2005. Introduction to Glass Science and Technology. Published by Royal Society of Chemistry, 2nd edition (March 29, 2005). Cambridge, United Kingdom. ISBN10:0854046399.

Tomporowski, A., Flizikowski, J, Opielak, M., Kasner, R., Kruszelnicka, W. 2017. Assessment of energy use and elimination of CO_2 emissions in the life cycle of an offshore wind power plant farm. Polish Maritime Research, 24(4): 93–101. DOI: 10.1515/pomr-2017-0140.

Tóth Szita, K. 2017. The Application of Life Cycle Assessment in Circular Economy. Hungarian Agricultural Engineering 31: 5–9. DOI: 10.17676/HAE.2017.31.5.

Tóth Kiss, K. 2007. Glass wastes as basic material in the waste management of glassworks. Építőanyag, 59(4): 114–117. http://epa.niif.hu/02200/02231/00012/pdf/EPA02231_Epitoanyag_200704_114-117.pdf

Viscosity calculator, (Accessed on 22/05/2019) https://glassproperties.com/viscosity/ .

Solutions for Sustainable Development – Szita, Jármai & Voith (eds.)
© 2020 Taylor & Francis Group, London, ISBN 978-0-367-42425-1

Circular economy solutions for industrial wastes

Klara Szita Tóthné
Association of LCA Center, University of Miskolc

Zs. István & R.S. Bodnárné
Bay Zoltan Applied Research Institute LTD

A. Zajáros
ÉMI Nonprofit ÉMI Ltd

ABSTRACT: The concept of circular economy (CE) has been in the centre of the strategical thinking since 2015 and numerous practical initiatives are in progress. It is a new way towards sustainability. This study is based on Hungarian case studies where circular solutions of industrial waste treatment have been implemented. One of them is a solvent regeneration from wastewater to reuse it after distillation process. The sustainability analysis which is mostly based on Life Cycle Sustainability Assessment resulted about 15 -20 percent benefit compared to linear technological process. The other studies are based on the best circular practices of two Hungarian waste flows in the CIRCE2020 project. This project aims to facilitate a larger uptake of integrated environmental management approach in five specific Central European industrial areas by changing patterns from single and sporadic company recycling interventions to an integrated redesign of industrial interactions based on the concept of circular economy. The scenarios have been analysed by LCA and LCC methods and EF-based indicators of circular scenarios of the donor side.

1   INTRODUCTION

The circular economy (CE) has been in the centre of the strategical thinking since 2015 and numerous practical initiatives are in progress. It is expected in Europe to promote economic growth by creating new businesses and job opportunities, saving cost of materials, dampening price volatility, improving security of supply while at the same time reducing environmental pressures and impacts. Many initiatives started to implement circular solution. Especially these focused on the waste treatment to the plastic industry, the chemicals, the critical raw materials and rubber value chain or construction demolition waste to reuse or recover their own materials. The CE means a new perspective to reach the UN's sustainability goals, while there is also economic growth if the linear production chain is replaced by closed loop solutions and industrial symbiosis.

2   CIRCULAR ECONOMY

2.1   *General overview*

The concept of Circular Economy (CE) takes an approach to embarking on a sustainable path and greening the economy from the perspective of waste. Its basic philosophy is that the use of the waste output of one system as the input to another system enables resource savings and reduces environmental stress. This approach is a combination of multiple theoretical concepts and practical solutions, such as" cradle to cradle", "blue economy" (Pauli 2010), 3R, 9R, waste mini-mization. The process of production is closed in the circular economy, where nearly 100% of all waste and by-products are reused or recycled (Szita 2017). When we are talking about a circular economy, not really a whole new conception the 21st century, but we need to talk about integrating economic, environmental and philosophical thoughts that have appeared in previous decades, which can trigger a paradigm shift at different levels of economic-social policy. Despite the idea of circular solution in the EU 2008/98 waste directive, the real breakthrough occurred when the European Commission (EC) supported the introduction of circular models in addition to the Ellen MacArthur Foundation (EMF 2015; EC 2015). The circular economy has become a new concept of global sustainability and development, and its political support is much stronger than it was before, and greater emphasis is placed on the unsustainability of the current linear economic model and the social advantage of circular solutions. The European model of a circular economy mostly approaches waste production from the sustainable aspect, as required by EU waste direct-ive 2008/98/EC. It seems that there has been a great need to incorporate waste-based models into the economic system, as is currently the case, in order to promote economic recovery. In the circu-lar economy model, the economy's metabolic processes flow in a closed system, waste and by-products are almost completely recycled. Their social and economic metabolism is organized in the same way as environmental metabolism processes, forming a typical industrial ecological system. It can appear at micro level (product, enterprise consumption), at meso level in eco-industrial parks, as industrial symbiosis or macro level, in regions, nations (Saidani et al. 2017). Some authors believe that CE combines 3R (reduce, reuse, recycle) and further 6R (repair, rema-nufacturing, redistribution, redesign, rethinking, recover), zero emission, LCA (life cycle analysis) and resource efficiency methods (Winans et al. 2017).

Several factors have contributed to making the circular economy a flagship for more sus-tainable, smarter solutions. These include economic challenges, uncertainties between individ-uals and across the economy, population growth, scarcity of resources, loss of biodiversity and climate change, or high unemployment and poor working conditions in some parts of the world. Of these, perhaps the scarcity of resources and the massive accumulation of waste were the strongest driving forces behind the move towards a circular economy. Meeting these chal-lenges is not tension-free and raises further questions about their impact on society and sus-tainability and well-being. It has become increasingly evident that the linear economic model of the consumer society (production - distribution - use - waste disposal) is incompatible with the model of sustainable economic growth. The circular economy means not just a new sus-tainability paradigm but also an economic one. Horvath (2019) quotes paper of Ramkumar et al. (2018), who says that adhering to the BAU (business as usual) philosophy - besides environmental threats - is a serious market risk that will have to be faced. Particular attention will be paid to policymakers, businesses and professionals in addressing global waste issues, the plastic crisis, resource depletion, water pollution, hygiene and climate change.

The European Union uses 13 tonnes of material per person per year, of which 4.9 tonnes of waste, 91% of which is already treated, but only 11.7% is recycled (Eurostat, 2019). The waste treatment is a key importance activity in all fields of economy (Mannheim & Siménfalvi 2018). It is in progress because earlier half of the waste has been deposited. However, one part of the waste is shipped outside Europe for treatment, as reflected in waste export statistics. The export trade of recyclable waste takes 35 thousand ton per year to outside Europe (Eurostat 2019).

The circular economy is based on three principles:

- The first is the preservation and development of natural capital through the regulated use of exhausted stocks and the balance of the flow of renewable resources. All this presupposes a massive reduction of material and creates conditions for, for example, soil regeneration.
- The second principle is to optimize resource yield through the circulation of products, parts, materials, maximizing their participation in the technical and biological cycle. This means that rebuilding, refurbishment, and maintenance are well designed to make materials as part of the economic process as possible.

- The third principle is to minimize negative externalities by eliminating, substituting or reducing toxic substances. With well-selected materials in the design phase, waste can be reduced, emissions can be reduced, and renewable energy can be used instead of fossil resources (EMF 2015).

2.2 *Circular economy in europe and hungary*

Institutional support for the circular economy is extensive and integrated into EU legislation. The EU has an ambitious circular economic package and determined an action plan and strategy (EC 2015, EP 2016). However, practical implementation can only be achieved with strong state support. Leading countries in circular programs are the Netherlands, Germany, Finland, Denmark, the United Kingdom and France. There are many good practices that can bring economic, environmental and social benefits. The advantages of a circular economy are proven to have significant potential in waste management, as a well-chosen technology reduces the need for raw material and thus reduces the cost of raw material but also reduces the environmental impact of recycling / recycling / recycling of waste. The regions, where CE models are integrated in economic development, and regional competitiveness can be enhanced by creating models for optimizing regional resources, innovative ways of waste management are creating new jobs and improving the quality of life for those living there by creating an environmentally friendly supply chain. The European circular economy action plan is not limited to waste management, but also covers areas of sustainability strategy such as eco-design, e-commerce, eco-labeling, green public procurement, cost-effective water recycling, and bio-based materials. Circular solutions can be linked to all areas of regional development (economic, social, environmental, rural and cultural), through several programs, including the Interreg program. It is illustrated the following few programs.

The R4R project is co-funded by the ERDF's INTERREG IVC programme. The aim of the project is to improve municipal waste recycling performance at regional level. The key actors implement a circular economy concept into waste management at regional level by local authorities in charge of waste management. Thus, also many legal, economic and educational instruments are decided at this decentralised level. The project covers: waste of electric and electronic, battery and hazardous waste collection; door to door selective collection; development of other collection systems; communication and advising initiatives; legal and economic instruments and bio waste collection. The R4R partners have agreed on a new notion, on the so called "DREC" (Destination RECycling) that includes: the municipal waste streams separated at source and collected separately with the purpose of recycling. The output from the sorting facilities, including bulky waste sorting centres, is going directly to facilities for recycling. The output from mechanical biological treatment installations is going directly to facilities for recycling. There 13 partners shared their experience in treatment and management of municipal waste. http://www.regions4recycling.eu/R4RTheProject

The SYMBI project aims to contribute to the improvement of the implementation of regional development policies and programmes related to the promotion and dissemination of Industrial Symbiosis and Circular Economy via legislative harmonisation to build sustainable economies. Participant countries: Finland, Greece, Hungary, Italy, Poland, Slovenia, Spain. The aim is to promote Industrial symbiosis to achieve resource efficient goals to share Materials to Minimize Waste: The project contains the following actions:

- Increase capacities for recycled materials
- Establish networks of selective waste selection
- Introduce 3rd generation waste treatment techniques
- Use waste as energy
- Create Ecoparks for waste transformation. https://www.interregeurope.eu/symbi/

The CIRCE2020 stands for Expansion of the Circular Economy concept in the CE local productive districts. It aims to introduce innovative cross-value chain waste governance models and transnational analytic tools to improve capacities of concerned waste public-private sector to reduce dependencies from primary natural resources within industrial processing. The project should also provide robust evidence about environmental and economic benefits from shifting to enhanced industrial symbiosis. https://www.interreg-central.eu/Content.Node/CIRCE2020.html

ReSiELP (2017-2020): Recovery of Silicon and other materials from End-of-Life Photovoltaic Panels. In order to keep the containing critical and precious metals in the loop, RESIELP brings together technologies from different fields to recover and purify critical and precious raw materials (Si, Ag) as well as by-product materials (glass, Al, Cu) in an environmentally friendly and circular economic process with a product-centric zero-waste approach. https://eitrawmaterials.eu/project/resielp

City symbiosis in Szentendre Industrial Park – PIMES Project (Szentendre, Hungary). The project aims to provide a community-wide construction demonstration for large-scale, integrated use of energy efficiency and renewable energy sources. The stakeholders developed solutions of energy saving possibilities, e.g.: ticking insulations and intelligent building management systems. Residential building and kindergarten rehabilitation and a new energy-efficient office building of ÉMI Nonprofit Ltd had been also completed. ÉMI was the project partner, national coordinator and team leader of the project. The ÉMI utilises the heat and cool energy from cleaned water of plant of sewage water treatment (Frey 2013)

In Hungary, a circular economy has also come to the force in recent years. The "Circular Hungary Program – the economy in cycles" started in 2018 by Circular Economy Foundation to promote the acceleration of the transition to circular economy in Hungary by identifying obstacles and gaps as well as improving conditions and environment at technological, regulatory, market, consumer behavior and financing levels for circular products, practices and projects. The Circular Economy Platform was officially established in Hungary as an initiative of the Business Council for Sustainable Development in Hungary (BCSDH), the Embassy of the Kingdom of the Netherlands, and the Ministry for Innovation and Technology in 2018 with more and more companies joining. In connection with international projects, several successful circular initiatives have been held in which the authors of the article have worked.

3   CASE OF DMSO PROJECT[1]

3.1   *The project and its aim*

The project called „DMSO contaminated industrial wastewater recycling by distillation" with identification number HU09-0090-A1-2013 is supported by the Norwegian Funds in the framework of Norwegian Financial Mechanism 2009-2014 Green Industry Innovative Programme and it connects to the Hungarian water quality improvement program. It started in 2015 and finished in 2017. The Project Promoter is a chemical plant, and they produce water filter to eliminate the arsenic, phosphorus, iodine and fluorine pollutants from drinking and technological waters. During the production process $1m^3$ high 20 w/w% dimethyl sulfoxide (DMSO) content hazardous wastewater is produced daily which is hazardous waste according to the Hungarian laws. The wastewater beside DMSO contains soluble polymer – ethylene-vinyl alcohol copolymer (EVOH) and minerals such as cerium-hydroxide. The main problem was that this wastewater needs to be collected and transferred to the incineration plant to be burned, so the transferred DMSO needs to be replaced with fresh solvent in the production process. This was expensive, and the waste treatment had high environmental impact. The aims of the project are to reduce the volume of the hazardous wastewater, to reuse the recovered solvent and water in the production process.

Based on professional literature DMSO (dimethyl sulfoxide) is frequently used as a solvent for chemical reactions and is also extensively used as an extractant in biochemistry and cell biology. DMSO can be efficiently recovered from aqueous solutions - even though contaminated with

---

1. The project called „DMSO contaminated industrial wastewater recycling by distillation" with identification number HU09-0090-A1-2013 is supported by the Norwegian Funds in the framework of Norwegian Financial Mechanism 2009-2014 Green Industry Innovative Programme.

volatile and/or non-volatile impurities - by distillation (Zajáros et al. 2017). The researcher studied the available technological solutions, and in cooperation with project promoter and experts of Budapest Technical University, and based on qualitative and quantitative analysis of the waste water had been chosen the optimal waste water treatment – which is the most adaptable for the existing manufacturing process in point of economic and efficiency. It was the Gaylord-methodology what suited the most from the development objectives.

3.2 *Applied method*

During the project the following methods have been applied:

- *analytical measurements:* qualitative and quantitative analysis of wastewater, and also the different separated fractions (water, solvent/DMSO/and other components), furthermore the comparing is mostly based on descriptive statistics (for example relative standard deviation);
- *sustainability analysis:* which is mostly based on Life Cycle Sustainability Assessment as with this approach we can analyse and compare the environmental, economical and social impacts of the current opened and the new closed technology.

3.3 *Results and conclusion*

Parallel with analytical measurement we started preparing and continuously updating with fresh results the sustainability analysis and we choose the Life Cycle Sustainability Assessment (LCSA) as evaluating method (UNEP, 2012). The assessment is based on ISO 14040 standard, we applied SimaPro 7.2 demo (Pre, 2010) and GaBi 4.4 (PE, 2010) programs furthermore we use Ecoinvent database and CML 2001 method to evaluate the environmental impacts. The goal of the LCSA is to evaluate the environmental, economic and social impact of the closed technology. The functional unit is 1 m^3 AsMet adsorber product. The system boundaries were determined from gate to waste treatment. First, we tried to rely on technological and experimental data, any other additional information is coming from Ecoinvent database or literary sources. We determined two scenarios: in the first one, the recovery of DMSO takes 98 % (DMSO_R'), in the second one beside this recovery it is used renewable resources too (DMSO_R'+PV). We analyzed the environmental impact of the solvent and we compared them with the results in the two different scenarios. Considering the results there were significant differences in each impact categories, so according to results of the life cycle assessment, we can say that the environmental effect significantly decreases if we recover the solvent and reuse it in the manufacturing process LCA. In case of the E-LCA (Environmental Life Cycle Analysis) the other component of the wastewater – for example the cerium – has a higher environmental effect than the DMSO, therefore, the impact of the solvent recovery is not so significant. The most significant differences are in the Abiotic Depletion (ADP), Global Warming Potential (GWP), Ecotoxicity (ETP/HT) impact categories.

In case of the Life-Cycle Cost (LCC) was a nearly 15% decrease – namely cost reduction – is caused mostly by the high price of the solvent, because if we reuse as much solvent as we can by distillation, we need to buy a little amount of fresh solvent. However, after the test running of the installed system, we should analyse the ratio of the cost reduction from the technology development (reusing the recovered water and solvent, less amount of hazardous waste) and the increased expenses (arisen energy and water demand).

In case of the S-LCA (Social Life Cycle Assessment), we can barely observe the effect of the new, closed technology in both two scenarios. In this view of the LCA, we supposed that the number of employees is not changed, and the comparison is based only on the rate of the salaries and total production costs. As the amount of the salaries is much smaller than the total production costs, the effects of the opened and closed technologies are almost the same. However, it is important to monetize the other impacts of the closed technologies, such as increased value added, increased professional knowledge/competence, better working circumstances but it is a big challenge due to these are subjective elements.

Application of the Life Cycle Sustainability Assessment (LCSA) we can analyze not only the environmental impacts (E-LCA), but also the economical (LCC) and social (S-LCA) effects too. According to these the criteria of sustainability is the following:

$$LCSA_{linear} (=E\text{-}LCA+LCC+S\text{-}LCA) > LCSA_{closed}(= E\text{-}LCA + LCC + S\text{-}LCA) \qquad (1)$$

4 CASE OF THE HUNGARIAN PILOT

4.1 *The CIRCE2020 project*[2]

The CIRCE2020 project, supported by the Interreg CENTRAL EUROPE Programme funded under the European Regional Development Fund, aims to widen the application of the circular economy (CE) concept in the Central-European region by introducing innovative solutions for the industrial waste management in order to reduce dependencies from primary natural resources within industrial processing. Business as usual (BaU) waste management is frequently based on a linear approach without considering their potential of valorisation within the production system or into another one (industrial symbiosis). In this sense, CIRCE2020 represents an important step to move to a closed loop system based on innovative reusing, remanufacturing and recycling products, thanks to the testing and implementation of CE solutions in 5 pilot areas in the following countries: Austria, Croatia, Hungary, Italy and Poland.

Within the project, life cycle approach and related Life Cycle Assessment (LCA) and Life Cycle Costing (LCC) methodologies have been adopted to the specific project goals.

4.2 *The applied methods of the CIRCE2020 project*

The Product Environmental Footprint (PEF) was applied to test the environmental sustainability of the pre-selected CE cases in the pilot areas, an LCA needs to be performed based on the latest Product Environmental Footprint (PEF) methodological requirements. For a consistent application of the PEF methodology, a guideline has been developed based on ISO 14040 2006 by ETRA (2018), what is not aimed to be PEF fully compliant, but it can be used for the studies aligned with the PEF methodology (EC, 2017). The most relevant impact categories which together contribute to at least 80% of the weighted environmental impact have to be investigated. (ETRA, 2018).

The economic impact of the BaU and CE solutions was investigated by Life Cycle Costing (LCC) method. The method was elaborated by Bay (2018), based on SETAC 2008 and Swarr et al. 2011.

It was a priority of the CIRCE2020 project to harmonize the development of the PEF-based LCA and the LCC guidelines.

4.3 *Case studies in Hungary*[3]

In Hungary waste tyre and plastic waste of stoma bag production are assessed. Both are currently treated in incinerators to produce energy while specific recycling solutions would lead to material recovery. Shredded recycled rubber is a multipurpose material and recycled plastic can be used to produce new products. (IFKA 2019 a,b).

---

2. This project is supported by the Interreg CENTRAL EUROPE Programme funded under the European Regional Development Fund.
3. The LCA and LCC analysis made by LCA Center Association on behalf of IFKA Public Benefit Non-Profit Ltd. for the Development of Industry in connection with CIRCE2020 project.

4.4 *Main conclusion of Hungarian case studies*

The examination of one of the waste streams was linked to the production of tyre. In the plant, where huge amount of rubber waste is produced beside the tyres, but a lot of them are reused in their materials after different waste treatment methods. However, the most significant part of the volume of waste is incinerated, which we considered as a BaU technology solution. The other waste flow was a mixed plastic, which was produced during production of stoma bag. Here the waste rate is very high related to input materials, and it has been burnt too.

The aim of the studies was to investigate other possibility what offers a better solution than burning; with applying circular solution (CE). These analyzes were carried out on the basis of the above-mentioned methodology elaborated by ETRA. In both cases, the functional unit was 1 kg of waste. The analysis did not include the production process, only the steps of transport and disposal were analyzed, in the case of the two-waste stream BaU and CE scenarios. In case of CE of tyre waste, the normalized and weighted environmental load decreased with 15%, but inside of this the climate change decreased with 30%, although some other impact categories increased.

The LCC showed the CE scenario is less "cost demanding" as its total cost is significantly lower – with 69% -than the total costs of the BaU scenario. In case of plastic waste, the environmental impact of the CE scenario is 31% of the BaU scenario. This means that it can reach environmental benefit with another/better waste disposal/recycling alternative. In a LCC perspective, there was not a significant difference between BaU and CE scenarios.

The environmental load of the circular solution has decreased in both selected waste streams comparing with the present situation, but it is not sure that these scenarios are the best solutions. The cause of it that the analysis focused on only one possible solution but there could be another potential alternative on circularity which environmental results are not known.

5 CONCLUSION

More regional research activities had verified, that both circular and bio-economy contribute to the achievement of the UN's Sustainable Development goals, support the sustainable use of renewable resources while conserves biodiversity and soil fertility and prevents the climate change, strengthening of innovation and competitiveness.

The institutional support of the circular economy is wide-ranging at international level: OECD, UNEP, WEF reports and programs can be monitored from 2011 and integrated into the EU legislation. The EU has an ambitious Circular Economy Package (2018) and has already defined its relevant Action Plans. However, the practical implementation can only be achieved with vigorous government support. In the EU Denmark, the Netherlands, the United Kingdom and Germany are leading. There are already many good practices in the EU, with large economic, environmental and social benefits, although we cannot really identify dedicated circular economy projects.

The economic benefits of circular economy have been proven to have significant potential in waste management because the lower cost of raw materials, with a well-chosen technology, reduces the environmental impact. It is true if a plant reuses the solvent after recovery from the waste, see the case of DMSO, but it is true even if industrial cooperation, symbiosis within an industrial park or within 50 km, is developed for recycling of rubber waste. However, there are cases where recovery of material does not result in an optimal, sustainable solution.

REFERENCES

Aguilar A., Wohlgemuth, R. & Twardowski, T. 2018. Perspectives on bioeconomy, New Biotechnology 40, 181–184.
Bay 2018. Life Cycle Costing guideline for the CIRCE2020 project, Bay Zoltán Nonprofit Ltd for Applied Research supported by Balázs Sára external expert, 2018.
CIRCE2020 https://www.interreg-central.eu/Content.Node/CIRCE2020.html.

EC 2013. European Commission Recommendation of 9 April 2013 on the use of common methods to measure and communicate the life cycle environmental performance of products and organisations. OJ L 124/4.

Ellen MacArthur Foundation -EMF 2015. *Towards A Circular Economy: Business Rationale for an Accelerated Transition*, https://www.ellenmacarthurfoundation.org/assets/downloads/TCE_Ellen-MacArthur-Foundation_9-Dec-2015.pdf.

European Parlament &European Commission 2008. Directive 2008/98/of the European Parliament and of the Council of 19 November 2008 on waste and repealing certain Directives, Official Journal of the European Union, 22.11.2008.

European Parlament 2016. Circular economy package Four legislative proposals on waste http://www.europarl.europa.eu/EPRS/EPRS-Briefing-573936-Circular-economy-package-FINAL.pdf.

ETRA 2018. Guidelines for the adaptation of LCA methodology to estimated environmental impact, ETRA supported by Ecoinnovazione Srl, 2018.

Európai Bizottság (EC) 2015. A körforgásos gazdaságról szóló jogalkotási csomag: kérdések és válaszok http://europa.eu/rapid/press-release_memo-15-6204_hu.htm.

European Commission 2017. PEFCR Guidance document, - Guidance for the development of Product Environmental Footprint Category Rules (PEFCRs), version 6.3, December 15 2017.

EUROSTAT 2019. Circular economy indicators, Trade in recyclable raw materials, https://ec.europa.eu/eurostat/data/database.

Frey L. 2013. Concerto communities in EU dealing with optimal thermal and electrical efficiencyof buildings and districts, based on microgrids, Del5-14_3rd Report.

Horváth B. 2019. Körforgásos gazdasági modellek és hatékonyságuk mérése Doktori értekezés, Szent István Egyetem, Gödöllő.

ISO 2006. ISO 14040 and ISO 14044: Environmental management, Life cycle assessment, 2006.

IFKA 2019a. LCA Report of Hungarian Pilot, made by LCA Center Association on behalf of IFKA Public Benefit Non-Profit Ltd. for the Development of Industry in connection with CIRCE2020 project.

IFKA 2019b. LCC Report of Hungarian Pilot, made by LCA Center Association on behalf of IFKA Public Benefit Non-Profit Ltd. for the Development of Industry in connection with CIRCE2020 project.

Mannheim, V. & Siménfalvi, Z. 2018. Life Cycle Assessment for Waste Management Processes. The 9th annual international experts conference Enviromanagement, Strebsko Pleso, 10–11 October 2018: 1–17. ISBN 978-80-85655-36-0. http://nmc.sk/wp-content/uploads/2018/09/ANOTACIA_MANNHEIM.pdf.

Pauli, G. 2010. Kék gazdaság – 10 Év 100 Innováció 100 Millió Munkahely, A Római Klub Jelentése, PTK KTK Kiadó, Pécs.

PE International Gmbh 2010. GaBi 4.4 software.

Pre 2010. SimaPro 7.2 www.pre.nl.

Ramkumar, S. F.& Kraanen, F. & Plomp, R. & Edgerton, B. & Walrecht, A. & Baer, I. & Hirsch, P. 2018. Linear Risks. Amsterdam: Circle Economy. 14. p.

ReSiELP 2017. Recovery of Silicon and other materials from End-of-Life Photovoltaic Panels, (2017–2020), https://eitrawmaterials.eu/project/resielp/.

Saidani M. & Yannou, B. & Leroy Y. & Cluzel F. 2017. How to Assess Product Performance in the Circular Economy? Proposed Requirements for the Design of a Circularity Measurement Framework, Recycling, 2, 6; doi:10.3390/recycling2010006.

Szita T.K. 2017. Application of life cycle assessment in Circular Economy, Hungarian Agricultural Engineering.

SETAC 2008. Life cycle costing 978-1-880611-83-8, 232 pp.

Swarr, T.E, Hunkeler, D., Klöpffer, W., Pesonen, H.L., Ciroth, A., Brent, A.C., & Pagan, R. 2011. Environmental Life Cycle Costing: A Code of Practice. Pensacola (FL): Society of Environmental Toxicology and Chemistry (SETAC). 98 p.

SYMBI project, http://hameenliitto.fi/en/symbi-project.

Ulmann L. ed. 2015. Circular Economy in Europe Towards a new economic model Growth Within: A Circular Economy Vision for a Competitive Europe. 2015. europeanfiles.eu/wp-content/uploads/issues/2015-september-38.pdf (2016.06.15.).

UNEP 2012. Social Life Cycle Assessment and Life Cycle Sustainability Assessment.

Zajáros, A., Szita T.K., Matolcsy K. & Horváth D. 2017. Life Cycle Sustainability Assessment of solvent recovery from hazardous wastewater. Periodica Politechnica Chemical Engineering 62(3), pp. 305–309.

Winans K., Kendall A. & Deng H. 2017. The history and current applications of the circular economy concept. Renewable and Sustainable Energy Reviews, 68, 825–833.

Solutions for Sustainable Development – Szita, Jármai & Voith (eds.)
© *2020 Taylor & Francis Group, London, ISBN 978-0-367-42425-1*

New approach for characterizing the "naturalness" of building materials

A. Velősy
ÉMI Nonprofit Kft, Szentendre, Hungary

ABSTRACT: About 87% of the life of a civilized population is spent in a built environment, so it is necessary and convenient to introduce an easy-to-use, quick evaluation process to characterize "naturalness" of building materials. With the help of the parameter "NAT-TECH" (NT_{grad}) we can try to give a more evident, more feasible tool for the designers, instead of complicated and expensive evaluation software. This is why the "weighted technical-ecological evaluation value" described in more details in this article, which can be deduced from the graphical representation of the "NAT-TECH" (NT_{grad}) and three other ecological and three relevant technical parameters. If the listed features - suitably - are shown in a web graph, then the area covered by the chart can define by using a single numerical value – "weighted technical-ecological evaluation value" – which material is the most appropriate for the decision maker in the given technical - ecological environment.

1 INTRODUCTION

As about 87% of the life of a civilized population is spent in a built environment or in community spaces surrounded by this environment (Karl et al. 2009), an expedient and obvious step to search the harmony between the highly technological and the nature-close construction in order to improve and protect the health of building users. Therefore it is necessary and convenient to introduce an easy-to-use, quick evaluation process that would make the job of architects and the situation of principal contractors easier. Nowadays, the research on natural building materials, including natural thermal insulation materials is in the public interest of ecological thinking and planning. The research of natural thermal insulation materials so is a popular and up-to-date topic both in international and domestic scientific and producer communities.

2 SIMPLIFIED PROCEDURE FOR TECHNICAL-ECOLOGICAL ASSESSMENT OF BUILDING MATERIALS

Global climate change is now perceived in our daily lives (Klepeis 2001). Without an exact aid for determining the concept of "naturalness" in European and overseas literature, we have no choice but to try to characterize "naturalness" inversely by the level of technology - for example, by the parameter of "NAT-TECH" (NT_{grad}) that represents the proportion of natural-technology. After that, the methodology of ecological assessment of building materials can be reconsidered in the knowledge of this parameter (Simon 2012, NyME 2011). With the help of this, we can try to give a more evident, more feasible tool for the designers instead of complicated and expensive evaluation software.

The recommended ecological weight value (w) of each technological intervention (B) (for the support of the evaluation principle in the working hypothesis phase) was determined as follows:

In the list above, the hypothetical threshold for the ecological weight of handicraft interventions up to A-J and those for industrialized interventions up to K-N are given.

	(B_w)
A, direct use without shaping (within one day of on-road transportation (45km))	0,00
B, simply shaping (splitting, rough shaping with hand tools)	0,01
C, breaking, crushing, shredding (by hand, with hand tools)	0,02
D, more exacting, more precise shaping (dry grinding, carving, fine forming by hand)	0,04
E, cutting, mechanical cleaning (with simple machine tools)	0,07
F, milling, grinding, cutting to size (with machine tools)	0,09
G, dissolving, glueing (chemical intervention with natural solvent adhesive)	0,11
H, high-demand shaping (mechanical cutting, wet grinding, polishing)	0,16
I, drying, heat treatment (natural and handmade)	0,50
J, mechanical forming by hot processes (pressing, rolling, forging, firing)	1,50
K, melting, steaming, hardening, maturing (technological processes, external energy demand)	5,00
L, industrial-scale chemical intervention, treatment (dissolving, glueing, mixing)	7,00
M, industrial scale alloying, firing, hot forming (serial production-forming processes)	10,00
N, large-scale production, composite blends (including the production of reinforced concrete, fiber reinforced materials, etc.)	15,00

Multiplying the above theoretical values (B_w) by the number of technological steps (L_n) finally, the NT_{grad} parameter describing the degree of technology being sought.

$$B_w \times L_n = NT_{grad}$$

Of course, this list needs to be expanded and refined on the basis of social conventions and ecological impact studies, but the aim at present is only to establish the principle to prove that it is possible to create a simple parameter that can describe the "naturalness" of building materials and processes.

The new created NAT-TECH (NT_{grad}) and three other ecological parameters (non-renewable primary energy, global warming potential, acidification potential) and three relevant technical parameters (body density, thermal conductivity and specific heat capacity) are used to achieve a single value, the TEL_{wn} (weighted number of Technical-Ecological Level) If the listed parameters - suitably - are shown in a web graph, then the area covered by the chart

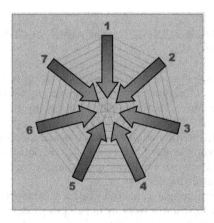

Figure 1. Meaning of the interpretation range of parameters of diagram-axes (favorable is the middle of chart, green colour).

can define by using a single numerical value – weighted Technical-Ecological Level TEL_{wn} – which material is the most appropriate for the decision maker in the given technical - ecological environment.

The interpretation range of parameters of diagram-axes for construction thermal insulation materials, are as follows:

Table 1. The interpretation range of parameters of diagram-axes.

Parameter	Property	Symbol É	Value range	Unit	Favorable, if the value is
The essential technical characteristics:					
1	body density	ρ	1 – 500	[kg/m^3]	low
2	thermal conductivity	λ	0.001 – 0.200	[W/mK]	low
3	specific heat capacity	c_p	2500 – 500	[J/kgK]	high
Ecological parameters:					
4	non-renewable primary energy	PET	0 – 25	[kWh/kg]	low
5	global warming potential CO_{2eq}	GWP	-2500 – +2500	[g/kg]	low
6	acidification potential SO_{2eq}	AP	0 – 25	[g/kg]	low
Degree of technology					
7	degree of technology	NT_{grad}	0 – 2500	[-]	low

The values of each product in the graph axis are the mathematical mean values reported in the research literature.

In this ranking order, we consider the most favorable material that has the lowest nominal TEL_{wn} value. For example – by staying at the topic of natural thermal insulation (University of Sopron 2012, Bauen mit nachwachsenden Rohstoffen 2008, Stoffliche Einsatzmengen nachwachsender Rohstoffe in Deutschland 2016) – a cork plate that was used in the immediate vicinity of the site of extraction and production.

Individual values of cork plate are shown in Table 2.

Based on these, the value of the "weighted technical-ecological evaluation value" (TEL_{wn}) of the natural untreated cork, described on the web-chart (Figure 2), is favorable: 29.134.

The practical values – in case of the natural and industrial thermal insulation materials (Bejó et al. 2014, Szalay 2016) discovered so far - ranged from the 29 (locally used cork) to 359 (vacuum cell thermal insulation). The difference between the extreme values is multiple of about 12.3, which shows that the process under development already allows for a fairly fine classification in the test phase.

In this respect, Figure 2–7 present web-charts and related TEL_{wn} values for the thermal insulating materials processed during the research. Calculations were based on data from international sources (CIPRA 2007, Climalp 2014, FÖK 2002, Institut Bauen und Umwelt e. V., VIP Panel data, Ökologische Wärmedämmstoffe im Vergleich 2.0 2010, Sprengard et al. 2012, Weiß et al. 2001).

Table 2. Individual values of cork plate.

Parameter	Property	Symbol	Value	Unit
1	body density	ρ	120	[kg/m^3]
2	thermal conductivity	λ	0.038	[W/mK]
3	specific heat capacity	c_p	1670	[J/kgK]
4	non-renewable primary energy	PET	1.05	[kWh/kg]
5	warming potential CO_{2eq}	GWP	-1220	[g/kg]
6	acidification potential SO_{2eq}	AP	1.89	[g/kg]
7	degree of technology	NT_{grad}	105.21	[-]

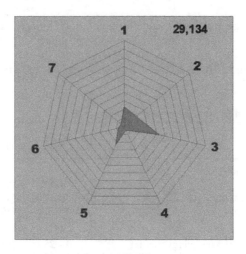

Figure 2.   Graphical interpretation of weighted technical-ecological evaluation value (TEL$_{wn}$) of the natural untreated cork. (29,134).

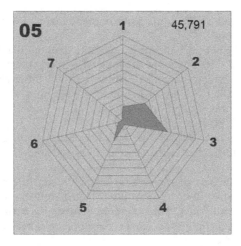

Figure 3.   TEL$_{wn}$ value of straw board. (45,791).

Table 3.   Individual values of straw board.

Parameter	Property	Symbol	Value	Unit
1	body density	ρ	90	[kg/m^3]
2	thermal conductivity	λ	0.055	[W/mK]
3	specific heat capacity	c_p	1600	[J/kgK]
4	non-renewable primary energy	PET	0,123	[kWh/kg]
5	warming potential CO_{2eq}	GWP	-1250	[g/kg]
6	acidification potential SO_{2eq}	AP	0,852	[g/kg]
7	degree of technology	NT_{grad}	8,04	[-]

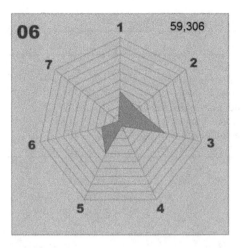

Figure 4. TEL_{wn} value of reed plank. (59,306).

Table 4. Individual values of reed plank.

Parameter	Property	Symbol	Value	Unit
1	body density	ρ	182,5	[kg/m³]
2	thermal conductivity	λ	0.058	[W/mK]
3	specific heat capacity	c_p	1600	[J/kgK]
4	non-renewable primary energy	PET	0.810	[kWh/kg]
5	warming potential CO_{2eq}	GWP	-1328	[g/kg]
6	acidification potential SO_{2eq}	AP	1.101	[g/kg]
7	degree of technology	NT_{grad}	72,54	[-]

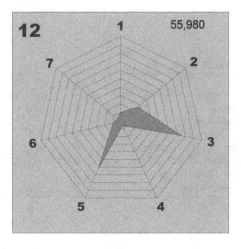

Figure 5. TEL_{wn} value of blown cellulose fiber. (55,980).

Table 5.   Individual values of blown cellulose fiber.

Parameter	Property	Symbol	Value	Unit
1	body density	ρ	50	[kg/m^3]
2	thermal conductivity	λ	0.042	[W/mK]
3	specific heat capacity	c_p	1900	[J/kgK]
4	non-renewable primary energy	PET	1.272	[kWh/kg]
5	warming potential CO_{2eq}	GWP	498,5	[g/kg]
6	acidification potential SO_{2eq}	AP	2,435	[g/kg]
7	degree of technology	NT_{grad}	81,31	[-]

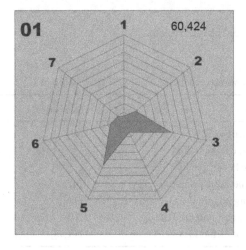

Figure 6. TEL_{wn} value of lamb wool. (60,424).

Table 6.   Individual values of lamb wool.

Parameter	Property	Symbol	Value	Unit
1	body density	ρ	25	[kg/m^3]
2	thermal conductivity	λ	0.0375	[W/mK]
3	specific heat capacity	c_p	1600	[J/kgK]
4	non-renewable primary energy	PET	3.51	[kWh/kg]
5	warming potential CO_{2eq}	GWP	380,5	[g/kg]
6	acidification potential SO_{2eq}	AP	3,46	[g/kg]
7	degree of technology	NT_{grad}	193.96	[-]

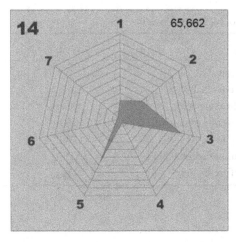

Figure 7.   TEL$_{wn}$ value of autoclaved aerated concrete thermal-insulation sheet. (65,662).

Table 7.   Individual values of blown cellulose fiber.

Parameter	Property	Symbol	Value	Unit
1	body density	ρ	115	[kg/m^3]
2	thermal conductivity	λ	0.045	[W/mK]
3	specific heat capacity	c_p	925	[J/kgK]
4	non-renewable primary energy	PET	0.995	[kWh/kg]
5	warming potential CO$_{2eq}$	GWP	351	[g/kg]
6	acidification potential SO$_{2eq}$	AP	1.036	[g/kg]
7	degree of technology	NT$_{grad}$	84,42	[-]

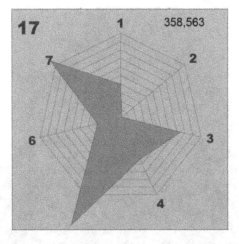

Figure 8.   TEL$_{wn}$ value of vacuum panel thermal-insulation board. (358,563).

Table 8. Individual values of blown cellulose fiber.

Parameter	Property	Symbol	Value	Unit
1	body density	ρ	200	[kg/m^3]
2	thermal conductivity	λ	0.007	[W/mK]
3	specific heat capacity	c_p	800	[J/kgK]
4	non-renewable primary energy	PET	12.96	[kWh/kg]
5	warming potential CO_{2eq}	GWP	3121,5	[g/kg]
6	acidification potential SO_{2eq}	AP	7.55	[g/kg]
7	degree of technology	NT_{grad}	2534.98	[-]

3 CONCLUSION

Technical-ecological evaluation among architects is now largely based on experience, market access and, to some extent, emotional, cultural attachments, which cannot be considered objective. At the same time, architectural designers are not expected to train at a high level in ecological assessment, purchase expensive software, databases, and learn how to handle them for their daily decisions. Hopefully, the described anomaly can be successfully mitigated by the technical-ecological evaluation principle what has been provided, for which an easy-to-use "desktop" decision-making kit or software can be developed. The ease with which the elaborated method and factor is applied is expected to make the process – currently in research phase – a popular decision-making tool. However an inspiring interest was experienced in the early stages of research from architect and also in the LCA's point of view, therefore the work will continue with the involvement of specialists.

Figure 9 shows our goal of separating high-tech (gray) and natural (green) processes along the red line.

Nevertheless the effort to draw such a sharp line between the two major areas, at least, may lead to failure because of the great differences in cultural and economic capabilities. On the other hand, a "blue zone" can be laid onto this "red line" that cleverly compromises between the two areas. This zone can be a path used by all of us. This article and the PhD thesis behind it deal with this attempt and create to build its foundations.

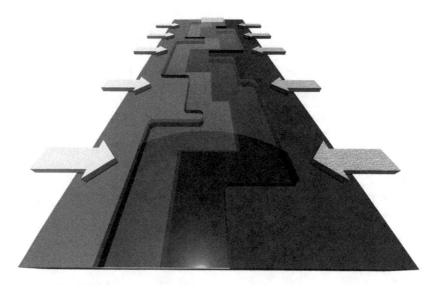

Figure 9. Separation of high-tech (gray) and natural (green) processes.

REFERENCES

Bauen mit nachwachsenden Rohstoffen/Entwurf,Konstruktion, Bauprodukte/2008/http://www.iwu.de/dateien/IWU_Bauen_nachwachs_Rohstoffen_081029.pdf.

Bejó L. & Kuzsner Á. & Hantos Z. & Karácsonyi Zs. 2014. A faanyag szerepe a környezetbarát építészetben 1. rész: A környezetbarát építőanyagok szerepe az épületek energiamérlegében és CO_2 kibocsátásában.

CIPRA 2007. Nachhaltiges Bauen und SANieren in den Alpen Modul 2: Energie und gebäude climalp, eine Informationskampagne der CIPRA/www.baubook.info/.

Climalp 2014. Energieinstitut Voralrbeg Climalp 1-5 issues/http://www.energieinstitut.at/wp-content/uploads/2015/07/140415_climalp_Modul2_DE.pdf/.

FÖK 2002. BauBioDataBank, Független Ökológiai Központ/www.fenntarthato.hu/.

Institut Bauen und Umwelt e. V. (vormals Arbeitsgemeinschaft Umweltverträgliches Bauprodukt e.V., AUB)/www.bau-umwelt.com/.

Karl T.R. & Melillo J. M. & Peterson T.C. (eds.) 2009. Climate Change Impacts in the United States - U.S. Global Change Research Program Cambridge University Press, http://www.iooc.us/wp-content/uploads/2010/09/Global-Climate-Change-Impacts-in-the-United-States.pdf/.

Klepeis N.E. 2001. The National Human Activity Pattern Survey (NHAPS) A Resource for Assessing Exposure to Environmental Pollutants/https://www.buildinggreen.com/blog/we-spend-90-our-time-indoors-says-who.

NyME 2011. Az ökológia tárgya, presentation http://www.nyme.hu/fileadmin/dokumentumok/emk/vad gazdalkodas/vadaszati_okologia/vadaszati_okologia.pdf/.

Ökologische Wärmedämmstoffe im Vergleich 2.0 2010. Leitfaden zur Dämmstoffauswahl für den normgerechten Einsatz mit Kapiteln zu Bauphysik, Planung, Qualitätssicherung und Ökobilanz sowie 24 detaillierten Dämmstoffbeschreibungen/www.muenchen.de/bauzentrum/.

Porextherm Dämmstoffe GmbH/www.innodaemm.de/daemmstoffe/vakuum-paneele.html/.

Simon E, 2012. Az Ökológia alapjai, Bevezetés az ökológiába,/http://ecology.science.unideb.hu/files/02-Bevezeto.pdf/.

Sprengard C. & Treml S, & Holm A. H. 2012. Technologien und Techniken zur Verbesserung der Energieeffizienz von Gebäuden durch Wärmedämmstoffe Metastudie Wärmedämmstoffe – Produkte – Anwendungen – Innovationen – FIW München Bericht FO/http://www.energie-experten.org/fileadmin/Newsartikel/Newsartikel _02/FIW_Metastudie_Wärmedäm&mstoffe.pdf/.

Stoffliche Einsatzmengen nachwachsender Rohstoffe in Deutschland 2016. https://mediathek.fnr.de/grafiken/stoffliche-einsatzmengen-nachwachsender-rohstoffe-in-deutschland.html/data downloaded: 17.07.2016 18:37/.

Szalay Zs. 2016. Az épületek teljes életciklusa, chapter in: Zöld András – Szalay Zsuzsa – Csoknyai Tamás, Energiatudatos Építészet 2.0, Terc Kiadó.

University of Sopron 2012. "Környezettudatos energiahatékony épület TÁMOP- 4.2.2.A-11/1/KONV-2012-0068" project catalog/2012/.

VIP Panel data,/www.innodaemm.de/daemmstoffe/vakuum-paneele.html/.

Weiß R.G. & Paproth O. 2001. Wärmedämmung für Wohngesundheit und Energieeinsparung Issue Leitfaden Ökologische Dämmstoffe 2 Publisher: NABU Bundesverband Naturschutzbund Deutschland Bonn, Krefeld.

Part F: Smart Manufacturing and Smart Buildings

Solutions for Sustainable Development – Szita, Jármai & Voith (eds.)
© 2020 Taylor & Francis Group, London, ISBN 978-0-367-42425-1

Grouping and analyzing PLC source code for smart manufacturing

O. Hornyák

Department of Information Engineering, University of Miskolc, Hungary

ABSTRACT: This paper deals with two approaches to use Group Technology in the field of Programmable Logic Controller grouping. The first approach analyses the functionality of the machines to control. A mathematical model is given, and a Genetic Algorithm Clustering approach is used. The second approach analyses the PLC codes and finds the longest matching sequences in them using the Smith-Waterman algorithm. Both approaches are explained by examples.

1 INTRODUCTION

1.1 *Motivation*

Shortening market time is key in modern and smart manufacturing (Kusiak, 2018). Engineering design processes require reusing the existing knowledge from similar projects (Erdélyi et al, 2009). Grouping the machines, the processes or project is in the engineer's interest for a long time. It is a challenging task to it in a smart, automated way. This requires computational intelligence and soft computing methods. Some tasks require simulation (Tamás, 2017) and visualization (Mileff et al, 2015)

The designing of manufacturing machines has undergone a dynamic change in recent days. In the past, the most important design parameters were the functionality and robustness of the machines.

Today we need to consider the customization cost of the source code of the machine controller. With the wide spread of various Information Technologies, it is now possible to use software-based solutions where conventional controller hardware such as mechanical actuators was applied.

Recent manufacturing machines are operated by more precise and smaller software driven controllers which are based on mathematical models. The side effect of this is that modern software controllers have more complex code. Therefore, its software architecture is even more complex.

The cost of creating a new project from an existing and running project is a very important factor.

Sometimes the new requested product variant can be very different in its functionality,

There are two approaches in practice:

In some cases, it is very helpful to have a 'max-project' that contains all the possible variants. The functionality which is not required for a certain project will be simply deactivated. Obviously, this approach does not result in real functional differences from a hardware or software point of view.

The other approach is to start from a minimum project, which contains the components that do not belong to any manufacturing machine. The variant requested could be achieved by adding or copying components. Both approaches are widely used by the industries.

The two approaches can be summarized as follows:

- Designers of manufacturing machines use modular build techniques including the mechanical parts, standardized communication and software modules.
- It is expedient to create controller software automatically for manufacturing machines produced in small batches or completely customized.

Modular machines design and reusing controller code have numerous advantages:

- repeated tasks can be executed more precisely,
- the risk of outage of senior developers can be reduced,
- the quality process and product documentation can be increased.

2 MATHEMATICAL MODEL

2.1 *The genetic algorithm clustering problem*

In (Hornyák & Sáfrány, 2009b) and (Hornyák & Sáfrány, 2010) the overview is given about the feasible coding systems for Group Technology (GT). Let's have a binary polycode (See Figure 1) coding system as follows:

- Each bit of the k bit long vector (code) represents a feature.
- 0 means that the controller does not implement the feature,
- 1 means that the feature is realized and activated.

Let us investigate n projects. The problem can be represented by an n x k matrix M, see for example (1).

$$
\begin{array}{c}
1 \\ 2 \\ 3 \\ 4 \\ 5 \\ 6 \\ 7 \\ 8 \\ 9 \\ 10
\end{array}
\begin{bmatrix}
0 & 0 & 0 & 0 & 0 & 1 & 0 & 1 & 0 & 0 \\
1 & 1 & 1 & 1 & 0 & 1 & 0 & 0 & 0 & 0 \\
0 & 1 & 1 & 1 & 1 & 1 & 0 & 0 & 0 & 0 \\
0 & 0 & 0 & 0 & 0 & 1 & 1 & 0 & 0 & 0 \\
0 & 0 & 0 & 0 & 0 & 0 & 0 & 1 & 1 & 0 \\
1 & 1 & 0 & 1 & 1 & 0 & 0 & 0 & 0 & 1 \\
0 & 0 & 0 & 0 & 0 & 1 & 1 & 0 & 0 & 0 \\
1 & 1 & 1 & 0 & 1 & 0 & 0 & 0 & 0 & 0 \\
0 & 0 & 0 & 0 & 0 & 1 & 1 & 1 & 0 & 0 \\
0 & 0 & 0 & 0 & 0 & 0 & 0 & 0 & 1 & 1
\end{bmatrix}
\tag{1}
$$

Let p_i represent the ith row. The hamming distance can indicate the count of positions at which the bits are different. The following matrix describes the Hamming distance between p_i and p_j.

The smaller the Hamming distance is the more similar the two projects are. It would be interesting to know, that based on project similarity how many groups can be formed.

The following chapter describes a Genetic algorithm clustering approach.

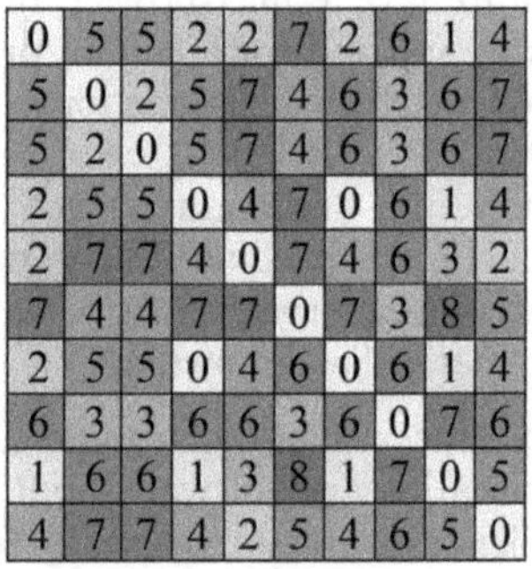

0	5	5	2	2	7	2	6	1	4
5	0	2	5	7	4	6	3	6	7
5	2	0	5	7	4	6	3	6	7
2	5	5	0	4	7	0	6	1	4
2	7	7	4	0	7	4	6	3	2
7	4	4	7	7	0	7	3	8	5
2	5	5	0	4	6	0	6	1	4
6	3	3	6	6	3	6	0	7	6
1	6	6	1	3	8	1	7	0	5
4	7	7	4	2	5	4	6	5	0

Figure 1. Heat map of hamming distances for *M*.

288

2.2 A clustering method using genetic algorithm

Machine cell formation is analogous to finding groups of machines. Cheng, 1992 described an approach that assigns machines to parts to manufacture, and form groups of those. De Lit et al focuses on minimizing the traffic of items between the cells. Gonçalves & Resende (2002) and James et. al. (2007) used a hybrid genetic algorithm for the cell formation problem. In Wang & Kusiak (2000) an overview is given for applying the computer intelligence in manufacturing, while Suresh & Kay (2012) describes a state of the art twelve years later.

The author of this paper also has some clustering/soft computing experience: Szabó et al. (2019), Hornyák & Barthal (2008), Hornyák & Sáfrány (2009a).

The new approach in this paper is that here group forming factor is the PLC code and not the part to manufacture.

The rows in matrix M need to be rearranged so that there will be a block of 1s near the p distance of the diagonal item. Gonçalves, & Resende (2002) defines two measures of the goodness of the grouping, called Grouping Efficiency and Grouping Efficacy.

Grouping Efficiency is calculated as:

$$\eta = q \cdot \eta_1 + (1 - q) \cdot \eta_2 \qquad (2)$$

where
η_1 is the proportion of 1s in the diagonal block,
η_2 is the proportion of 0s in the off-diagonal block,
q is a weight coefficient.

Grouping Efficacy is calculated as:

$$\mu = \frac{N_1 - N_1^{out}}{N_1 + N_0^{in}} \qquad (3)$$

where
N_1 is the total count of 1s in the matrix,
N_1^{out} is the count of 1s outside the diagonal blocks,
N_0^{in} is the count of 0s inside the diagonal blocks.

Genetic Algorithm Clustering, GAC is a method that uses clustering capabilities of genetic algorithm, see Maulik & Bandyopadhyay (2000). Both the Grouping Efficiency and Grouping Efficacy were implemented as the fitness function.

The n number of rows and k number of columns had been split to y and x number of blocks $w_1, w_2, \ldots, w_y$ and $u_1, u_2, \ldots, u_x$ so that

$$\sum_{1}^{y} w_i = n \ and \ \sum_{1}^{x} u_j = k \qquad (4)$$

If the GAC creates an instance which does not meet (4) then the fitness function will be 0. Otherwise, the fitness function is calculated as described above.

The pseudo code of the GAC is as follows:

Create the initial set of the population.
Evaluate each of them
While not to stop do
    Select two random elements
    Crossover random genes of those two
    Mutate the new specimen randomly
    Evaluate the new specimen

The new population will be the top elite of the previous population plus the best ones of the new specimens
End while

After the block diagonalization process, the (1) matrix should look like (5).

$$\begin{array}{c}8\\6\\3\\2\\4\\1\\9\\7\\5\\10\end{array}\begin{bmatrix} 1 & 1 & 1 & 0 & 1 & 0 & 0 & 0 & 0 & 0 \\ 1 & 1 & 0 & 1 & 1 & 0 & 0 & 0 & 0 & 1 \\ 0 & 1 & 1 & 1 & 1 & 1 & 0 & 0 & 0 & 0 \\ 1 & 1 & 1 & 1 & 0 & 1 & 0 & 0 & 0 & 0 \\ 0 & 0 & 0 & 0 & 0 & 1 & 1 & 0 & 0 & 0 \\ 0 & 0 & 0 & 0 & 0 & 1 & 0 & 1 & 0 & 0 \\ 0 & 0 & 0 & 0 & 0 & 1 & 1 & 1 & 0 & 0 \\ 0 & 0 & 0 & 0 & 0 & 1 & 1 & 0 & 0 & 0 \\ 0 & 0 & 0 & 0 & 0 & 0 & 0 & 1 & 1 & 0 \\ 0 & 0 & 0 & 0 & 0 & 0 & 0 & 0 & 1 & 1 \end{bmatrix} \tag{5}$$

You can see three groups identified by the algorithm here:

Group 1 consists of (8, 6, 3, 4),
Group 2 consists of (1, 9, 7),
Group 3 consists of (5, 10).

3 COMPUTER ALGORITHM FOR CODE SIMILARITY DETECTION

3.1 *Software metric-based approach*

Many computer algorithms were specified for detecting software similarity, mainly for detecting plagiarism. (Parker & Hamblen, 1989) defined seven layers of code modification in a plagiarism spectrum. Some software metrics can be used to detect similarity. Halstead (1997) was among the pioneers, the suggested metric used the following quantities:

n_1 = number of unique or distinct operators.
n_2 = number of unique or distinct operands.
N_1 = total usage of all the operators.
N_2 = total usage of all the operands.

Using these metrics, we can calculate:

$$V = (N_1 + N_2)\log_2(n_1 + n_2) \tag{6}$$

$$E = (n_1 N_2 (N_1 + N_2)\log_2(n_1 + n_2))/2n_2 \tag{7}$$

where V refers to the volume of the program and E refers to the efforts to create the program.

You may find the logarithmic function odd at first sight. However, in this metric, a correlation between the number of bugs in the program, programming or debugging time of the program and the complexity of the program was found.

This approach offers good measures for calculation algorithms. In PLC programs, however, there are not too many operators to count.

3.2 *Local alignment detection using Smith-Waterman algorithm*

Smith & Waterman (1981) developed their algorithm to detect common molecular subsequences, see also Pearson (1991). Two molecular sequences were compared:

$A = a_1 a_2 \ldots a_n$ and $B = b_1 b_2 \ldots b_m$.

The algorithm works as follows: 1. Set up a H_{n+1} x $m+1$ matrix whose first row and column have the 0 indexes and are zeroed.

$H_{k0} = H_{0l} = 0$ for $0 \le k \le n$ and $0 \le l \le m$

Then H_{ij} is the maximum similarity of two segments ending in a_i and b_j respectively. It is calculated as:

$$H_{ij} = \max \left\{ \begin{array}{l} H_{i-1,j-1} + s(a_i, b_j), \\ \max_{k \ge 1}\{H_{i-jk} - W_k\}, \\ \max_{l \ge 1}\{H_{i,j-1} - W_l\} \end{array} \right\} \tag{8}$$

where $s(a_i, b_j)$ is a score function for similarity, W_k is a weight (cost) function for a k-long deletion and W_l is a cost function for inserting l length of the new sequence.

The highest score in the matrix indicates the maximum local alignment of the two sequences.

Once the matrix elements are calculated the maximum element must be found. That refers to the end of the maximum alignment. The traceback algorithm will find the way back. Find the next highest score. There can be three types of movement:

1. A diagonal jump implies there is an alignment (either a match or a mismatch).
2. A top-down jump implies there is a deletion.
3. A left-right jump implies there is an insertion.

3.3   *Checking similarity of PLC programs*

Look at the following two PLC programs that are in Structured Text format:

```
INTERFACE
USEPACKAGE CAM;

PROGRAM StartUp;
PROGRAM Movement;
END_INTERFACE
IMPLEMENTATION
PROGRAM startup

HMI_rotaryknife_show_position: = 200;

END_PROGRAM
PROGRAM movement
VAR backcurrpos:INT: = 0;
END_VAR

LABEL1:;

HMI_rotaryknife_show_position: =
HMI_rotaryknife_show_position - 5;

IF HMI_rotaryknife_show_position =
50 THEN
current_offset[0]: = 95;
backcurrpos: = current_offset[0];
END_IF;

IF HMI_rotaryknife_show_position =
0 THEN
HMI_rotaryknife_show_position: =
200; END_IF;

GOTO label1;
END_PROGRAM
END_IMPLEMENTATION
```

```
INTERFACE
USEPACKAGE CAM;

PROGRAM Movement;
END_INTERFACE
IMPLEMENTATION
PROGRAM movement
VAR backcurrpos:INT: = 0;
END_VAR

LABEL1:;

HMI_rotaryknife_show_position: =
HMI_rotaryknife_show_position - 5;

IF HMI_rotaryknife_show_position =
50 THEN
current_offset[0]: = 95;
backcurrpos: = current_offset[0];
END_IF;

IF HMI_rotaryknife_show_position =
0 THEN HMI_rotaryknife_show_position: = 200;
END_IF;

GOTO label1;
END_PROGRAM
END_IMPLEMENTATION
```

A computer program was written which tokenizes the program, then finds the similar sequences among the tokens. The following keywords were found:

INTERFACE
A USEPACKAGE
B CAM
C PROGRAM
D STARTUP
E MOVEMENT
F END_INTERFACE
G IMPLEMENTATION
H HMI_ROTARYKNIFE_SHOW_POSITION
I END_PROGRAM
J VAR
K BACKCURRPOS
L INT
M END_VAR
N LABEL1
O IF
P THEN
Q CURRENT_OFFSET[0]
R END_IF
S GOTO
T END_IMPLEMENTATION

The computer program then built the matrix as discussed before. In this example the longest matching sequence was:

BCDDGHDJDKMNOPQSPQSTJU

We can introduce a new metric that compares the longest matching sequence with the overall program length:

$$s = \frac{l_l}{l_t} \tag{9}$$

where s is the relative similarity, l_l is the longest matching sequence, l_t is the total program length.

4 CONCLUSIONS

In this paper, various aspects of finding similarity in PLC codes were investigated. Local alignment detection algorithm can identify those coding blocks that should be generalized. So that PLC software developers can create coding libraries of their content. This can speed up development time thus contribute to smart manufacturing.

ACKNOWLEDGEMENTS

This research was supported by the European Union and the Hungarian State, co-financed by the European Regional Development Fund in the framework of the GINOP-2.3.4-15-2016-00004 project, aimed to promote the cooperation between the higher education and the industry.

REFERENCES

Cheng, C.H. 1992. Algorithms for grouping machine groups in group technology. *Omega*, 20(4), 493–501.

De Lit, P., Falkenauer, E., & Delchambre, A. 2000. Grouping genetic algorithms: an efficient method to solve the cell formation problem. *Mathematics and Computers in simulation*, 51(3-4), 257–271.

Erdélyi, F., Tóth, T., Kulcsár, G., Mileff, P., Hornyák, O., Nehéz, K., & Körei, A. 2009. New Models and Methods for Increasing the Efficiency of Customized Mass Production. *J. Mach. Manuf*, 49, 11–17.

Faidhi, J.A., & Robinson, S.K. 1987. An empirical approach for detecting program similarity and plagiarism within a university programming environment. *Computers & Education*, 11(1), 11–19.

Gonçalves, J.F., & Resende, M.G. 2002. A hybrid genetic algorithm for manufacturing cell formation. *AT&T Labs Research Technical Report* TD-5FE6RN, 1–30.

Halstead, M.H. (1977). Elements of software science (Vol. 7, p. 127). New York: Elsevier.

Hornyák, O., & Barthal, G. 2008. Soft Computing Methods for Behaviour Based Control. In *Micro-CAD'2008 International Scientific Conference*.

Hornyák O. & Sáfrány, G. 2009a. A csoporttechnológia alkalmazásának vizsgálata moduláris gyártógépek vezérlőkódjának automatikus generálására, *MicroCAD'2009 International Scientific Conference*, Miskolc, pp. 2009, pp. 43–49.

Hornyák, O., & Sáfrány, G. 2009b. Group technology for automated generation of machine controller code. In 5th International Symposium on Applied Computational Intelligence and Informatics (pp. 17–22). IEEE.

Hornyák, O., & Sáfrány, G. 2010. Models and methods to detect similarity of Manufacturing machines. In *Hungarian Journal of Industrial Chemistry*, Vol38(2) (pp. 149–153).

James, T.L., Brown, E.C., & Keeling, K.B. 2007. A hybrid grouping genetic algorithm for the cell formation problem. *Computers & Operations Research*, 34(7), 2059–2079.

Kang, H.S., Lee, J.Y., Choi, S., Kim, H., Park, J.H., Son, J.Y, Do Noh, S., 2016. Smart manufacturing: Past research, present findings, and future directions. *International Journal of Precision Engineering and Manufacturing-Green Technology*, 3(1),111–128.

Kusiak, A. 2018. Smart manufacturing. *International Journal of Production Research*, 56(1-2), 508–517.

Maulik, U., & Bandyopadhyay, S. 2000. Genetic algorithm-based clustering technique. *Pattern recognition*, 33(9), 1455–1465.

Mileff, P., Nehéz, K., & Dudra, J. 2015. Accelerated Half-Space Triangle Rasterization. *Acta Polytechnica Hungarica*, 12(7).

Parker, A., & Hamblen, J.O. (1989). Computer algorithms for plagiarism detection. *IEEE Transactions on Education*, 32(2), 94–99.

Pearson, W.R. 1991. Searching protein sequence libraries: comparison of the sensitivity and selectivity of the Smith-Waterman and FASTA algorithms. *Genomics*, 11(3), 635–650.

Smith, T.F., & Waterman, M. S. (1981). Identification of common molecular subsequences. Journal of molecular biology, 147(1), 195–197.

Szabó, N.P., Nehéz, K., Hornyák, O., Piller, I., Deák, C., Hanzelik, P.P., Kutasi, Cs. & Ott, K. 2019. Cluster analysis of core measurements using heterogeneous data sources: An application to complex - Miocene reservoirs. *Journal of Petroleum Science and Engineering*, 575–585.

Suresh, N.C., & Kay, J.M. (Eds.). 2012. *Group technology and cellular manufacturing: a state-of-the-art synthesis of research and practice*. Springer Science & Business Media.

Tamás, P. (2017). Decision Support Simulation Method for Process Improvement of Intermittent Production Systems. *Applied Sciences*, 7(9), 950.

Wang, J., & Kusiak, A. 2000. *Computational intelligence in manufacturing handbook*. CRC Press.

Environmental assessment of buildings in Sudan

S.A.A. Ismail & Zs. Szalay
Department of Construction Materials and Technologies, Budapest University of Technology and Economics, Budapest, Hungary

ABSTRACT: Buildings are responsible for 40% of the greenhouse gas emission in the world. Sustainability in building design and construction is an ideal solution to mitigate climate change and solve housing shortage issues. Due to the high costs of the building industry in Sudan compared to the low GDP per capita, sustainable building materials and technology must be applied. The main objective of the research paper is to provide an overview of the sustainability concept and apply the method of Life Cycle Assessment for analysing building construction and materials in Sudan. The environmental impact of two typical buildings is compared. The first case study represents a sustainable, low cost, traditional building technology, while the second one represents a modern building type.

1 INTRODUCTION

About 14% of the world population lives in a slum, without adequate hygiene, sewage systems, using drugs and exposed to abuse (World Housing 2019). Like many other developing countries, Sudan suffers from a dire economic situation. There is a huge gap between the GDP per capita and the cost of building materials. The building materials industry is insufficient both in terms of quantity and quality, due to the lack of infrastructure, financing, production facilities, education, and training programs (Elkhalifa 2011). The sustainability of buildings is an important aspect to deal with both in the developed as well as in the developing world. The aim of this research paper is to introduce the importance of sustainability in building construction, and show how we could quantify it with the help of Life Cycle Assessment (LCA).

A comparison study of the greenhouse gas emissions associated with the 3 building types (traditional, semimodern, and modern) in Nepal, found that traditional buildings release about one-fourth of the greenhouse gas emissions released by semimodern buildings and less than one-fifth of the emissions of modern buildings (Bhochhibhoya et al. 2017). A study presented an environmental impact comparison of four different building structural systems widely used in Central Europe revealed that there is a significant potential for improving the environmental potential of low-rise buildings (Žigart et al. 2018). Bin Marsono and Balasbaneh (2015) compared global warming potential (GWP) for seven combinations of exterior and interior walls with various building construction materials for residential buildings in Malaysia.

1.1 *Sustainability and sustainable development goals*

"Sustainable development is a development that meets the needs of the present without compromising the ability of future generations to meet their own needs." (UN-WCED 1987)

The 17 Sustainable Development Goals (SDGs) of the 2030 Agenda for Sustainable Development adopted by world leaders in 2015 officially came into force in 2016. Many of the SDGs are directly related to buildings, like affordable and clean energy, industry innovation and infrastructure, sustainable cities and communities, responsible consumption and production, and climate action (United Nations 2019).

1.2 Sustainable architecture

"... instead of trying to invent ever more and different types of technologies, we should try to understand why existing technologies are not more widely adopted. Only when we understand this better we will be in a position to know whether it is the materials which are at fault or the method by which it is introduced". (Martin 1983)

Norton (1999) suggests the following criteria for the assessment of sustainable architecture and building technologies: makes substantial use of locally available materials and local means of transport; does not depend on equipment that is not easily available; uses skills that can be realistically developed in the community; can be afforded within the local socio-economic context; produces a durable result; responds to and resists the effects of the local climate; provides flexibility to adapt to local habits and needs; and can be replicated by the local community (Elkhalifa 2011).

1.3 Life cycle assessment and its purpose

Life Cycle Assessment (LCA) is a method to evaluate the environmental effects of a product or process throughout its entire life cycle. An LCA entails examining the product from the extraction of raw materials for the manufacturing process, through the production and use of the item, to its final disposal, and thus encompassing the entire product system (ISO 14040:2006). LCA has proven to be an efficient method for analyzing and comparing buildings and building constructions.

1.4 Goal and structure of the paper

The goal of this paper is to compare the environmental impact of two case study buildings in Sudan with the help of LCA. After a brief overview of Sudan, the two case studies and the methodology are introduced. The first case study is a building made with traditional techniques, while the second one is a typical modern building.

2 SUDAN OVERVIEW

Sudan is located in North-East Africa. It is bounded by the Red Sea and Egypt from the North, Ethiopia, and Eritrea from the East, South Sudan from the South, Central African Republic, Chad and Libya from the West (Figure 1).

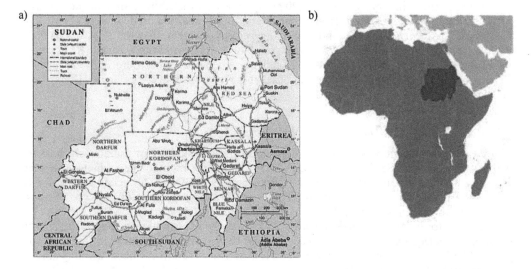

Figure 1.   a) Sudan map.   b) Location of Sudan in Africa,
Source: Food and Agriculture Organization (FAO) (2005).

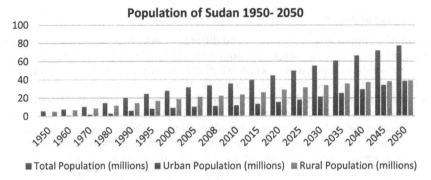

Population of Sudan 1950- 2050

■ Total Population (millions) ■ Urban Population (millions) ■ Rural Population (millions)

Figure 2.   Total, urban and rural population of Sudan 1950 – 2050.
Source: Based on United Nations (2015).

Before 2011 Sudan was the largest country in Africa, but due to the secession of South Sudan, its current area is about 1.886 million km², making it number three in Africa and number 16 in the world (FAO 2005).

2.1   *Population and urbanization in Sudan*

Due to the development process, considerable expansion is expected in the urban population. Urbanization has accelerated due to the war in the South and the drought that hit the Eastern, Western and Kordofan regions during the 1980s.

The urban population has increased from 6.8% of the total population in 1950 to 32.9 in 2008 and estimated to approach 49.8% by 2050 as shown in Figure 2 (United Nations 2015).

2.2   *Availability of raw materials for building material production in Sudan*

Most durable materials used in construction in Sudan are imported at present. There are a number of high-quality raw materials available. However, there is a misuse of natural resources due to the weak economic infrastructure and the enormous size of the country. According to the Ministry of Industry (2005), quarries are distributed all over the country: black cotton soil (semi clay material) (central and South), can be used as brick after treatment; granite (central, North and East); aggregates (central, North, East and West); marble (central, North and East); lime (central, North, East and West); basalt (Khartoum); limestone (north and east); sandstone (Khartoum); quartz (Khartoum and east); stone (Khartoum and east); sand (Khartoum, east, and Kordofan); and aggregates (Khartoum and east).

2.3   *Access to finance and housing supply in Sudan*

The cost of construction is extremely high compared to salaries in Sudan, which range between SDG 1000 (US$22.124) and SDG 4023 (US$89) per month. The exchange rate used for trade by the Central Bank of Sudan (CBoS) is US$1 = SDG45.20). The time needed to raise the funds to build a house ranges from 20 to 83 years depending on the type of house and interest rates, assuming that the saving rate is about 25% of income (CAHF 2018).

According to Sudan's Report for Habitat III 2016, the contribution of public agencies in the supply of adequate housing is limited. Sudanese citizens satisfy their need for housing with great difficulty by using their own savings and transfers from family members working abroad (UN- HABITAT & MEPD 2016).

2.4 Building materials for different residential areas classes

In the building regulation in Sudan, houses are classified according to the specification of construction materials. The residential areas are classified into three categories: 1st and 2nd class areas, 3rd class areas and 4th class areas.

The building materials in Sudan are classified into three types: (a) modern materials: i.e. concrete, red bricks with cement mortar, cement bricks and corrugated iron sheets; (b) traditional permanent materials: red bricks combined with mud bricks for wall construction, mud construction for walls and roofing made from sticks, thatch and mud; and (c) traditional materials: i.e. thatch used for roofing and for walls. In first class areas, residential buildings are made of red bricks with clay or cement mortar, reinforced concrete ceilings, and roofs or corrugated iron sheets for roofing (Elkhalifa 2011, NCR 2003).

3 METHODOLOGY

The method used in this research paper is to measure the environmental impact of two case study buildings using life cycle assessment. The first case study represents an accurate, sustainable, low cost building technology, while the second one represents a modern building type.

3.1 Case studies

3.1.1 The first case study: "earthbag building"

The first case study represents an eco- friendly solution which was constructed in October 2018 in the southern part of Khartoum. This is a new building where traditional earth technology (called earthbags or Superadobe) was used as a construction method. The goal of this project was to revive the old building techniques used in different parts of Sudan in the past. Local earth was utilized mainly with some amount of stabilizer for durable, economical and environmentally friendly building material. The structure was a dome shape (2 domes) of 4- 5 diameter. Plaster was applied to the external and internal parts of the dome (Figure 3).

The construction materials are locally available: earth, sand, lime, cement, barbed wires, and polypropylene bags. The polypropylene bags were filled onsite with the earth mixture

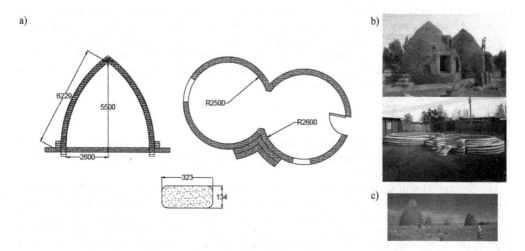

Figure 3. a) Building cross-section, plan and earthbag dome's dimension. b) Photos of the building. c) The ancient building's photo.
Source: Own drawings are based on the information from Home for Sudan (HS4S) (2018), Terra Form (2019), and Updated by the Author's Calculations.

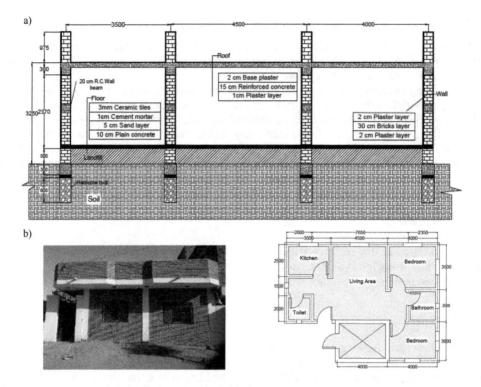

Figure 4.   a) Construction details (cross-section). b) Plan and photo of the building.

(85% earth + 15% lime/cement). The earthbags were laid in vertical rows. Barbed wires were used between earthbags for the shear strength. Windows are single glazed wooden type. The floor area is 41 m^2.

3.1.2 *The second case study: "modern building"*
The second case study is a single-family house with a floor area of 96 m^2. This building is typical for the urban areas in Sudan. Usually, it has a single floor with two or more rooms inside. The roof is wooden or reinforced concrete flat roof. The foundation is a reinforced concrete grade beam on hardcore. The wall is constructed from red clay bricks. Sometimes a thin wall beam is laid in the middle of the wall to support the load in case of adding a new floor in the future. Mortar is used to hold bricks together and also as a plaster layer. Windows are single glazed with aluminum frame (Figure 4).

3.2 *Life cycle assessment*

The assessment process includes identifying and quantifying energy and materials used and wastes released to the environment, assessing their environmental impact and evaluating opportunities for improvement.

  The functional unit of the study is 1 m^2 of building floor area for 50 years. The following life cycle stages were considered: product stage (A1-3), transport (A4), replacement (B4) and end-of-life (C2-C3-C4) (Figure 5). Operational energy use was not considered, as typically these buildings are not equipped with heating nor cooling systems. The reference study period is 50 years. Replacement of plaster assumed during this period. OpenLCA software with the aid of an excel worksheet and ecoinvent v2.2 database was used for the calculation for the impact assessment. Global warming potential (GWP) and cumulative energy demand (CED) was calculated for the buildings.

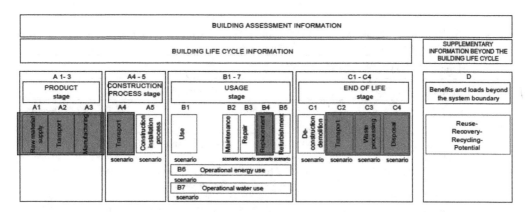

Figure 5. Life cycle assessment process,
Source: EN 15978 (2011).

3.3 *Quantity of materials*

In order to check the environmental impact of a building, the first step is to estimate the type and quantities of the building materials. The bill of quantities was calculated using the collected information (plans, estimations and dome calculator) for both cases.

The calculation of the earthbags mixture materials was done according to the collected information and updated by the author's estimations (HS4S 2018, Terra Form 2019).

Polypropylene bags thickness is measured by mil (1 mil = 0.0254 mm). In 1 m^2 there are 7.5 bags (thickness = 0.134 m) (Terra Form 2019). Barbed wire thickness is 2.5 mm. The total amount of the barbed wire has been calculated using the cross-sectional area of the wire and the different parameters of the dome shape (Just fence 2019).

Table 1 illustrates the amount and types of raw materials (cement, sand, clay, gravel, bricks, concrete, steel, and wood and aluminum for doors and windows for the modern and earthbag building. Based on the author's estimations and different resources, the total cost of the modern building is about SDG 2,100,000 (equivalent to US$ 46,460) while the total cost of the earthbag building is about SDG 350,000 (equivalent to US$ 7,743) (Elkhalifa 2011).

4 RESULTS AND DISCUSSION

As shown in Figure 6a and b, the amount of GWP and CED are higher for 1 m^2 building element in the modern building than in the earthbag building.

According to Figure 7, the largest amount of CO_2 in the modern building and also in the earthbag building is coming from the wall and foundation. In the modern building, the share of the wall and foundation is 54% and 19% in the GWP, respectively. In the earthbag building, the share of the wall and foundation is 61% and 31%, respectively. There is no roof in the earthbag building because of the dome shape, so there is no GWP effect from the roof element (Figure 7).

The total absolute values of the GWP are 113928.97 (kg CO_2-eq) for the modern building and 12468.25 (kg CO_2-eq) for the earthbag building. The floor area of the modern building is larger than the earthbag building, so the direct comparison is misleading.

As shown in Table 1, the modern building contains a larger quantity of materials than the earthbag building. A comparison between the two buildings' materials was done. In the modern building for the wall element, the highest effect is coming from bricks, plaster and concrete, while for the foundation element the highest effect is from bricks, concrete and crushed stone. In the earthbags building the highest effect for the wall element is from clay, cement, and barbed wire, while for the foundation element the highest effect is from clay, cement and gravel (Figure 8).

Table 1. Building materials and type of construction work for the modern and earthbag building, calculated from the plans and construction details (Figures 3, 4).

Building Materials and Type of Construction Work

Item	Modern Building			Earthbag Building		
	Material	Density (kg/m^3)	Quantity (kg)	Material	Density (kg/m^3)	Quantity (kg)
Wall	acrylic dispersion, 65% in H2O, at plant	1700	95,20	base plaster, at plant	2200	1437,98
	base plaster, at plant	2200	10348,80	polypropylene, granulate, at plant **	900	14,23
	brick, at plant	1920	144559,30	Barbed steel wire **	7850	128,65
	cement mortar, at plant	2080	4892,16	portland calcareous cement, at plant	1440	2895,37
	concrete, normal, at plant	2420	31877,86	sand, at mine	1600	5578,42
	reinforcing steel, at plant	7850	1044,09	gravel, crushed, at mine	1800	6,48
				expanded clay, at plant	1089	7593,62
Roof	concrete, normal, at plant	2420	35893,44			0,00
	base plaster, at plant	2200	6652,80			
	reinforcing steel, at plant	7850	822,93			
	lime, hydrated, packed, at plant	560	2688,00			
Foundation	limestone, crushed, washed	1600	47980,80	gravel, round, at mine	2300	6224,41
	concrete, normal, at plant	2420	32061,07	sand, at mine	1600	3633,68
	concrete, sole plate and foundation, at plant	2200	8327,55	gravel, crushed, at mine	1800	1362,63
	brick, at plant	1920	35985,60	expanded clay, at plant	1089	4946,34
	reinforcing steel, at plant	7850	1050,09	Portland calcareous cement, at plant	1440	1885,99
Floor	cement mortar, at plant	2080	1878,24	base plaster, at plant	2200	1505,20
	sand, at mine	1600	7017,60			
	ceramic tiles, at regional storage	2790	734,22			
	concrete, sole plate, and foundation, at plant	2200	19487,60			
	sand, at mine	1600	21052,80			
	gravel, crushed, at mine	1800	15789,60			
	expanded clay, at plant	1089	23881,77			
Windows	aluminum sheet	2739	23.71	sawn timber, hardwood, planed, kiln dried	500	14.26
	flat glass, uncoated, at plant	2530	82,73	flat glass, uncoated, at plant	2530	25,76
Door	door, outer, wood-aluminum, at plant	2739	15,20	door, inner, wood, at plant	500	2,88
Lintel	concrete, normal, at plant	2739	5,97			

** Estimated with the aid of special sources or calculator (Terra Form 2019, MILLER SUPPLY INC 2019, Just fence 2019).

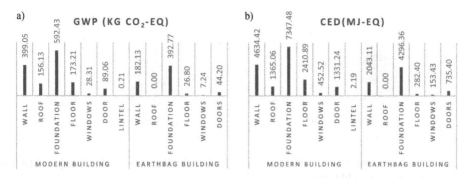

Figure 6.   LCA result for 1 m^2 building element in the modern and earthbag building for:
a) Global Warming Potential. b) Cumulative Energy Demand.

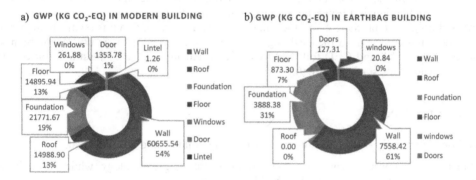

Figure 7.   Share of the building elements in the GWP of the building for the: a) Modern Building.
b) Earthbag Building.

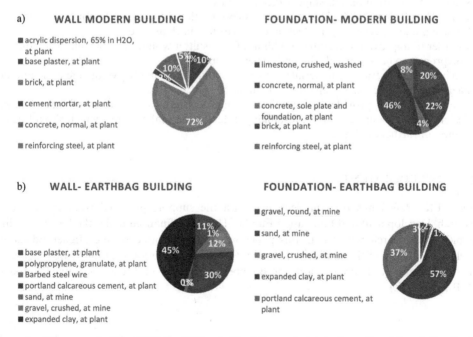

Figure 8.   The amount of the GWP (kg CO2-eq) of building materials in the wall and foundation for the
a) modern building. b) earthbag building.

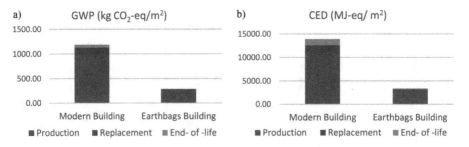

Figure 9. LCA Result (Production, Replacement and End of life) of Modern and Earthbag Buildings for: a) Global Warming Potential b) Cumulative Energy Demand.

Figure 9 presents the GWP and CED of the building through the different life cycle stages for the modern and earthbag buildings for the functional unit. The amount of GWP and CED in the modern building is 4 times that in the earthbag building.

Also, during the replacement and the end of life, the amount of GWP is relatively higher in the modern building than in the earthbag building. The production stage utilizes the highest amount of the GWP and CED followed by the replacement and the smallest amount occurs during the end of life stage (Figure 9).

5 CONCLUSIONS

Based on the analysis of the case studies the following points could be concluded:

- Earthbag building represents an eco-environment building technology, while the environmental impact related to the modern building is higher.
- The amount of GWP and CED per m^2 floor area are 4 times higher in the modern building than in the earthbag building. The highest value of CO_2 is produced during the production stage, concentration on this stage is recommended.
- For both buildings, the highest amount of GWP is related to wall and foundation.
- Some materials have a larger effect on the environment than others.
- In order to improve the earthbag building LCA result, it would be important to modify the polypropylene bags, barbed wire and cement properties.
- Environmentally friendly materials must be sustainable during the whole life cycle and not hazardous to our environment and human health.
- The two buildings provide different living standards (floor area, number of rooms, proper design for bathroom, etc.) and presumably different thermal comfort, which should be the subject of further analysis.

ACKNOWLEDGEMENT

Project FK 128663 has been implemented with the support provided from the National Research, Development and Innovation Fund of Hungary, financed under the FK_18 funding scheme. The research reported in this paper was also supported by the Higher Education Excellence Program of the Ministry of Human Capacities in the frame of the Water sciences & Disaster Prevention research area of the Budapest University of Technology and Economics (BME FIKP-VÍZ).

REFERENCES

Bhochhibhoya. S, Zanetti. M, Pierobon. F, Gatto. P, Maskey. R.K, and Cavalli. R. 2017. The Global Warming Potential of Building Materials: An Application of Life Cycle Analysis in Nepal. *International Mountain Society*. DOI: 10.1659/MRD-JOURNAL-D-15-00043.1.

Bin Marsono A.K., Balasbaneh, A.T. 2015. Combinations of building construction material for residential building for the global warming mitigation for Malaysia. *Construction and Building Materials* (85): p. 100-108. DOI: 10.1016/j.conbuildmat.2015.03.083.

CAHF 2018. *Housing Finance in Africa "A review of Africa's housing finance markets"*. Johannesburg: Centre for Affordable Housing Finance in Africa (CAHF).

Elkhalifa, A.A. 2011. The Construction and Building Materials Industries for Sustainable Development in Developing Countries. Ph.D. Thesis. University of Camerino, Camerino, Italy.

EN 15978: 2011. *Sustainability of construction works - Assessment of environmental performance of buildings - Calculation method*. May. 2019.

FAO 2005. *AQUASTAT Survey-Irrigation in Africa in figures*. Rome: Food and Agriculture Organization (FAO).

HS4S 2018. *Homes for Sudan*, http://homesforsudan.blogspot.com/.

ISO 14040: 2006. *Environmental management – Life Cycle Assessment – Principles and Framework*.

Just fence 2019. *GI barbed wire 12 X 12*. https://www.justfence.in/gi-barbed-wire-12x12.html. May 2019.

Martin, R. 1983. Symbols, Security and Salesmanship as Factors in Technology Rejection. In CIB (Ed.), *Proceedings of International Symposium "Appropriate Building Materials for Low-Cost Housing" held in Nairobi 7-14 November*: pp. 291-301. London: E. & F. N. Spon.

MILLER SUPPLY INC. 2019. *Poly Bag Dimension*, https://www.millersupplyinc.com/poly-bag-dimensions-i-24.html. May 2019.

Ministry of Industry 2005. *Comprehensive Industrial Survey*. Khartoum: Ministry of Industry.

NCR 2003. *The Current Situation of Housing in Khartoum*. (A.M. Mohamed, & A.A. Adam, Eds.) Khartoum: National Center for Research.

Norton, J. 1999. Sustainable Architecture: A Definition, *Habitat Debate*, (5):2.

Terra Form 2019. *Super Adobe Dome Materials Cost Calculator*, http://terra-form.org/tools/earthbagdomecalc.html, May. 2019.

UN-HABITAT & MEPD 2016. *Sudan's Report for United Nations' Third Conference on Housing and Sustainable Urban Development, (Habitat III)*. Development and United Nations Human Settlements Program (UN-HABITAT) - Ministry of Environment, Forestry and Urban Development National Council for Physical Development (MEPD).

United Nations 2015. *World Urbanization Prospects: The 2014 Revision*. New York. https://esa.un.org/unpd/wup/Publications/Files/WUP2014-Report.pdf.

United Nations 2019. *Sustainable Development Goals*, May 2019 (https://sustainabledevelopment.un.org).

UN-WCED 1987. *The Brundtland Report "Our common future"*. p. 43. Oxford: Oxford Press.

World Housing 2019. *A home for everyone*. (https://worldhousing.org).

Žigart M, Lukman RebekaKovačč, Premrov M, Leskovar VesnaŽ. 2018. Environmental impact assessment of building envelope components for low-rise buildings. *Energy*. doi: 10.1016/j.energy.2018.08.149.

Part G: Innovation and Efficiency

Solutions for Sustainable Development – Szita, Jármai & Voith (eds.)
© 2020 Taylor & Francis Group, London, ISBN 978-0-367-42425-1

The innovation shell, and barriers of disruptive innovation

Z. Bartha & A.S. Gubik
University of Miskolc, Miskolc, Hungary

ABSTRACT: If we look at the technological possibilities offered by breakthrough results in computer sciences, materials sciences or biotechnology, we expect robust changes to take place in our everyday life in the near future. According to the expectations, these changes should reshape our society and economy in a major way. The possibilities offered by the breakthrough results, however, may only affect our lives if they are put to practical use. In our study, we investigate the characteristics of the innovation process that creates the bridge between technological/theoretical possibilities, and practical innovations. We detect the main influencing factors that form the so called innovation shell. The innovation shell has three components: the corporate core (made up of ownership interests, managerial motivations, corporate culture and structure, and people); the innovation ecosystem (made up of research and financial infrastructure, and regulating institutions); and finally the values of customers and stakeholders. Some of these factors create barriers others create incentives for innovation. In this study, we focus on the former ones.

1 INTRODUCTION

The following quote by Harari characterizes well the zeitgeist regarding our expectations of the future: "A thousand years ago [...] there were many things people didn't know about the future, but they were nevertheless convinced that the basic features of human society were not going to change. [...]In contrast, today we have no idea how [...] the world will look in 2050. We don't know what people will do for a living, we don't know how armies or bureaucracies will function, and we don't know what gender relations will be like (Harari 2018, pp. 226-227)." Most of those who are at least somewhat interested in future developments expect revolutionary changes that will transform the fundamentals of the society in the very near future.

The idea that our environment changes at an increasing rate is not new. The representativeness bias identified by Tversky and Kahneman (Thaler 2015) makes us believe that current and fresh memories describe the most influential trends. We might feel that the current changes are very fast, and have a deep impact on the society, but the real reason we feel that way maybe because we do not have any memories of past changes. Greenwood et al. (2005) have shown that the weekly housework has decreased from 58 hours in 1900 to 18 in 1975 on the average in the USA thanks to the introduction of new household durable goods. These changes can safely be called revolutionary, but current generations have basically no memories of them.

The ignorance of our minds about past changes seems to be compensated by our tendency to be very imaginative about the future. Starting from 1972 experts on artificial intelligence have been surveyed many times, asking them about the expected amount of time needed to develop a human level artificial intelligence. The median of responses has been constantly fluctuating around 50 years (Bartha&Gubik 2018). Expectations about future changes are generally based on our conception about technological possibilities, but we tend to forget that the pace and direction of technological development are also influenced by layers of the society that are very stable and only change at a very slow rate.

In this study, we focus on the social influencing factors that can reroute and slow down the pace of technological development, and so they have a clear impact on our future as well. These social factors are institutional in nature, meaning that they change very slowly, and the change itself is influenced by the community through the political system and public debates.

2 POSSIBILITIES AND REALITIES

Technological possibilities become a reality through institutions. Institutions affect the inception and diffusion of new ideas. Economics describes this process through the concept of innovation that has many models, but one of the traditional approaches, often attributed to Schumpeter, is the linear model, that is built on the invention-innovation-diffusion triangle (Stoneman 1995). These three steps are often separated by a long time from each other. Fuel cells, for example, have been perceived as a promising technology for more than a century. The slow diffusion of new techno-logies may be due to technological reasons, but it is also influenced by socioeconomic factors featured in Figure 1. This innovation shell summarizes the factors that are generally regarded as innovation bottlenecks in the literature: cost, know-ledge, market, institutional and other barriers (OECD 1997).

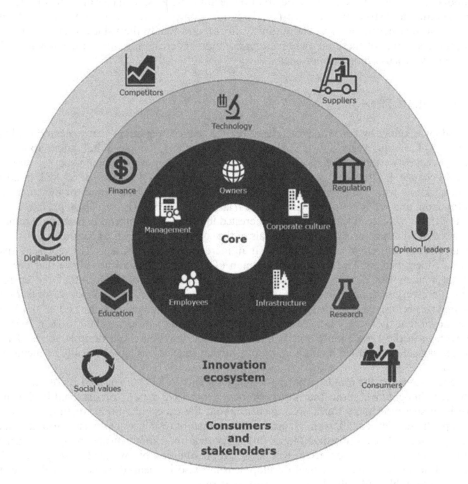

Figure 1.   Innovation shell.
Source: own elaboration.

The corporate core consists of the internal factors that contribute to the conversion of technological possibilities and new ideas into marketable products. Ownerships interests and managerial motivations are crucial since the Schumpeterian creative destruction process requires a long-term commitment. The corporate culture and the organisational structure are components that can both encourage and suppress new ideas. Finally, the role of employees is essential too, as they can be the source of creative ideas, and they are the ones who follow through the innovation process characterised by distortions and inertia.

The corporate core functions as a part of the innovation ecosystem formed around the company. The extent to which the innovation processes of the firm are supported by the ecosystem depends on the quality of the ecosystem, and also on the level of embeddedness of the firm. Organisations doing basic research, and generating new technologies are one of the components of the innovation ecosystem. So are the market and government regulations, as well as the financial infrastructure that funds the innovation process. The incentives of the business/ economic environment influence the creation and introduction of innovation ideas. The capitalist system that is based on the private ownership of businesses, and regulations guaranteeing market competition favours innovation because it creates the incentives to generating a large number of innovative ideas(Erixon&Weigel 2016). A profound analyses also reveal that the strength of the incentives depends on firm size, ownership structure (these two being part of the corporate core), and on the actual rules of market competition (which are part of the innovation ecosystem).

The outer layer of the innovation shell consists of the consumers and other stakeholders. The diffusion and market success of innovation depends on the openness of the market and the consumers, and also on the rules that regulate the marketability of new products and ideas. The diffusion of some ideas can be slow because of the resistance of the community, others may be slowed down by the barriers created by rules and regulations. The latter barriers can also be explained by social uncertainty. When disruptive technologies appear, one typical reaction of the society is to try to limit uncertainties and feared negative consequences through creating prohibitively strict regulations. The regulation lag is also an issue, as the time needed to change the rules can well exceed the pace at which new technologies are developed.

3 RESEARCH AND/OR INNOVATION?

Even though the importance of the innovation process is unquestionable, the methods used to measure progress in this field are extremely limited. One of the common ways to measure innovation performance is to use research and development (R&D) expenditures as a proxy, but there is a fairly large difference between the two concepts. According to the Oslo Manual "An innovation is the implementation of a new or significantly improved product (good or service), or process, a new marketing method, or a new organisational method in business practices, workplace organisation or external relations" (OECD 2005, p 46). This definition reflects the Schumpeterian model.

While statistics on the innovation activity are scarce, R&D statistics are collected and distributed based on a standardised methodology. International and longitudinal comparisons are generally made using the expenditures per GDP indicator. R&D expenditures have slightly increased in the 1985-2015 period, the OECD average rose to 2.36% in 2015 (OECD 2018a). The number of patents has also grown, in 2015 142,714 patterns were filed, which is a 22.2% rise compared to 2000 (OECD.stat).

4 INNOVATION AND INNOVATION

Innovations can occur in many different forms, and this is the primary reason why no standardised measurement method has been developed so far. There is a clear difference between radical (or disruptive) and incremental innovation: incremental innovation is an improvement that is made within the framework of an already existing solution; radical innovation, on the other hand, breaks the old framework (Norman &Verganti 2014).

Most innovations are incremental in nature. They only bring a small amount of novelty but are fairly safe and cost efficient. Modifications made to an already existing product that improve on the quality, or reduce prices represent a typical example to incremental innovation. The larger and the more dominant a firm is, and the bigger control it has over its markets, the more its development strategy is shifted towards incremental innovation. This makes its financial performance more predictable, but it obviously slows down the pace of technological change.

Radical innovation has a number of barriers. One of these barriers is the firm itself, as most organisations strive to be stable and secure. These goals can be achieved if only those innovative ideas get the go ahead sign that fit the competencies, the profile, the manufacturing systems, the marketing methods, and the distribution channels of the firm(Norman, 2010). For these reasons radical innovations are often rejected by investors, but customers can also reject radically new solutions. As a result of these barriers, the vast majority of radical innovation projects fail. The chance of failure in case of radical innovations is 96% (Norman & Verganti 2014).

There is a distinctive ambiguity to the modern innovation activity. On the one hand, statistics measuring innovation performance suggest a constant improvement and an increasing rate of change. On the other hand, radical innovation is extremely scarce. This scarcity means that the speed at which technological possibilities can be transformed into real products is actually very slow. The following sections introduce some of the components of the innovation shell and highlight the barriers that slow down the innovation process.

5 THE CORPORATE CORE

The corporate core consists of components that can be directly influenced by the innovator, but the socioeconomic effects impact the core as well through trending business models and patterns. Narrow framing within the organisation in order to limit risk.

or dispersed ownership are examples to such trends.

5.1 *Ownership structure*

While the majority of innovation efforts is attributed to large corporations (around two thirds of the R&D expenditure is spent by large enterprises in the OECD – OECD 2017a), the ownership structure of these firms has undergone major changes in the last few decades. Managerial and ownership tasks were separated as the share of financial investors as well as the company size has been rising. Managers and owners have different target functions, and the level and quality of information they have access to is also different. Institutional investors and small shareholders typically treat their ownership stake as a financial asset, and so their main expectation towards the firms they own is stable and predictable business operation. Part of the predictability is the steady and calculated rise in firm revenues and the predictability of the return on investments. If these indicators are stable, it is likely that share values and dividends paid will be predictable as well. The latter two factors determine the return rate of the company as a financial asset – the only indicator that institutional investors and small shareholders are really interested in.

While an increasing share of owners prefers companies with predictable financial results, radical innovation involves a great amount of uncertainty, meaning that companies involved in radical innovation projects cannot have predictable financial results. "Real" capitalist owners, the ones who have a clear picture in their minds about the firm's vision, are happy to take risks and embrace uncertainty, but the number and share of these owners is on the decline. This is how the ownership structure can act as a barrier to radical innovation(Erixon&Weigel 2016).

5.2 Management targets

As owners prefer predictability, managers are overly cautious in development decision. The most common approach taken by managers is described as narrow framing by Thaler (2015). Narrow framing is a decision making method that is rooted in excessive risk avoidance and a strong present bias. Managers using narrow framing break up the development efforts of the firm into different projects. The projects are treated and evaluated independently. Let's assume that the firm has 10 quite risky projects (the chance of failure is around 50%), but they all have positive expected values! The rational choice would be to undertake all projects as the positive expected value would increase the overall value and profits of the company. Narrow framing, the fact that all projects have different leaders, and are evaluated individually, leads to the rejection of most of these projects.

A survey conducted by Innovation Leader and KPMG (2018) among 270 managers, confirms that narrow framing is indeed fairly common in development projects. The position of people responsible for innovation projects is also characterised by a high level of insecurity. 70% of the respondents agreed that support coming from top management is the most important incentive, but even if that is provided, intra-organisational struggle for corporate resources is very common. There are some specific characteristics based on firm size, type of innovation, and industry. Some firms have a dedicated department for innovation (this is typical in the pharmaceutical industry in case of radical innovations). In other cases, the business units are involved in the development process directly (typical for incremental innovation). A common barrier is that the innovation process is not assigned to the proper unit or department, and there is no clear division of labour and responsibilities among the different units.

5.3 Incentives of the corporate culture

The IL-KPMG survey (2018) revealed that managers believe that cultural problems are the second biggest barrier in innovation processes (after intra-organisational struggles). Large firms that have a stable market position developed a corporate culture that discourages from risk taking, and according to many respondents, corporate culture specifically punishes and stigmatises failure. The larger the firm, the bigger a challenge it is to create incentives for risk taking, and to encourage people to adopt an entrepreneurial mind-set in general. It is no surprise that few are willing to take on risky projects in such an environment.

Most companies do not have a formal compensation plan for innovation projects. Around 50% of the respondents (IL-KPMG 2018) reported that there is some sort of compensation scheme available (that can take many forms, e.g. financial incentives, rewarding contributions to projects, corporate awards etc.), but 35% claimed that there is no incentives in place at all. No wonder employees often feel that they do not receive sufficient support for their efforts. Shortcomings were reported in the area of trying/testing new ideas, and tolerating mistakes, too (Accenture 2013).

5.4 Employees

People are just as important for the success of the innovation process, as physical resources or incentives. Surveys actually reveal that the main barrier of innovation is related to the human resources and culture. IBM compiled a report about the future of corporations that is based on 1130 interviews. The top three barriers identified were the following: 1) lack of supporting corporate culture; 2) financial limitations; and 3) problems with human resources (IBM 2008).

The attitude of employees is obviously crucial for the success of the innovation project. Innovation may require additional training for the employees, could lead to higher performance requirements, and in some cases can lead to layoffs as well. These are all possible explanation to why employees may not be enthusiastic about innovation. A survey conducted among 2500 German firms showed that employees are especially hostile towards innovation projects that 1) endanger jobs, 2) have high adaptation costs, and 3) increase the work load. The level of hostility is also dependent on the corporate strategy, innovation goal, the external opportunities of employees, and the market position of the firm. The survey ultimately concludes that employee resistance is larger against innovation projects targeted at enhancing the performance of the employees, than projects undertaken in order to increase customer satisfaction (Zwick 2000).

6 INNOVATION ECOSYSTEM

The innovation ecosystem is made up of elements related to the environment of the firm. The ecosystem may be actively influenced by the firms. The ecosystem approach is based on the older idea of regional innovation systems (Cooke et al. 1998), and it is widely accepted by experts that it is a primary influencer of the innovation process (see the summary prepared by Rücker Schaeffer 2018). We highlight three components of the ecosystem in this study.

6.1 *Technology and research*

Technological possibilities are created by organisations specialised in basic research (universities, research agencies). Several megatrends affect these organisations. The financing of universities has been changing for the past decades, and these changes represent a major threat for most of them. Financing uncertainties are coupled with negative demographic trends in the West, and these trends have led to an increasing globalisation in the higher education sector. Globalisation also means that basic research is often conducted in international networks.

Concentration is also a dominant trend is basic research. Big Data, the processing of extremely large databases with algorithms using machine learning, has been gaining ground in basic research projects (OECD 2016, 2018b). Big Data is highly resource intensive: there are only a limited number of large databases on the one hand, and the processing of the data requires a lot of processing power. But Big Data has an impact on businesses as well: digitally mature companies are also more innovative (Shawn 2018a). One of the significant effects of these megatrends is that knowledge is increasingly concentrated, and so it creates a geographical barrier to innovation in many regions.

6.2 *Regulations*

In some markets idea phrasing and devoted resources influence the pace of innovation; in other markets the regulatory environment is backward, most stakeholders have a vested interest in sustaining the status quo, and new technologies can only trickle down very slowly. The education sector is a good example of such a market: teaching methods and materials usually lag way behind the state of the art technology.

In health care, the centralised nature of both the financing system and the service providers is the barrier that slows down the diffusion of new technologies. The development of a new drug cost \$1.3 billion on average (Avik 2012), and the high development costs discourage innovation on the one hand and lead to a further market concentration on the other hand. Clinical tests and licensing regulations put a huge time cost on the development process. The time required to introduce a new drug to the market after filing the first patent application is 15.1 years on the average (Gingrich 2013).

The transport market is also an area where regulations significantly influence innovation. Self-driving cars are expected to bring revolutionary changes in this field, the number of accident could be reduced by 90% (Litman 2018). They could also reduce the time cost of commuting, the stress level, and the emission levels. The resistance of some stakeholders (transport service businesses, car manufacturers), and the uncertainty around responsibilities, the lack of proper regulatory background may delay the introduction of the technology until 2030, and it could be as late as 2050 when self-driving cars are widely used (Litman 2018).

The regulatory system has been becoming more complex due to natural needs, which leads directly to the outer layer of the innovation shell. Before we proceed, we highlight two further points.

- As people naturally want to avoid uncertain situations, we tend to create more and more rules to increase the predictability of human interactions in society. This increases the comfort level in the short term but creates barriers to radical innovation in the long term.
- In certain cases, experts call for global regulations. Technologies like artificial intelligence could potentially create unforeseen threats of large magnitudes. In an extreme scenario, they could create damages that could not be put right by ex post regulation. In such cases setting the principles up ex ante could be especially important (Tegmark 2017).

6.3 *Financing*

The megatrends discussed in the previous sections have changed the R&D and innovation funding priorities. Social challenges that go beyond the traditional growth and employment goals, security challenges created by the new technologies have led to new innovations objectives, such as data privacy, access to data and data sharing, creating incentives for cooperation and encouraging open innovation in general, and updating the framework for competition and intellectual property rights (OECD 2018b).

Starting from 2010-11 public funding of R&D projects (in per GDP terms) has been decreasing throughout the OECD (OECD 2018b), and the use of tax incentives has been on the rise (OECD 2017b).

7 CONSUMERS AND SOCIAL VALUES

Customers and other stakeholders form the outer layer of the innovation shell. Considerations regarding the role of the human factor were already discussed in case of the employees; this time we concentrate on the attitude of consumers and stakeholders toward market innovations. When new services or products are introduced to the market, the key to success is whether or not customers will understand and accept them. This is especially important in radical innovations, because destructive innovations can change existing habits, and can change beliefs as well. Resistance to change is part of human nature (Gatington-Robertson 1989); its extent is influenced by factors related to the customers and to the innovation. The personality, attitudes, value orientation, previous experience with innovation, perception and motivation are customer-related factors (Ram 1987). Relative advantage, compatibility, risk, complexity, and expectations about a better product are innovation-related factors.

Customer decisions are influenced by switching costs: these costs do not have to be paid if the old solution is chosen instead of the new one. A new word processor or a new keyboard layout may lead to more efficiency in the long run, but they increase the switching cost in the short term (Shapiro-Varian 1999). Innovations that create short term costs, and only offer long term benefits typically fail during the market introduction.

8 CONCLUSION

Technological possibilities impact the socioeconomic environment and are introduced to our everyday life through the innovation process. Although the current zeitgeist prompts us to believe that our socioeconomic environment will be fundamentally changed in the next decades, some innovation barriers work against this expectation. More and more money is spent on R&D, the number of patents has been rising, but the positive trends seem to only affect incremental innovation. Incremental innovation only generates slow change; it incrementally introduces new versions of the original product that are somewhat better looking or performing, or a bit more efficient. If we only had incremental innovation during this century, we can tell for sure that our socioeconomic environment will not be fundamentally different 100 years from now.

Innovation that has a destructive effect on our environment is called radical innovation. This latter type of innovation is slowed down by a number of socioeconomic factors. These factors can be identified in all three layers of the innovation shell. Dispersed ownership, growing firm size and hierarchy, and the managerial approach shaped by the values of the institutional investors all contribute to the disappearance of classical innovators/leaders from the top of the corporate hierarchy. If innovator leaders are crowded out, the corporate culture also discourages from taking risks, and innovators lose out in the intra-organisational struggles fought over corporate resources. Further problems can be created by the attitude and the knowledge level of the employees.

Some elements of the innovation ecosystem can act as a barrier to radical innovation as well. As research becomes increasingly global and concentrated, many regions are left without basic research organisations. While the rules and regulations of the society and the economy make the environment more predictable, and some areas even call for a global system of rules, they also make radical innovation projects extremely expensive and drag them out for a very long time. Deficiencies of the financing infrastructure form another barrier.

Finally, social values, negative attitude towards change is yet again a serious barrier. Employees fear the loss of their jobs and the rise in their workload; customers resists new products because of the high switching costs.

REFERENCES

Accenture 2013. Corporate Innovation Is Within Reach: Nurturing and Enabling an Entrepreneurial Culture. A 2013 study of US companies and their entrepreneurial cultures. http://www.fintech-ecosystem.com/assets/study-corp-innovat-n-entrepreneur-l-culture—accenture-fall-2015.pdf (Retrieved March 4, 2019).

Avik S. & Roy, A. 2012. STIFLING NEW CURES: The True Cost of Lengthy Clinical Drug Trials. Project FDA Report, No. 5 March 2012.

Bartha, Z. & S. Gubik. 2018. Oktatásikihívások a technikaiforradalomtükrében. *Észak-magyarországi-StratégiaiFüzetek*, 15(1),15–29.

Cooke P., Uranga M. & Etxebarria G. 1998. Regional systems of innovation: An evolutionary perspective. Environment and Planning, vol. 30, no. 9, pp.1563–1584.

Erixon, F. & Weigel, B. 2016. *The Innovation Illusion. How so little is created by so many working so hard.* Yale University Press, London.

FuelCellToday 2018. http://www.fuelcelltoday.com/history. (Retrieved March 4, 2019).

Gatignon, H. & Robertson, T.S. 1989. Technology diffusion: An empirical test of competitive effects. *Journal of Marketing*, 53(9): 35–49.

Gingrich, N. 2013. B*reakout. Pioneers of the future, Prison Guards of the Pat and the Epic Battle That Will Decide America's Fate.* Regnery Publishing Inc. Washington.

Greenwood, J., Seshadri, A. & Yorukoglu, Y. 2005. Engines of Liberation. *The Review of Economic Studies*, 72(1),109–133, doi: 10.1111/0034-6527.00326.

Harari, Y.N. 2018. *21 Lessons for the 21st Century.* Spiegel &Grau, New York.

IBM 2008. The enterprise of the future. Life sciences industry edition. IBM GLOBAL CEO STUDY https://www-935.ibm.com/services/us/gbs/bus/pdf/gbe03080-usen-ceo-ls.pdf (Retrieved March 4, 2019).

IL-KPMG 2018. Benchmarking Innovation Impact 2018. https://cdn2.hubspot.net/hubfs/2711843/06_14_2018_FINAL_Linked_BenchmarkingInnovationImpact2018.pdf (Retrieved March 4, 2019).

Litman, T. 2018. Autonomous Vehicle Implementation Predictions Implications for Transport Planning. Victoria Transport Policy Institute https://www.vtpi.org/avip.pdf (Retrieved March 4, 2019).

Norman, D.A. 2010. Technology First, Needs Last. *Challenges to Design Research* XVII.2 March + April 2010.

Norman, D.A. & Verganti, R. 2014. Incremental and radical innovation: Design research versus technology and meaning change. *Design Issues*, 30(1),78–96.

OECD 1997. Oslo Manual. Proposed Guidelines for Collecting and Interpreting Technological Innovation Data. European Commission Eurostat http://www.oecd.org/dataoecd/35/61/2367580.pdf (Retrieved March 4, 2019).

OECD 2005. Oslo Manual, Guidelines for Collecting and Interpreting Innovation Data. (Retrieved October 13, 2018). A joint publication of OECD and Eurostat. DOI: 10.1787/9789264013100-en.

OECD 2015. Frascati Manual Guidelines for Collecting and Reporting Data on Research and Experimental Development OECD Publishing. (Retrieved October 13, 2018). doi: 10.1787/9789264239012-en.

OECD 2016. OECD Science, Technology and Innovation Outlook 2016, OECD Publishing, Paris. (Retrieved October 13, 2018). doi: 10.1787/sti_in_outlook-2016-en.

OECD 2017a. Business R&D, in OECD Science, Technology and Industry Scoreboard 2017: The digital transformation, OECD Publishing, Paris. (Retrieved October 13, 2018). DOI: 10.1787/sti_scoreboard-2017-21-en.

OECD 2017b. OECD Time-Series Estimates of Government Tax Relief for Business R&D Summary report on tax expenditures, 20171 2 http://www.oecd.org/sti/rd-tax-stats-tax-expenditures.pdf (Retrieved October 13, 2018).

OECD 2018a. Gross domestic spending on R&D (indicator). (Retrieved May 7, 2019). doi: 10.1787/d8b068b4-en.

OECD 2018b. OECD Science, Technology and Innovation Outlook 2018 Adapting to Technological and Societal Disruption. OECD Publishing, Paris. (Retrieved May 7, 2019). doi: 10.1787/sti_in_outlook-2018-en.

Ram, S. 1987. A model of innovation resistance. *Advances in Consumer Research*, 14(4): 208–213.

Rücker Schaeffer, P., Fischer, B. &Queiroz, S. 2018. Beyond Education: The Role of Research Universities in Innovation Ecosystems. *Foresight and STI Governance*, vol. 12, no. 2, pp. 50–61. (Retrieved May 7, 2019). DOI: 10.17323/2500-2597.2018.2.50.61.

Sandberg, B. 2011. *Managing and Marketing Radical Innovations*. New York: Routledge.

Shapiro, C. & Varian, H.R. 1999. *Information Rules. A Strategic Guide to the Network Economy*. Harvard Business School Press, Boston.

Shawn, R. 2018a. 2018 TIBCO CXO Innovation Survey. https://www.apexofinnovation.com/wp-content/uploads/2018/12/2018-TIBCO-CXO-survey.pdf (Retrieved March 4, 2019).

Stoneman, P. 1995. *The Handbook of Economics of Innovation and Technological Change*. Blackwell, Cambridge MA.

Tegmark, M. 2017. Life 3.0: *Being Human in the Age of Artificial Intelligence*. Penguin Random House LLC, New York.

Thaler, R.H. 2015. *Misbehaving: The Making of Behavioral Economics*. W.W. Norton & Company, New York.

Zwick, T. 2000. Empirical Determinants of Employee Resistance Against Innovations. Centre for European Economic Research (ZEW), Mannheim.

Solutions for Sustainable Development – Szita, Jármai & Voith (eds.)
© 2020 Taylor & Francis Group, London, ISBN 978-0-367-42425-1

Lifetime analyses of S960M steel grade applying fatigue and fracture mechanical approaches

Z.S. Koncsik

Institute of Materials Science and Technology, University of Miskolc, Miskolc, Hungary

ABSTRACT: The structural integrity analysis of an engineering component is based on information belonging to loading condition of the structure, to properties of engineering materials and their changes due to loading and environment, furthermore to material discontinuities resulting from production and/or operation. Beside this one of the driving intentions in the industry is weight reduction, which can be achieved by applying high strength materials. High strength steels produced by innovative processes, like thermomechanical rolling are increasingly applied in different engineering structures. The application of these metal grades enables longer life time, improved performance and thinner wall thickness, which require less welding activity. Besides these advantageous properties, these steel grades can contain material discontinuities, for example cracks, in their microstructure. Aiming to get information about their response to mechanical cyclic loads and static loads, low cycle fatigue and crack tip opening displacement investigations were carried out.

1 INTRODUCTION

The appearance and propagation of weldable high strength steels is the result of their certain advantageous properties, which is, beside their high strength, owing to their appropriate toughness (Lukács et al. 2018; Lukács 2019). However, their application, propagation is not simple, due to the limited application experience based on their novelty, and in some cases the incomplete knowledge about their properties. These last, the incomplete information about their properties, is nowadays also a researched area (Kuzsella et al. 2012; Kuzsella et al. 2013; Tisza & Lukács 2014; Tisza & Lukács 2015). The producers generally publish the results of standard investigations (chemical composition, tensile- and impact test). There is substantially less information for the structure design engineer to carry out fatigue or fracture mechanical controls (Glodez et al. 2009; Lukács 2005; Dobosy et al. 2018; Lukács et al. 2018). For fatigue control, the standard EUROCODE 3, which deals with production of welded steel structures and which is widely applied, contains information until 460 MPa guaranteed yields strength. The 12. chapter, published in 2007, is relevant up to 700 MPa yield strength, with constraints. The document published by the International Institute of Welding contains fatigue data up to 960 MPa yield strength.

The measures, material characteristics needed for fracture mechanical control of these above mentioned steels in their brittle condition are only in small quantities available, or only for grades with lower strength (Hagedorn & Eckel 1992). The microstructural changes due to welding heating processes and the property changes based on this are even less known in this field. Therefore, it is reasonable to determine the fatigue and fracture mechanical properties of welded joints, for which a good initiation basic is the collection of data about the base material.

The aim of this paper is to characterize the low cycle fatigue (LCF) properties, focused on Manson-Coffin relationship, furthermore to determine and publish the critical crack tip opening displacement (CTOD) parameter of the thermomechanical rolled high strength steel S960M.

Table 1. Chemical composition of the investigated steel grade in wt %, according to the producer's certificate.

C	Si	Mn	P	S	Al	Cr	Ni	Mo	Cu	V	Ti	Nb	N	B
0.084	0.329	1.65	0.011	0.0005	0.038	0.61	0.026	0.29	0.016	0.078	0.014	0.035	0.006	0.0015

Table 2. Tensile properties of the investigated material.

Properties	$R_{p0.2}$ MPa	R_m MPa	A_5 %
Value	1051	1058	16.9

Figure 1.   The microstructure of the investigated S960M steel grade, N = 500x, Acid: 2% HNO_3.

2   INVESTIGATED MATERIAL

The chemical composition of the investigated ALFORM 960M material, produced by Voestalpine Anarbeitung GmbH, is given in Table 1.

The tensile properties of the investigated steel grade according to the producer's certificate are given in Table 2.

The investigated specimens have a 15 mm wall thickness. Aiming to analyze the microstructure of the material longitudinal section was prepared, which is shown in Figure 1. In Figure 1 elongated grains, as a consequence of forming, and microstructure appropriate to fast cooling can be seen.

3   LOW CYCLE FATIGUE TEST

3.1   *Investigation parameters*

From the sheet with a nominal thickness of 15 mm the low cycle fatigue specimens, shown in Figure 2 were machined.

Low cycle fatigue tests (ASTM E606, 2012) were carried out on MTS 810.23 type electrohydraulic materials testing equipment. During fatigue, the equipment was controlled due to computer, applying microconsol type MTS 458.20. Data collection and evaluation was performed by LCF program of computer program group MTS 759.20. For control and measurement of strain 632.13C-20 type axial strain-gouge was applied, with L_e = 10 mm. LCF tests

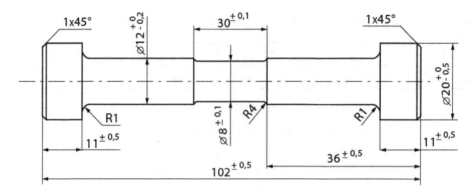

Figure 2.   The geometry of low cycle fatigue specimens.

were carried out at room temperature, on air. The stress ratio of controlled total strain ampli-tude was R = -1, the geometry of loading function was triangle. The test frequency was selected to 0.1 Hz. As damage criteria, the 35% reduction of maximal load at the tensile side was selected. MTS 680.01 type hydraulic grip was applied for gripping the specimens.

The computer program stores the hysteresis loop for the following cycle numbers: the first 10 cycle, following that 20; 30; 40; 50; 60; 70; 80; 90; 100; 200; 300; 400; 500; 600; 700; 800; 900; 1000; 2000; and so on. The appearance of cracks produces new situation, the form of hysteresis curve changes again, therefore it is reasonable to increase the frequency of their records. The pro-gram stores back from the N_t number of cycles to failure the hysteresis curves regarding to cycle numbers of $N_t - 2000$, $N_t - 1000$, $N_t - 900$, $N_t - 800$, ..., $N_t - 200$, $N_t - 100$, $N_t - 90$, $N_t - 80$, ..., $N_t - 20$, $N_t - 10$, $N_t - 9$, $N_t - 8$, ... $N_t - 2$, $N_t - 1$, N_t, with 100-100 cohesive set of points.

As the program stored the necessary data during the investigation, with their application the different measures on certain specimens, and their changes can be illustrated on diagrams. Following the fatigue test of each specimen the stored data was analyzed. The most important conclusions are as follows:

- During the investigations the change of controlled variable, so the maximum and minimum values of total strain amplitude in accordance with cycle number was smaller, then 1%, which is demanded in standards as testing uncertainty.
- The maximum/minimum value of stress amplitude depending on cycle number was during the first some cycles slightly increasing/decreasing, following that, until damage it was monotonic decreasing/increasing.
- The total strain – time diagrams belonging to the representative hysteresis curves followed almost perfectly the selected triangle function.
- On the representative hysteresis curves, no signs of buckling could be observed. On the hys-teresis curves, just before damage of specimens tested with larger strain amplitude, there could be definitely seen the inflexion on pressed side denoted to presence of crack.

The data needed for illustration of diagrams are available in digital form.

3.2   Results of low cycle fatigue tests

The macroscopic determination of characteristics during low cycle fatigue of different materials is crucially carried out by empirical correlation. One part of them determine connection between plastic strain amplitude (plastic strain range) and number of cycles to failure. Nowadays, besides that the model was first announced in 1954, the Manson-Coffin empirical equation is the most widely applied by low cycle fatigue tests at room temperature (Czoboly et al. 1984):

$$\varepsilon_{ap} = \varepsilon_f' \cdot (N_t)^c \tag{1}$$

where:
ε_{ap}: plastic strain amplitude;
N_t: number of cycles to failure;
ε_f', c: material constants determined from the investigation.

Illustrating the data in double logarithmic scale it will give a straight line. Beside the measured values the data concerning elastic strain are also added. Since in case of elastic strain Hooke's-law is valid, there exists the following connection between stress amplitude and flexible strain amplitude:

$$\varepsilon_{ae} = \frac{\sigma_a}{E} \tag{2}$$

where:
ε_{ae}: elastic strain amplitude;
σ_a: stress amplitude;
E: modulus of elasticity, GPa.

The connection between elastic strain amplitude and number of cycles to failure can be given as follows:

$$\varepsilon_{ae} = \frac{\sigma_f'}{E} \cdot N_t^b \tag{3}$$

where σ_f' and b are material constants determined from investigation results.

The total strain amplitude can be determined as a sum of the elastic and plastic strain amplitude:

$$\varepsilon_a = \varepsilon_{ae} + \varepsilon_{ap} = \frac{\sigma_f'}{E} \cdot N_t^b + \varepsilon_f' \cdot N_t^c \tag{4}$$

In practice the value of elastic strain amplitude, which is relatively small and its determination from the measured hysteresis curves is inexplicit, will be calculated from the Equation (2), while the plastic strain amplitude from the following equation

$$\varepsilon_{ap} = \varepsilon_a - \varepsilon_{ae} \tag{5}$$

The results of low cycle fatigue give also information about fatigue hardening/softening of the material. Presenting the stress amplitude in accordance with cycle number applying constant total strain amplitude for fatigue tests the hardening/softening curve can be observed. Since this graph is changing during fatigue test, the connection between stress and plastic strain amplitude is determined, based on the agreement, at 50% of stress amplitudes of the life time, which is called cyclic yield curve. Based on experience the measured values can be narrowed with the following equation:

$$\sigma_{a50} = K \cdot \varepsilon_{ap}^n \tag{6}$$

where
σ_{a50}: the value of stress amplitude at the 50% of life time;
K and n: material contants determined during the investigation.

In case of every specimen tested with different total strain amplitude until damage the following data were collected:

- number of cycles to failure: N_t;
- stress amplitude at the 50% of life time: σ_{a50};

Table 3. Results of low cycle fatigue tests.

Specimen Id.	Number of cycles to failure N_t, cycle	Stress amplitude σ_a, MPa	Total strain amplitude ε_a, -	Flexible strain amplitude ε_{ae}, -	Plastic strain amplitude ε_{ap}, -
M24	548	784.6	0.008	0.003886	0.004114
M25	959	756.7	0.0075	0.003717	0.003783
M33	1008	763.7	0.0065	0.003770	0.002730
M26	1136	732.2	0.0060	0.003591	0.002409
M23	2046	713.9	0.0055	0.003512	0.001988
M30	3099	707.8	0.0045	0.003454	0.001046
M28	3731	710.7	0.0045	0.003474	0.001026
M29	4206	699.9	0.0040	0.003424	0.000576
M27	6939	699.9	0.0037	0.003411	0.000289

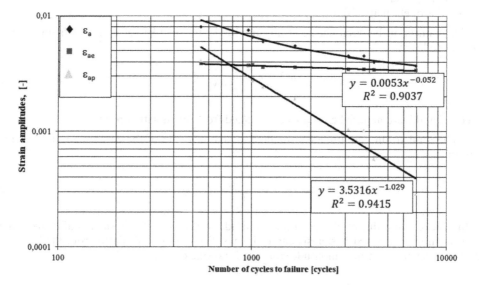

Figure 3. Connection between strain amplitude – number of cycles to failure.

- elastic strain amplitude: ε_{ae};
- plastic strain amplitude: ε_{ap}.

Furthermore, all the controlled total strain amplitudes (ε_a) were available. All data are presented in Table 3.

Based on the results given in Table 3 the connection between strain amplitudes and number of cycles to failure (Figure 3.), and plastic strain amplitudes – stress amplitudes (Figure 4.) are illustrated. In Figure 3 the connection between elastic strain amplitude – number of cycles to failure, and plastic strain amplitude –number of cycles to failure are also illustrated and given with the appropriate parameters of the connection and the correlation index square, which can prove the closeness of the approaching.

The figure illustrates well, that both functions fit appropriate to the measured values, which is also proven by the correlation index of 95.06% and 97.03%, respectively. The third line is approached by the equation (5).

Figure 4 presents the connection between plastic strain amplitude – stress amplitude. On this diagram is also available the equation describing the relationship between the data and the square of correlation index.

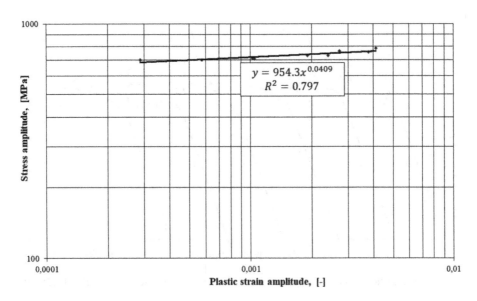

$$y = 954.3x^{0.0409}$$
$$R^2 = 0.797$$

Figure 4. Relationship between plastic strain amplitude – stress amplitude.

The approaching function fits decently to the measured values, however the correlation index value is lower, 89.27%.

Based on the investigation it can be stated, that it would be reasonable to complete the test range with some more specimens. However, the available results already initiate the possibility for designers to determine the low cycle fatigue resistance of different welded joints, and to determine the different behavior compared to the base material.

4 FRACTURE MECHANICAL INVESTIGATION

4.1 *Investigation parameters*

Advanced high strength steels have the advantage, that they possess also high toughness. Common fracture toughness values are not simple to determine, due to their ductile behavior (Zhu & Joyce 2012). Aiming to determine the crack tip opening displacement character of the investigated high strength steel 5-5 three-point bending specimen were worked out, with L-T and T-L orientation. The CTOD/δ_c value is depending on the specimen geometry. Specimens were machined according to the standard (ISO 12135, 2016). The initial nominal dimensions of the specimens are given in Table 4.

The notched specimens were precracked by fatigue with R = 0.1 asymmetry factor. The initial maximum load was F_{max} = 8560 N, which was gradually reduced as the crack arose and started to propagate to the smallest value of F_{min} = 3230 N. The fatigue precracking and later the fracture mechanical tests were carried out on MTS 810.23 type electro-hydraulic testing machine.

Completing the fatigue precracking, the length of cracks on specimen surface changed between a = 12.73 – 12.92 mm. The real crack lengths were measured after opening the specimens according to the requirements of the standard (ISO 12135, 2016).

Table 4. Initial dimensions of three-point bending specimens.

Thickness, B	Width, W	Support distance, S	Notch depth, a
mm	mm	mm	mm
13	26	104	11.46 – 11.52

321

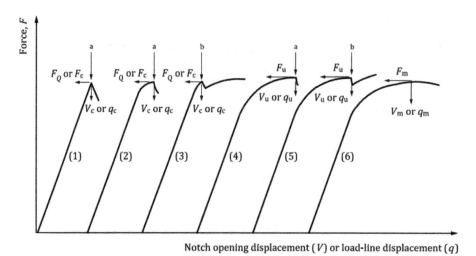

Figure 5. Characteristic types of force versus displacement records in fracture tests (ISO 12135, 2016).

Aiming to determine the critical crack tip opening displacement (CTOD/δ_c) load-displacement diagrams were recorded during the static three-point bending tests. Different types of diagrams could be observed, according to the standard (ISO 12135, 2016), see Figure 5. To measure the opening of the crack mouth displacement gauge type MTS 632.02C was applied with measurement range of ± 1,5 mm. For the investigation own program was applied in the frame program of MTS TestWare, the controlled variable was the piston displacement. During investigation the piston displacement and crack tip opening data were registered. Drawing diagrams from the data in most cases a sudden springing, the so-called pop-in could be observed.

After loading, the specimens were marked by heat tinting in a chamber at 250 °C for an hour. Following that, they were broken open and the fracture surface was examined in order to determine the original crack length a_0, and any stable crack extension Δa that may have occurred during the test.

4.2 Determination of fracture toughness in terms of δ

Analyzing the recorded load-notch opening displacement diagrams two different types of diagrams could be observed, in Figure 5 they are type (2) and type (6). In case of diagram type (2) the F and V values were taken at fracture, in case of type (6) F and V values were taken when the test record exhibited a maximum force plateau prior to fracture with no significant pop-ins. In the first case, F_c and V_c values are recorded, while in the last case F_m and V_m values.

δ_0 value is calculated from the following relationships:

$$\delta_0 = \left[\frac{S}{W} \cdot \frac{F}{(BB_N W)^{0.5}} \cdot g_1 \left(\frac{a_0}{W} \right) \right]^2 \cdot \frac{1 - \nu^2}{2R_{p0.2}E} + \frac{0.4(W - a_0)}{0.6a_0 + 0.4W + z} \cdot V_p \tag{7}$$

where
F: applied load, N;
B_N: specimen net thickness between side grooves, mm;
g_1: geometry function given in standard (ISO12135, 2016);
ν: Poisson's ratio;
$R_{p0.2}$: yield strength, MPa;
z: the initial distance of the notch opening gauge measurement position from the notched edge of the specimen, mm;
V_p: plastic component of V, mm.

Table 5. CTOD results

Specimen Id.	Pop-in information	Δa measured mm	Δa calculated mm	F_{max} N	Type of F and V	Type of δ at B = 13mm mm	Value of δ_0 mm
M11	no pop-in.	0.503	0.303	29402.9	F_m and V_m	δ_m	0.20
M12	no pop-in	0.264	0.283	30102.5	F_m and V_m	δ_m	0.16
M13	2. pop-in	0.131	0.258	29908	F_c and V_c	δ_c	0.11
M21	1.pop-in	0.166	0.272	27400.5	F_c and V_c	δ_c	0.14
M23	no pop-in	0.262	0.317	27805	F_m and V_m	δ_m	0.23
M24	no pop-in	0.613	0.348	29103	F_m and V_m	δ_m	0.29
M25	2. pop-in	0.133	0.280	29101	F_c and V_c	δ_c	0.16

When F and V correspond to crack instability after stable crack extension, so that

$$\Delta a < 0.2 \ mm + \frac{\delta}{1.87} \cdot \frac{R_{p0.2}}{R_m} \tag{8}$$

they shall be recorded as values F_c and V_c.

When F and V correspond to crack instability after stable crack extension, so that

$$\Delta a \geq 0.2 \ mm + \frac{\delta}{1.87} \cdot \frac{R_{p0.2}}{R_m} \tag{9}$$

they shall be recorded as F_u and V_u.

Qualifying the δ_0 fracture toughness data conformed to the standard (ISO 12135, 2016), 3 results for L-T direction (M11, M12, M13) and 4 results for T-L direction (M21, M23, M24, M25) could be analyzed. In 4 cases there are no pop-in phenomena on the diagram, in 1 case the first pop-in is significant, while in two cases the second pop-in is significant. The results of fracture mechanical investigation and the qualification of δ_0 fracture toughness value is given in Table 5.

The requirements concerning to initial crack length, crack and notch relationship, surface roughness, measuring equipment, precracking conditions (ISO 12135, 2016) were observed, measured and controlled before, during and after the investigation, respectively.

According to Table 5, δ_c values can be observed, which belong to brittle behavior. In cases, where δ_0 has a higher value, so it becomes δ_m, the specimens show more plastic behavior, especially in the vicinity of crack tip.

5 CONCLUSION

In this paper the application possibility of a thermomechanical rolled high strength steel is introduced, with the presentation of two, not common, like yield strength, designing and controlling parameters. In case of high loads, furthermore on stress concentrating areas, the parameters of Manson-Coffin relationship are determined, which is necessary for lifetime estimation of structures loaded by low cycle fatigue.

In addition, results of crack tip opening displacement measurement are presented for the control of these steel grade, operating in brittle fracture environment, at room temperature.

In both areas, the damage of welded joints and especially their microstructures have to be considered. For the control and lifetime prediction of these welded structures and their base material both, the low cycle fatigue parameters and the crack tip opening displacement data are essential. Aiming to receive reliable results, further investigations are necessary, focused on inhomogeneous microstructures, as for example welded joints.

The described article was carried out as part of the EFOP-3.6.1-16-2016-00011 *"Younger and Renewing University – Innovative Knowledge City – institutional development of the University of Miskolc aiming at intelligent specialisation"* project implemented in the framework of the Széchenyi 2020 program. The realization of this project is supported by the European Union, co-financed by the European Social Fund.

REFERENCES

Czoboly, E., Ginsztler, J. & Havas, I. 1984, Ismeretek a kisciklusú és termikus fáradásról. *GÉP*, 36: 241–253. ISSN 0016-8572

Dobosy, Á., Gáspár, M. & Lukács, J. 2018, The Influence of Mismatch Effect on the High Cycle Fatigue Resistance of High Strength Steel Welded Joints. *Advanced Materials Research* 1146: 73–83. doi: 10.4028/www.scientific.net/AMR.1146.73

Glodez, S., Knez, M., Jezernik, N. & Kramberger, J. 2009, Fatigue and fracture behavior of high strength steel S1100Q, *Engineering Failure Analysis* 16: 2348–2356, doi:10.1016/j.engfailanal.2009.03.023

Hagedorn, K.E. & Eckel, M. 1992, Investigations in fracture mechanics failure concepts with thermomechanically rolled steels, *Nuclear Engineering and Design*, 137: 343–353.

International Standard ISO 12135, 2016, Metallic materials — Unified method of test for the determination of quasistatic fracture toughness.

Kuzsella, L., Lukács, J. & Szűcs, K. 2012, Fizikai szimulációval végzett vizsgálatok S960QL jelű, nagyszilárdságú acélon; Gépipari Tudományos Egyesület, *GÉP* LXIII.: 35–40, ISSN 0016-8572

Kuzsella, L., Lukács, J. & Szűcs, K. 2013, Nil-Strength Temperature and Hot Ten-sile Tests on S960QL High-Strength Low-Alloy Steel; *Production Processes and Systems*, 6: 65–76.

Lukács, J. 2005. Dimensions of lifetime management. In *Materials Science Forum* 473–474: 361–368. doi: 10.4028/www.scientific.net/MSF.473-474.361.

Lukács, J., Dobosy, Á. & Gáspár, M. 2018. Fatigue Crack Propagation Limit Curves for S690QL and S960M High Strength Steels and their Welded Joints. *Advanced Materials Research* 1146: 44–56. doi: 10.4028/www.scientific.net/AMR.1146.44.

Lukács, J. 2019. Fatigue crack propagation limit curves for high strength steels based on two-stage relationship. In *Engineering Failure Analysis* 103: 431–442. doi:10.1016/j.engfailanal.2019.05.012

Standard ASTM E606/E606M – 12, 2012, Standard Test Method for Strain-Controlled Fatigue Testing

Tisza, M. & Lukács, Zs. 2014. Theoretical and experimental investigation of large strain cyclic plastic deformation in dual-phase steels. *Production Processes and Systems* 7: 87–102.

Tisza, M. & Lukács, Zs. 2015. Formability Investigations of High Strength Dual-Phase Steels. *Acta Metallurgica Sinica English Letters* 28: 1471–1481.

Zhu, X-K. & Joyce, J.A. 2012, Review of fracture toughness (G, K, J, CTOD, CTOA) testing and standardization, *Engineering Fracture Mechanics* 85:1–46, doi: 10.1016/j.engfracmech.2012.02.001.

Solutions for Sustainable Development – Szita, Jármai & Voith (eds.)
© 2020 Taylor & Francis Group, London, ISBN 978-0-367-42425-1

Efficient application of S690QL type high strength steel for cyclic loaded welded structures

H.F.H. Mobark & J. Lukács

Institute of Materials Science and Technology, Faculty of Mechanical Engineering and Informatics, University of Miskolc, Miskolc, Hungary

ABSTRACT: The mass reduction of engineering structures is one of the basic trends nowadays, which can be attained by applying high strength steels. During the manufacturing of these structures, the main joining technology is the welding, and gas metal arc welding (GMAW) is the most frequently used fusion welding process. Furthermore, cyclic loads are the most dangerous loads, which occur in case of different engineering structures. Accordingly, the determination and application of different design or limit curves is a global trend for the innovative and economical design and operation of the structures. The purpose of the paper is to present the determination of high cycle fatigue and fatigue crack propagation design or limit curves for S690QL type high strength steels and their GMAW joints, under different matching conditions. The determined design limit curves can be applied for engineering critical assessment and structural integrity calculations of cracked, cyclic loaded structures or structural elements.

1 INTRODUCTION

Protecting the environment (greenhouse gas emissions and more particularly CO_2 emissions) and increasing resource efficiency (including energy) both constitute major transversal issues in the universe of the research technology development (RTD) programs that are proposed by European Steel Technology Platform (ESTEP). Security and safety represent the third very important objective to be addressed, not only in the relevant industries but also in customers' everyday lives as users of steel solutions (surface transport, buildings, energy production, etc.) by developing new intelligent and safer steel solutions (ESTEP 2013, ESTEP 2017). These can be approached from a materials science of view by applying different high strength steels (Peters et al. 2015). In case of different steel structures, the main manufacturing and joining technology is the welding, conventional and advanced methods of fusion and pressure welding processes can be applied (AWS 2019).

Welded parts have been used in the majority of engineering applications (Pijpers et al. 2007, Baluch et al. 2014) such as engineering structures, power generation, onshore and offshore structures, furthermore in vehicle industry (thick plates) and in automotive industry (thin plates). Welded joints are very sensitive parts of the structures because the welded regions are in complex microstructural and stress conditions. Steel and consumable producers currently provide a diversified spectrum of high strength base materials and filler metals. Specific design solutions and economic aspects of modern steel constructions can be found in the practice; extensive weight and production costs reduction can be achieved with increasing the steel strength (Schroepfer & Kannengiesser 2016). High strength structural steels with nominal yield strengths from *690 MPa* upwards are applied in a growing amount of different applications.

During the welding process, the joining parts are affected by heat and mechanical loads, which cause inhomogeneous characteristics. These specialities appear in deflections, which are basically acceptable, or rather in failures, which are basically unacceptable, and influence both the complex behavior and the load-bearing capability of the welded joints (Gáspár & Sisodia

2018, Sisodia et al. 2019, Varbai et al. 2019). Discontinuities in base materials and their welded joints have especially high danger in case of cyclic loading conditions (Balogh et al. 2015), which are frequent in case of different structures and structural elements (e.g. bridges, industrial and transporting pipelines, vehicles).

There are different documents containing both high cycle fatigue and fatigue crack growth limit or design curves and rules for the prediction of the crack propagation (e.g. BS 7910 2015, Berger et al. 2018). The background of the design or limit curves consists of two basic parts: statistical analysis of numerous investigations (high cycle fatigue and fatigue crack propagation tests) and description of behavior of the investigated base materials and their welded joints under different conditions.

Based on the above mentioned facts, the aim of our paper is the determination of limit or design curves for S690QL type base materials and their differently matched welded joints under high cycle fatigue (HCF) and fatigue crack growth (FCG).

2 INVESTIGATIONS

2.1 Base materials and filler metals

30 mm thick RUUKKI Optim 700QL plates and 15 mm thick SSAB Weldox 700E plates were used for the welding, and for the investigations of base materials and welded joints. INE INEFIL NiMoCr, Thyssen UNION X85 and UNION X90 solid wires were used for gas metal arc welding (GMAW), applying M21 type mixed shielding gas with 18% CO_2 + 82% Ar content. The chemical composition of the base materials and the filler metals and the mechanical properties can be seen in Table 1 and Table 2, respectively. The sources of the data are as follows: Optim 700QL, Weldox 700E and Union X85 – inspection certificates of suppliers

Table 1.   Chemical composition of the investigated base materials and used filler metals, wt%.

Material	C	Si	Mn	P	S	Cr	Ni	Mo
RUUKKI Optim 700QL	0.16	0.31	1.01	0.010	0.001	0.61	0.21	0.205
SSAB Weldox 700E	0.14	0.30	1.13	0.007	0.001	0.30	0.04	0.167
INE INEFIL NiMoCr	0.80	0.50	1.60	0.007	0.007	0.30	1.50	0.250
Thyssen Union X85	0.07	0.68	1.62	0.010	0.010	0.29	1.73	0.60
ThyssenUnion X90	0.1	0.8	1.8	N/A	N/A	0.35	2.3	0.6

Material	V	Ti	Cu	Al	Nb	B	N	Zr
RUUKKI Optim 700QL	0.010	0.016	0.015	0.041	0.001	0.0015	0.003	N/A
SSAB Weldox 700E	0.011	0.009	0.01	0.34	0.001	0.002	0.003	N/A
INE INEFIL NiMoCr	0.09	N/A	0.120	N/A	N/A	N/A	N/A	N/A
Thyssen Union X85	< 0.01	0.08	0.06	< 0.01	N/A	N/A	N/A	< 0.01
Thyssen Union X90	N/A	N/A	N/A	N/A	N/A	N/A	N/A	N/A

Table 2.   Mechanical properties of the investigated base materials and used filler metals.

Material	$R_{p0.2}$ MPa	R_m MPa	A %	CVN impact energy J
RUUKKI Optim 700QL	809	850	17.0	-40°C: 106
SSAB Weldox 700E	791	836	17.0	-40°C: 165
INE INEFIL NiMoCr	750	820	19.0	-40°C: 60; -20°C: 90; 20°C: 120
Thyssen Union X85	≥ 790	≥ 880	≥ 16.0	-50°C: ≥ 47; 20°C: ≥ 90
Thyssen Union X90	890	950	15.0	-60°C: 47; 20°C: 90

(EN 10204 – 3.1); INEFIL NiMoCr – typical all weld metal analysis and mechanical properties; Union X90 – typical composition of solid wire, mechanical properties of all weld metal.

2.2 *Welding circumstances and parameters*

Width and length dimensions of the welded workpieces were *300 mm* and *125 mm*, respectively. For the equal stress distribution symmetrical *X*-grooved (double *V*-grooved) welding
joints were used, with *80°* opening angle, *1 mm* root face height, and *2 mm* gap between the
two plates. The welding equipment was a Daihen Varstroj Welbee Inverter P500L (WB-
P500L) MIG/MAG power source; *1.2 mm* diameter solid wires were applied. The applied
strength mismatch characteristics, in other words the base materials and filler metals pairing
can be found in Table 3.

The root layers (*2 layers*) were made by a qualified welder, while the filler layers (*18 layers*
for *30 mm* thick plates or *6 layers* for *15 mm* thick plates) by an automated welding car, in all
cases. During the welding process, a welding monitoring system (HKS 2019) was used, and
the workpieces were rotated systematically, after each layer. Figure 1 shows the structure of
the Optim 700QL welded joints (Dobosy 2017).

The welding parameters were selected based on both theoretical considerations and real
industrial applications (Gáspár & Balogh 2013, Májlinger et al. 2016, Dobosy et al. 2017,
Kalácska et al. 2017). Applying the registered data, the welding parameters for both base
materials and root layers/filler layers were as follows:

Table 3. Mismatch characteristics – base materials and filler metals paring during our investigations.

Base material	Filler metal	Mismatch condition
RUUKKI Optim 700QL	INE INEFIL NiMoCr	matching (M)
SSAB Weldox 700E	Thyssen Union X85	matching (M)
	Thyssen Union X90	overmatching (OM)
	Thyssen Union X85/Union X90	matching/overmatching (M/OM)

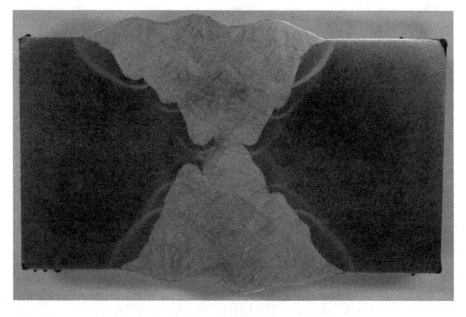

Figure 1.   The structure of the Optim 700QL welded joints (Dobosy 2017).

- preheating temperature: *150°C/180 °C*;
- interpass temperature: *150°C/180 °C*;
- welding current: *130–140 A/280–300 A*;
- voltage: *19.0–20.5 V/28.5–29.5 V*;
- welding speed: *20 cm/min/40 cm/min*;
- linear energy: *700–750 J/mm/1000–1100 J/mm*;
- calculated critical cooling time ($t_{8.5/5}$): *7-8 s/9-11 s*.

2.3   High cycle fatigue (HCF) tests

HCF tests were performed on base materials and on their welded joints, with an MTS 810 type electro-hydraulic testing equipment, at room temperature and in laboratory environment. Flat test specimens and constant load amplitude were applied during the tests, with $R = 0.1$ stress ratio, $f = 30\ Hz$ loading frequency, and sinusoidal loading wave form. Staircase method was used during both the preparation and the evaluation of the HCF test, based on the JSME S 002-1981 prescription (Nakazava & Kodama 1987).

The measured values and the basic lines ("*Mean*") of the determinable HCF design or limit curves for the Optim 700QL and the Weldox 700E base materials (see Mobark, Dobosy & Lukács (2018) too) and their welded joints are presented in Figure 2 and Figure 3, respectively. In Figure 2 and Figure 3 *WJ* means and designates the tested flat specimens cut from the welded joints, *BWJ* means and designates the specimens tested with the whole welded joint, x/y = center line of the specimen/crack growth direction, h = parallel to the rolling direction, k = perpendicular to the rolling direction, v = thickness direction, $1W$ = centerline direction of the welded joint, $3W$ = thickness direction in the welded joint. Flat specimens (*WJ*) were fractured characteristically in the weld metal; while whole butt welded joint specimens (*BWJ*) were fractured both in the weld metal and in the HAZ.

Table 4 summarizes the basic parameters of the determinable HCF design limit curves ("*Mean*" S-N curves), the N_k value is the number of cycles for the break point of the S-N curve, the $\Delta\sigma_D$ is the fatigue limit, and the $\Delta\sigma_{1E07}$ is the stress value belonging to $1*10^7$ cycles in the

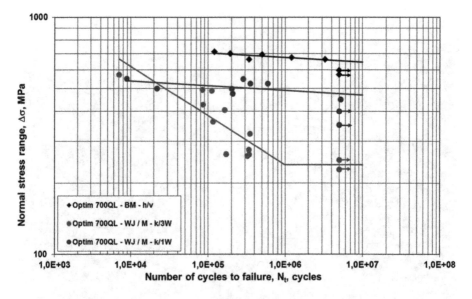

Figure 2.   Measured values and the basic lines ("*Mean*" S-N curves) of the determinable HCF design or limit curves for Optim 700QL base material (BM) and their welded joints (WJ).

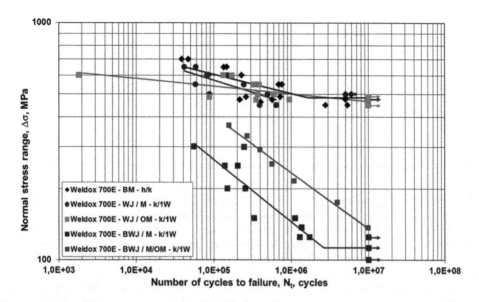

Figure 3. Measured values and the basic lines ("*Mean*" S-N curves) of the determinable HCF design or limit curves for Weldox 700E base material (BM), their welded joints (WJ) and butt welded joints (BWJ).

Table 4. Calculated results of high cycle fatigue (HCF) tests ("*Mean*" S-N curves).

		m	lg(a)	N_k	$\Delta\sigma_D$	$\Delta\sigma_{1E07}$
Base material	Mismatching – orientation	–	–	Cycle	MPa	MPa
Optim 700QL	BM – h/v	51.282	151.109	–	–	646
	WJ/M – k/3W	4.826	17.476	9.893E05	239	–
	WJ/M – k/1W	50.251	141.260	–	–	470
Weldox 700E	BM – h/v	12.453	39.650	1.677E06	483	–
	WJ/M – k/1W	9.960	32.469	–	–	–
	WJ/OM – k/1W	31.250	90.415	–	–	467
	BWJ/M – k/1W	3.831	14.284	2.660E06	113	–
	BWJ/M/OM – k/1W	4.207	15.966	–	–	136

cases, when the horizontal (endurance limit) part of the curves cannot be determined. The used equation, the Basquin equation, is as follows:

$$N * \Delta\sigma^m = a \qquad (1)$$

where N = number of cycles to failure, $\Delta\sigma$ = normal stress range, furthermore m and a are material constants.

2.4 Fatigue crack growth (FCG) tests

The FCG tests were executed on three-point bending (TPB) specimens, nominal W values were *26 mm* ($t = 15$ *mm*), and *28 mm* ($t = 30$ *mm*) and *13 mm* ($t = 15$ *mm*) for the base materials and the welded joints, in the *21* and *23* directions, respectively. The position of the notches correlated with the rolling direction (T-L, L-T and T-S). The positions of the cut specimens from the welded joints are shown in Figure 4, *21* and *23* directions (*21W* and *23W*) were used.

The notch locations, the notch distances from the centerline of the welded joints, were different, therefore the positions of the notches and the crack paths represent the most important

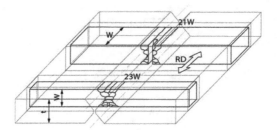

Figure 4. TPB specimen locations in the welded joints with the notch directions (*RD* = rolling direction).

and the most typical crack directions in a real welded joins. Post-weld heat treating was not applied after welding on GMAW joints (investigations in as-welded condition).

The FCG tests were performed with tensile stress, $R = 0.1$ stress ratio, sinusoidal loading wave form, at room temperature, and in laboratory environment, using MTS 312 type electro-hydraulic testing equipment. The loading frequency was $f = 20\ Hz$ for two-thirds of the growing crack's length, approximately, and it was $f = 5\ Hz$ for the last third. The propagating crack was registered with optical method, using video camera and hundredfold magnification.

Figures 5–6 and Table 5 shows selected experimental results on Optim 700QL welded joints. The crack length vs. the number of cycles curves can be seen in Figure 5, and the calculated stress intensity factor range vs. fatigue crack growth rate values are shown in Figure 6.

Secant method (ASTM E 647) was used to evaluate the fatigue crack growth data, using Paris-Erdogan (Paris & Erdogan 1963) relationship:

$$\frac{da}{dN} = C * \Delta K^n \qquad (2)$$

where da/dN = fatigue crack growth rate, ΔK = stress intensity factor range, furthermore C and n are material constants. The constants were calculated using the least squares regression method and the fatigue fracture toughness (ΔK_{fc}) values were determined using the crack length on the crack front measured by stereo microscope. The data belonging to stage II of the kinetic diagram of fatigue crack propagation have been eliminated during the least square

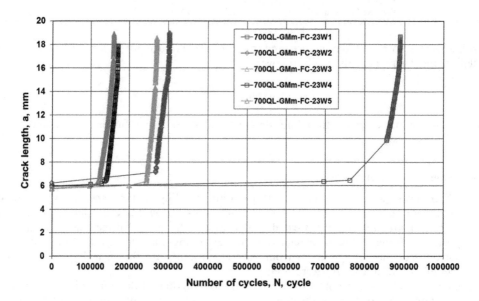

Figure 5. Crack length vs. number of cycles curves (T-S/23W, Optim 700QL/INEFIL NiMoCr).

330

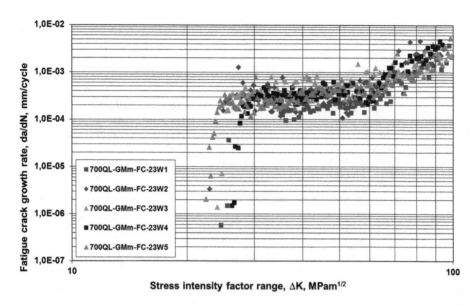

Figure 6.   Results of FCG tests on Optim 700QL/INEFIL NiMoCr welded joints (altogether 5 specimens).

Table 5.   Fatigue crack propagation test results executed on specimens cut from Optim 700QL/INEFIL NiMoCr welded joints.

Specimen ID: short and long		C	n	Correlation coefficient	ΔK_{fc}
		$MPam^{1/2}$, mm/cycle	–		$MPam^{1/2}$
6221	700QL-GMm-FC-23W1	1.260E-08	2.534	0.8666	99.32
6222	700QL-GMm-FC-23W2	7.993E-07	1.647	0.6938	94.54
6223	700QL-GMm-FC-23W3	1.232E-06	1.592	0.8402	97.42
6224	700QL-GMm-FC-23W4	2.266E-07	1.974	0.8320	93.80
6225	700QL-GMm-FC-23W5	2.573E-06	1.310	0.8190	98.60

regression analysis, for each specimen, systematically. Table 5 summarizes the FCG test results executed on specimens cut from Optim 700QL/INEFIL NiMoCr welded joints.

3   DESIGN OR LIMIT CURVES

3.1   *High cycle fatigue (HCF) design or limit curves*

The "*Mean*" S-N curves (see Figures 2–3) determined based on JSME prescription (Nakazava & Kodama 1987) can be completed with standard deviation (*SD*) values and these curves ("*Mean-2SD*" S-N curves) can be used as design or limit curves. The calculated parameters of these design or limit curves are summarized in Table 6; the "*Mean*" and the "*Mean-2SD*" S-N curves for the Optim 700QL and the Weldox 700E base materials and their welded joints are presented in Figure 7 and Figure 8, respectively.

3.2   *Fatigue crack growth (FCG) design or limit curves*

Kinetic diagram of fatigue crack growth can be simplified and described using both simple and two-stage crack growth relationships (BS 7910 2015). According to the main aim of this paper, which is the determination of design or limit curves, the simple crack growth

Table 6. Parameters of the high cycle fatigue (HCF) design or limit curves ("*Mean-2SD*" S-N curves).

Material grade	Mismatching – orientation	m	lg(a)	N_k	$\Delta\sigma_D$	$\Delta\sigma_{1E07}$
		–	–	cycle	MPa	MPa
Optim 700QL	BM – h/v	51.282	150.186	–	–	620
	WJ/M – k/3W	4.826	16.762	9.893E05	239	–
	WJ/M – k/1W	50.251	138.732	–	–	418
Weldox 700E	BM – h/v	12.453	38.557	1.677E06	483	–
	WJ/M – k/1W	9.960	31.739	–	–	–
	WJ/OM – k/1W	31.250	88.311	–	–	400
	BWJ/M – k/1W	3.831	13.752	2.660E06	113	–
	BWJ/M/OM –k/1W	4.207	15.822	–	–	126

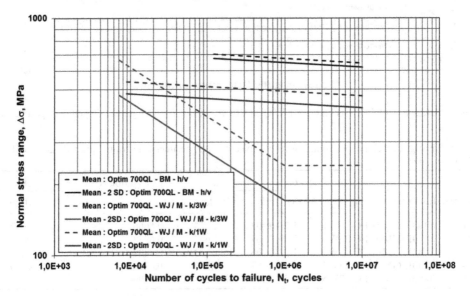

Figure 7. Determined high cycle fatigue (HCF) design or limit curves ("*Mean-2SD*" S-N curves) for Optim 700QL base material (BM) and their welded joints (WJ).

relationship was selected and used. Generally, the determination of the fatigue crack growth design or limit curves consists of six steps, as follows (Lukács 2003):

- determination of measuring values;
- classification of measured values into statistical samples;
- selection of the distribution function;
- calculation of the parameters of the selected three-parameter Weibull-distribution functions using the

$$F(x) = 1 - \exp\left[-\left(\frac{x - N_0}{\beta}\right)^{\frac{1}{\alpha}}\right] \tag{3}$$

equation, where N_0 is the threshold parameter, α is the shape parameter and β is the scale parameter;

- selection of the characteristic values of the distribution functions, which can be seen in Figure 9 schematically;
- calculation of the parameters of the design or limit curves.

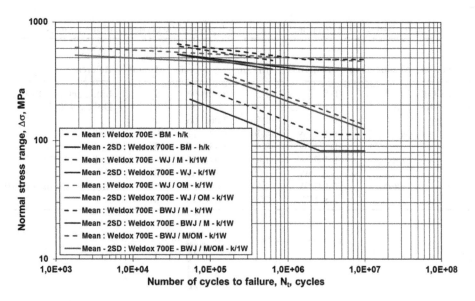

Figure 8. Determined high cycle fatigue (HCF) design or limit curves ("*Mean-2SD*" S-N curves) for Weldox 700E base material (BM), their welded joints (WJ) and butt welded joints (BWJ).

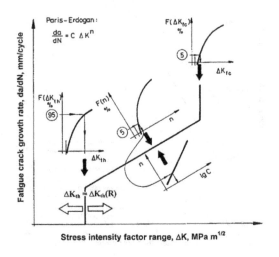

Figure 9. Method for determination of fatigue crack growth (FCG) design or limit curves (Lukács 2003).

Table 7 summarizes the characteristics of the determined fatigue crack growth (FCG) design or limit curves, and Figure 10 shows the curves.

In those cases, when the orientation and/or the path of the propagating crack is known, the values in Table 7 can be directly used. In those cases, when n and ΔK_{fc} values calculated in different directions are significantly different, and the orientation and/or the growing crack path is not known, the lowest value should be considered from the related ones. The unambiguous determination of the design or limit curves in the near threshold region (near ΔK_{th}) is difficult. On the one hand, if the ΔK_{th} value is not known, values from literature are usable; on the other hand, the ΔK_{th} value must be reduced by tensile residual stress field and may be increased by compressive residual stress field.

Table 7. Characteristics of the determined fatigue crack growth (FCG) design or limit curves.

| Base material | Mismatching | Orientation | n | C | ΔK_{fc} | Reference |
				MPam$^{1/2}$, mm/cycle	MPam$^{1/2}$	
Optim 700QL	M	T-S	1.20	6.52E-06	93	Balogh et al. (2015)
Weldox 700E	BM	T-L and L-T	1.70	8.09E-07	101	Dobosy (2017)
		T-S	1.50	2.06E-06	75	
	M	T-L/21W	4.10	1.12E-11	105	Dobosy (2017)
		T-S/23W	2.30	4.93E-08	80	
	OM	T-L/21W	1.85	4.02E-07	96	Dobosy (2017)
		T-S/23W	1.90	3.19E-07	61	
	M/OM	T-L/21W	2.67	8.88E-09	90	Mobark & Lukács (2019)
		T-S/23W	2.85	3.87E-09	67	

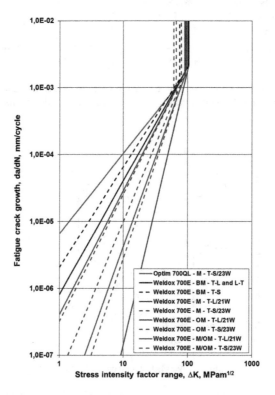

Figure 10. Determined fatigue crack growth (FCG) design or limit curves for Optim 700QL and Weldox 700E base materials and their welded joints.

4 SUMMARY AND CONCLUSIONS

Based on our investigations and their results, the following conclusions can be drawn.

The applied gas metal arc welding (GMAW) process and technological parameters are suitable for production welded joints of the investigated S690QL type high strength steels with eligible quality, where the appropriate quality contains the eligible resistance to high cycle fatigue (HCF) and fatigue crack growth (FCG). The welding causes unfavorable effects both on the basic mechanical properties and the fatigue resistance (HCF and FCG) of the investigated high strength steels. This statement has good correspondence with different experiments, which can be found in the literature (e.g. Pijpers et al. 2007).

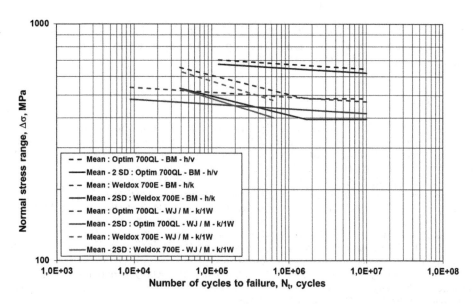

Figure 11. Comparison of the high cycle fatigue (HCF) design or limit curves for the investigated base materials and their GMAW joints.

The results of the executed HCF and FCG investigations justified the necessity of statistical approaches, especially referring to the determination of the number of the tested specimens, and the directions of the base materials (h, v and k) and the welded joints (k, $1W$ and $3W$).

Based on the Basquin equation, calculated curves ("*Mean*" S-N curves) can be used for the determination of HCF design or limit curves, applying "*Mean-2SD*" philosophy.

The resistance of the base materials to HCF is more advantageous than the resistance of the welded joints. An interesting observation that the base material type has characteristic influence on the HCF resistance of the base material and the GMA welded joint (see Figure 11).

Based on these results and the used methods (Lukács 2003) FCG design or limit curves can be determined for the investigated high strength base materials and their GMAW joints, applying simple crack growth relationship (BS 7910 2015). The determined design or limit curves correctly reflect the fatigue crack growth characteristics of both the base materials and their welded joints.

The mismatch phenomenon (matching (M), overmatching (OM) and matching/overmatching (M/OM)) has characteristic influence both on the HCF and on the FCG resistance. The mismatch phenomenon has different influence on various crack paths or crack propagation directions. The reliability of the test results is different in case of different matching conditions.

Further examinations required in order

– to draw statistically better established conclusions (e.g. increasing of the tested specimens in a given sample);
– to draw more accurate conclusions about the mismatch effect;
– to calculate the parameters of the design or limit curves more reliable;
– to study the effects of the welding residual stresses (stress fields);
– to determine HCF design or limit curves for different type of whole welded joints;
– to measure threshold stress intensity factor range (ΔK_{th}) values for base materials and welded joints.

ACKNOWLEDGEMENTS

The presented work was carried out as part of the EFOP-3.6.1-16-2016-00011 "*Younger and Renewing University – Innovative Knowledge City – institutional development of the University of Miskolc aiming at intelligent specialization*" project implemented in the framework of the Széchenyi 2020 program. The realization of this project is supported by the European Union, co-financed by the European Social Fund.

REFERENCES

ASTM E 647 2015. *Standard Test Method for Measurement of Fatigue Crack Growth Rates.*
AWS 2019. *Vision for Welding Industry.* American Welding Society https://app.aws.org/research/vision.pdf.
Balogh, A., Lukács, J. & Török, I. (eds.) 2015. *Weldability and the properties of the welded joints.* Miskolc: University of Miskolc. (In Hungarian.).
Baluch, N., Udin, Z.M. & Abdullah, C.S. 2014. Advanced High Strength Steel in Auto Industry: an Overview. *Engineering, Technology and Applied Science Research* 4(4): 686–689.
Berger, C. et al. 2018. *Bruchmechanischer Festigkeitsnachweis für Maschinenbauteile, FKM-Richtlinie.* VDMA Verlag.
BS 7910 2013 + A1:2015. *Guide to Methods for Assessing the Acceptability of Flaws in Metallic Structures.*
Dobosy, Á. 2017. *Design limit curves for high strength steel structures under cyclic loading condition.* PhD thesis, Miskolc: University of Miskolc. (In Hungarian.).
Dobosy, Á., Gáspár, M. & Jámbor, P. 2017. Weldability of S960M thermo-mechanically treated advanced high strength steel. In *Proceedings of the Third Young Welding Professionals International Conference (YPIC2017), Halle (Saale)* 16–18 August 2017.
ESTEP 2013. *Strategic Research Agenda – A vision for the future of the steel sector.* Brussels: European Steel Technology Platform. https://www.estep.eu/assets/SRA-2013.pdf.
ESTEP 2017. *Strategic Research Agenda (SRA).* Brussels: European Steel Technology Platform. https://www.estep.eu/assets/SRA-Update-2017Final.pdf.
Gáspár, M. & Balogh, A. 2013. GMAW experiments for advanced (Q+T) high strength steels. *Production Processes and Systems* 6(1): 9–24.
Gáspár, M. & Sisodia, R. 2018. Improving the HAZ toughness of Q+T high strength steels by post weld heat treatment. *IOP Conference Series: Materials Science and Engineering* 426, Paper: 012012. doi:10.1088/1757-899X/426/1/012012.
HKS 2019. https://hks-prozesstechnik.de/en/about-hks/.
Kalácska, E. et al. 2017. MIG-welding of dissimilar advanced high strength steel sheets. *Materials Science Forum* 885: 80–85.
Lukács, J. 2003. Fatigue crack propagation limit curves for different metallic and non-metallic materials. *Materials Science Forum* 414–415: 31–36.
Lukács, J. & Mobark, H.F.H., in press. *Mismatch effect on fatigue crack propagation limit curves of S690QL, S960QL and S960TM type base materials and their gas metal arc welded joints.* 72nd IIW Annual Assembly and International Conference, Bratislava, 7–12 July 2019.
Májlinger, K., Kalácska, E. & Spena, P.R. 2016. Gas metal arc welding of dissimilar AHSS sheets. *Materials and Design* 109: 615–621.
Mobark, H.F.H., Dobosy, Á. & Lukács, J. 2018.. Mismatch Effect Influence on the HCF Resistance of High Strength Steels and their GMA Welded Joints. In Jármai, K. & Bolló, B. (eds.), *Vehicle and Automotive Engineering 2, Proceedings of the 2nd VAE2018, Miskolc*, 23–25 May 2018. Lecture Notes in Mechanical Engineering. Springer International Publishing AG, part of Springer Nature.
Nakazawa, H. & Kodama, S. 1987. Statistical S-N testing method with 14 specimens: JSME standard method for determination of S-N curves. In Tanaka, T., Nishijima, S. & Ichikawa, M. (eds.), *Statistical research on fatigue and fracture. Current Japanese Materials Research* 2. London: Elsevier Applied Science.
Paris, P. & Erdogan F. 1963. A critical analysis of crack propagation laws. *Journal of Basic Engineering, Transactions of the ASME* 85: 528–534.
Peters, K. et al. 2015. The European steel technology platform faces resource efficiency in its strategic research agenda. *International Journal of Sustainable Energy Development* 4: 187–195.
Pijpers, R.J.M. et al. 2007. Fatigue experiments on very high strength steel base material and transverse butt welds. *Advanced Steel Construction* 5(1): 14–32.
Schroepfer, D. & Kannengiesser, T. 2016. Stress Build-Up in HSLA Steel Welds Due to Material Behavior. *Journal of Materials Processing Technology* 227: 49–58.
Sisodia, R., Gáspár, M. & Guellouh, N. 2019. HAZ Characterization of Automotive DP Steels by Physical Simulation. *International Journal of Engineering and Management Sciences* 4(1): 478–487.
Varbai, B. et al. 2019. Optimization of resistance spot welded UHSS sheets. In Barabás, I. (ed.), *Proceedings of 27th International Conference on Mechanical Engineering (OGÉT2019)*, Oradea, 25–28 April 2019. Hungarian Technical Scientific Society of Transylvania. (In Hungarian.).

Solutions for Sustainable Development – Szita, Jármai & Voith (eds.)
© 2020 Taylor & Francis Group, London, ISBN 978-0-367-42425-1

Improvement of the modern house's energy efficiency in the region of In Saleh

S. Oukaci, A. Hamid, D. Semmar & A. Naimi
Faculté de technologie, Laboratoire LTSM, Université Blida 1, Blida, Algerie

S. Sami
Renewable Energies Development Center, Algiers, Algeria

ABSTRACT: The construction materials used in the contemporary house of the arid regions of southern Algeria have generated thermal discomfort and energy consumption. In this context, a thermal study was made on a modern house in the region of In Saleh, which is known for the aridity of its climate. The purpose of the thermal study is to evaluate the house's thermal behaviour and to contribute to the improvement of its energy efficiency. The study includes an experimental study and a dynamic thermal modelling using the software PLEAIDES on several variants. Experimental results have shown that the composition of the constructive elements allowed reaching 5°C. The results obtained from the dynamic thermal modelling have demonstrated that the reinforcement of the roof allows improving the energy efficiency of the modern house, which leads to reach a gap of 7°C during the over-heating period, and allows reducing the energy consumption up to 38%.

The traditional habitat in Algeria has been abandoned and replaced by the modern typology, in which it has been found the use of new construction techniques. This abandonment has resulted in the loss of architectural heritage, discomfort and excessive energy consumption, especially in regions of the arid climate. This state of affairs has actually been noted by the experts as well as the occupants.

New materials are currently being used for construction in the arid regions of southern Algeria, including cinder block, concrete and hollow brick. Building materials that do not adapt to the climatic conditions of the region, despite the fact that the region has a variety of local materials: Adobe clay, natural insulators including petiole, bunch and trunk of palm wood. These local materials have been the subject of several studies whose purpose is to approve their effectiveness.

The aim of the research was to characterize local materials in the arid regions of southern Algeria. A study was done by (Oukaci et al.) on the characterization of the traditional earthen brick of the In Saleh region. Another research was done by (Nefidi et al) on local materials from the region of Oued Souf.

Other studies have aimed to improve the thermo-physical and mechanical characteristics of local materials. A study was made by (Benmancour et al), whose goal is the improvement of the classical earth mortar, by adding 5% of palm wood fibre. Another study has been done on the improvement of the earth brick by (Mekhermeche et al. 2011). Other researchers have focused on the characterization of different parts of the palm tree (Agoudjil et al, 2011). Through an experimental study, it has been approved that the petiole part of the palm is the most effective part in the case of insulation in a building. The thermal conductivity has been estimated at 0.058 W/mK. Another research has been undertaken on integrating plaster and natural fibres composite of palm wood by (Chikhi, 2016) to integrate it into the building. The study of the impact of the choice of building materials on the evolution of indoor temperatures has also been the aim of several studies. (A-Mokhtari 2008 et al.), through their study of a local located in Bechar, one of the arid regions of southern Algeria, have deduced that red brick responds better to climatic conditions compared to cinderblock. The temperature difference was estimated at 7°C during the overheating period.

The typology of the traditional house also has been the subject of research. (Fezzioui et al. 2014) have studied the impact of the patio on thermal comfort and energy consumption. The results approved the energy efficiency of the patio house compared to the modern house.

Through the studies cited, it has been found that the traditional house is more energy-efficient efficient thanks to its typology and building materials. The abandonment of the latter has led to excessive energy consumption and frequent shedding especially during the period of overheating. Within the framework of this problem, passive solutions have been proposed, which aim to improve the energy efficiency of the modern house of southern Algeria. Integrating traditional building materials in the composition of the building elements of the modern house can improve its energy efficiency and also contribute to the energy rehabilitation of existing modern homes.

Also, through the results of the literature, it was deduced that the integration of PCM phase-change materials also improves the thermal comfort in the habitat. Through the study done by (Boussaba et al. 2018), it has been shown that integrating phase-change materials can improve the energy efficiency of the building. The results of the literature have led us to propose several variants whose aim is to improve the energy efficiency of modern homes in arid regions.

1 METHODOLOGY

A thermal study was done to assess the impact of the use of building materials, and the improvement of the energetic efficiency use of the modern house of southern Algerian arid regions.

Our choice was focused on a modern typology house whose walls were built in breezeblock and roof in concrete. The chosen house has undergone changes in the composition of the building elements. In order to improve its efficient energy use, local traditional materials (natural insulators, adobe wastes and earth bricks) have been added, either in the roof or in the walls.

The chosen sample represents a variety of compositions of the constructive elements. This variety allowed us to do a thermal behaviour study of different spaces according to their composition. The research question of the undertaken work is part of an energy rehabilitation of houses built in this area. After abandoning use of traditional materials, there was evidence of energy overconsumption in its regions. İn order to reach our goal, which is the energy rehabilitation of the chosen case study, a thermal study, has been undertaken, which includes the following steps:

Steps of study:

1.1 *An experimental study*

In order to assess the evolution of temperatures during the summer period, measurements were taken inside and outside the chosen house, from 14 to 15 June 2018. The measurements were taken using a mercury type liquid thermometer, which was placed at 1.2m from the high floor and in the middle of the space to be measured; we proceeded with this method of measurement because the house does not contain any furniture.

1.2 *A dynamic thermal modelling*

A parametric study was made on the studied case, in order to assess the thermal behaviour of the chosen house on the long term and to study the impact of adding local construction materials and phase change materials in the roof on the efficient energy use. The modelling was done on three variants. The study was done using Pleiades software, which allows calculating in a precise way the thermal flows taking into consideration the climatic conditions of the studied region.

1.2.1 *Studied variants*

1.2.2 *Variant A*

State of play variant. Depending on the composition of the constructive elements of the studied case, the integration of traditional construction materials (Adobe, earth mortar) was made only in the roof of the east-facing spaces and in the walls of the west-facing spaces (see attached plan).

1.2.3 *Variant B*

In order to improve the thermal comfort of our case study, we opted for the integration of the insulation in concrete house roof. For this variant, we opted for the use of the composition of the floor of the traditional house in this region. The traditional composition was characterized by the integration of palm cluster, palm petiole and palm wood trunk. Also in this variant, the standard earth mortar was replaced by an improved earth mortar by adding 5% of palm wood fibre.

1.2.4 *Variant C*

The solution being proposed in this variant is to integrate in concrete roof, a layer of palm petiole insulation as well as the treatment of the interior surface of the roof with plaster with phase change materials. The variant C proposal can be an appropriate solution in the context of energy rehabilitation of modern homes already built with concrete roof.

In order to consider the weekly operation of the house, scenarios have been proposed: an Occupancy scenario, the number of occupants is estimated at five people. Dissipated power scenarios, the purpose of which is to determine the home appliance power input according to their operation. Scenarios of ventilation, for this scenario, the flow of rate of the ventilation was calculated according to the climatic data of the site among others the dry temperature and its relative humidity.

When integrating the ventilation scenarios, it was taken into consideration, the simulation season as well as the simulation type, for example; during the summer period in the case of the simulation without temperature setpoint, night ventilation was proposed. For our modelling, four ventilation scenarios have been proposed.

Thermostat setpoint scenario: the purpose of which is to determine the heating and cooling requirements, the setpoint was set at 20°C during the winter period and 27°C during the summer period.

2 STUDY CASE PRESENTATION

The chosen house is located in the centre of the city of In Saleh in southern Algeria, the region is classified to be in zone D '(DTR), the region is known for the aridity of its climate. The temperature may exceed 48°C during the overheating period.

2.1 Presentation of house plan

The house includes the following spaces: entrance hall, living room, women's living room, parents 'room, girls 'room, boys' room, kitchen, sanitary spaces, an area for animals and a courtyard (see Figure 1).

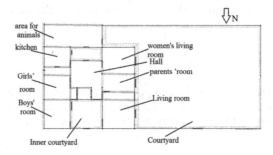

Figure 1.   House plan.

The chosen model is a modern typology in which traditional processes such as the integration of the inner courtyard and the vocation of traditional spaces have been integrated.

3   THERMO-PHYSICAL CHARACTERISTICS OF CONSTRUCTIVE ELEMENTS

For the thermo-physical characteristics of the materials used during the simulation, we have relied on the results in literature.

4   RESULTS AND DISCUSSION

4.1   Results of the experimental study

The measurements made in the different spaces allowed us to have the following results:

4.1.1   Results obtained in spaces with roof type 01

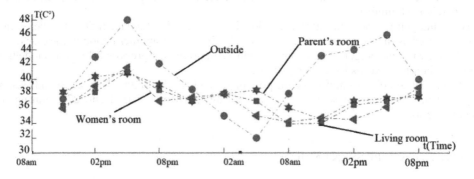

Figure 2.   Results obtained in spaces with roof type 01.

4.1.2 Interpretation of results obtained in spaces with roof type 01

According to the obtained results of the carried out measurements, outdoor temperatures vary from 32°C to 46°C and indoors can reach up to 44°C. Also, it was found that Indoor temperatures follow the fluctuations of outdoor temperatures despite the fact that the walls are built in breezeblock, with adding earth bricks (see figure 02), and the gap has been estimated at 2°C.

340

Table 1. Details of the constructive elements.

Composition	Thickness(cm)	R (m².K)/W Resistance
Low floor	14	0.13
Roof type 01	24	0.2
Roof type 02	36	0.33
Improved roofing 01*	40	2.13
Improved roofing 02*	34	1.54

* Improved roofing 01: integration of traditional local insulators in concrete roof.
*Improved roofing 02: integration of layer of palm petiole insulation and plaster
with phase change materials in concrete roof.

4.1.3 Results obtained in spaces with roof type 02

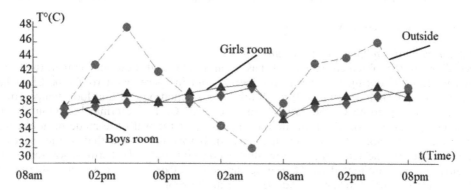

Figure 3. Results obtained in spaces with roof type 02.

According to the results obtained of the carried out measurements, outdoor temperatures varies from 32°C to 46°C. While indoor temperatures are almost constant and estimated from 37°C to 42°C. Indoor temperatures do not follow the fluctuations of outdoor temperatures (see figure 03), the gap was estimated at 5°C, and it is due to the use of materials with high thermal inertia (addition of adobe and earth mortar in Concrete roof).

5 SYNTHESIS

From the experimental results obtained in this study, it was found that the temperatures in spaces, whose the roof is built in Concrete, with the addition of adobe and earth mortar in the roof respond better to outdoor temperature fluctuations with a gap that can reach up to 5°C.

The integration of traditional local materials (local natural insulators, Adobe) in modern houses roof made of concrete, is an adequate passive solution to improve thermal comfort.

5.1 Results of dynamic thermal modeling

The simulations were made without setpoint and with thermostat setpoint on the three variants (A, B, C) during the winter and summer period. This allowed us to obtain the following results:

Simulation results during the summer period without thermostat setpoint:

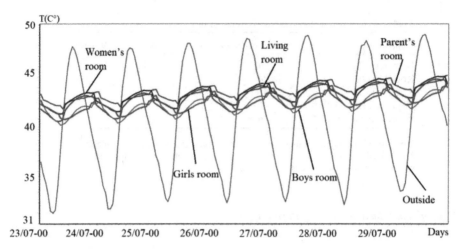

Figure 4.   Results of variant A.

5.1.2 *Interpretation of the variant A results*

According to the obtained results, it has been found that outdoor temperatures can reach up to 48°C during the overheating period, but in spaces with roof type 01, it has been found that indoor temperatures follow the fluctuations of outdoor temperatures and can reach up to 45°C, the gap is estimated at 3° C. This thermal behaviour is due to the use of concrete on the roof, despite the fact that the walls are made of adobe and breeze block (see table of construction elements). It was also found that in the spaces in the roof type 02, the temperatures are almost constant and vary between 42 and 43°C, the difference was estimated at 5°C during the period of overheating and it is due to the Addition of the earth mortar and adobe waste in the roof.

5.1.3 *Results of variant B*

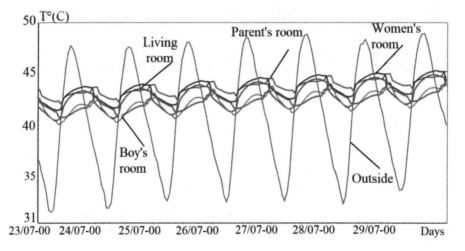

Figure 5.   Results of variant B.

From the obtained results, it was found that the temperatures became almost constant in the different spaces of the studied case; the difference was estimated from 6°C to 7°C in the different spaces. The integration of natural insulating materials, as well as the improved earth mortar, made it possible to respond in a better way to temperature fluctuations.

5.1.4 *Results variant C*

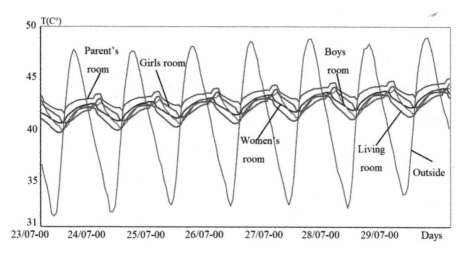

Figure 6.   Results of variant C.

According to the obtained results, it was found that the temperatures are almost constant in the different spaces after adding a layer of 7 cm of palm tree petiole insulation and plaster with phase change materials, which allowed a difference of 7°C compared to external temperatures.

5.2   *Results of the simulation during the winter period without thermostat setpoint*

5.2.1 *Results variant A*

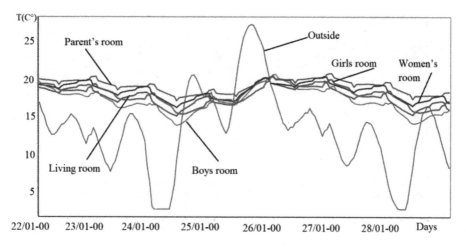

Figure 7.   Results of variant A.

According to the obtained results, we could figure out that outside temperatures can drop to 3°C during the coldest week. While the temperatures in the different spaces vary between 16°C and 18°C (see figure.).

343

5.2.2 *Simulation results of variant B*

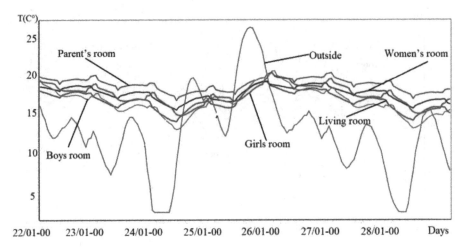

Figure 8. Results of variant B.

According to the results, it was noted that indoor temperatures have reached a comfortable range after strengthening the roof; temperatures vary from 20 to 22°C in all spaces. Comfort is guaranteed without the need for a heating system.

5.2.3 *Results of variant C*
Interpretation of the results of variant C:

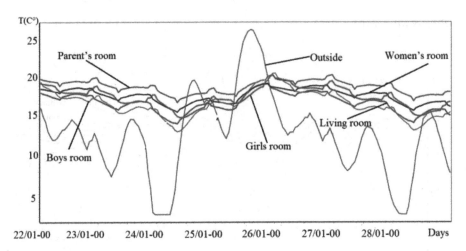

Figure 9. Results of variant C.

According to the obtained results, it was found that indoor temperatures have reached the comfortable range after strengthening the roof by insulation in palm tree petiole and interior treatment with a phase change material plaster coating, indoor temperatures vary from 19°C to 21°C in all spaces.

344

6 SYNTHESIS

The simulation results of the variant C have shown that we can have the same thermal behaviour as the variant B; the proposed variant C can be an alternative solution for strengthening the concrete roof, which can be an easy solution to execute, for the energy rehabilitation of the modern houses of the regions of the arid climate.

6.1 *Simulation results with thermostat setpoint*

The purpose of which is to determine the heating and cooling requirements: Interpretation of results with thermostat setpoint:

6.1.1 *Interpretation of the results obtained from the simulation with thermostat setpoint*

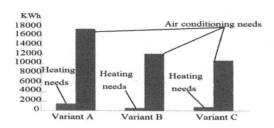

Figure 10.   Results of results with thermostat setpoint.

According to the obtained results, we could figure out that the reinforcement of the roof contributed to the improvement of the energetic efficiency of the studied case.

The simulation results of variant B have shown that the rate of reduction of consumption is estimated at 60% for heating and 30% for air conditioning compared to the initial case (variant A). In addition, the results of simulation of the variant(C) have shown that the proposed solution allows having satisfactory results and that improve the energy efficiency of the case studied, the reduction rate is estimated at 45% for heating and 38% for air conditioning.

7 CONCLUSION

The results obtained during this study approved the need for reinforcing the roof of modern houses built in concrete. The passive solutions proposed can also improve the energy efficiency of similar cases existing in the arid regions of southern Algeria. The proposed integration of phase change materials can also contribute to the energy rehabilitation of existing houses. The passive solutions proposed during this study made it possible to reach a temperature difference of 7°C during the period of overheating, and to enhance the comfort during the winter period without resorting to an active system. Although the results are satisfactory, summer comfort is not achieved. This problem is part of our research perspective, in which the integration of an adiabatic humidification cooling system is planned.

REFERENCES

B. Agoudjil, A. Benchabane, A. Boudenne, L. Ibos, M. Fois, 2011. Renewable materials to reduce building heat loss: Characterization of date palm wood, Energy and building.

Benmancour, N. 2015. Développement des composites naturels et caractérisation des composites naturels locaux adaptée à l'isolation thermique dans l'habitat. Thèse de Doctorat. Université Hadj lakhdher de Batna.

Chikhi, M. 2016. Young's. modulus and thermo-physical performances of bio-sourced materials based on date palm fibres. Energy and Buildings Vol 129. 589–597.

Naïma Fezzioui, Maatouk Khoukhi, Zohra Dahou, Karim Aït-Mokhtar et Saleh Larbi, 2014. Bioclimatic Architectural Design of Ksar de Kenadza South-west Area of Algeria Hot and Dry Climate.

H. Nefidi, S. OUKACI, A. Hamid. Evaluation thermique de l'habitat traditionnel en Algerie cas de la region de Souf. 2019. ISBN10:6138456130.

Mekhermeche, A et al. 2012. Contribution à l'étude des propriétés mécaniques et thermiques des briques de terre en vue de leur utilisation dans la restauration des Ksours Sahariennes. Mémoire de Magister. Université d'Ouargla.

L Boussaba, A.A Foufa, S. Makhlouf, G. Lefebvre, L. Rayon.. Elaboration and properties of a composite bio –based PCM for an application in building envelopes construction and building naturels.

Oukaci, S et al. 2018. Impact du choix des matériaux de construction dans un habitat du climat aride, cas d'In Saleh. ICEMAEP'18 Fourth International Conference on Energy, Materials, Applied Energetics and Pollution. Constantine. Algeria.

Solutions for Sustainable Development – Szita, Jármai & Voith (eds.)
© *2020 Taylor & Francis Group, London, ISBN 978-0-367-42425-1*

Innovative and efficient production of welded body parts from 6082-T6 aluminium alloy

R.P.S. Sisodia & M. Gáspár

Institute of Materials Science and Technology, University of Miskolc, Miskolc, Hungary

ABSTRACT: In this paper, the effect of special heat treatment (pre ageing, post ageing) of the recently developed Hot Forming and in-die Quenching (HFQ) process is used to compensate the Heat affected zone (HAZ) softening. Investigation was performed on 1 mm AA6082-T6 sheet by a Gleeble 3500 physical simulator with specifically selected peak temperatures using Rykalin-2D model (thin plate) and with two linear heat input (100 J/mm and 200 J/mm). Then the most critical subzone (T_{peak} = 440°C) was selected for further investigations. Three different heat-treating routes were programmed on the simulator: HAZ cycle + post ageing; solution annealing + HAZ cycle +post ageing; solution annealing + pre ageing + HAZ cycle +post ageing). Finally, the properties of the investigated HAZ subzone were examined by optical microscope and microhardness test.

1 INTRODUCTION

Aluminium production has increased worldwide over the past decades and is considered to be the most important alloy in terms of use. It is of great importance ranging from domestic industry to vehicles industry. Aluminium alloys are also used to achieve massive reductions, which are not only a weighty aspect of manufacturing in the automotive industry but also produce positive results for users later. The use of aluminium alloys is advantageous in terms of fuel consumption, it also meets the consumer requirement for more economical operation, and the reduction in vehicle weight also reduces emissions. An outstanding feature of aluminium is that it can be recycled under excellent conditions. On the basis of all these properties it can be said that due to its recoverability, it is an environmentally friendly element (Balogh et al., 2015). The increasing use of aluminium is based on the positive benefits of its properties, such as low density, good electrical and thermal conductivity, good formability, excellent corrosion resistance, strength and toughness comparable to the steels by the application of alloying elements. These properties are widely used in industry in a number of sectors, such as electronics, space technology, packaging industry, structural construction and the vehicle industry (Balogh et al., 2015). A wide variety of alloys are available for the production of welded structures and finished products made using aluminium and its alloys. Each of the alloys and alloy types has different properties that are important to both the designer and the manufacturer. The practical marking system used by Aluminium Association (AA), which groups the alloys according to their main alloying elements, was taken over by EN 573. The marking system denotes the alloys with a four-digit numerical sign, in the main alloy grouping (Balogh et al., 2015). as shown in Figure 1.

The aluminium alloy used in this paper is 6082-T6, whose main alloying constituents are magnesium and silicon, T6 represents solution heat-treated and artificially aged. It has similar properties to steel in terms of strength but after welding, welded and HAZ get soften, to analyse this softening effect and to improve the hardness we proposed to use the method of post ageing after welding.

Aluminium alloys of group 6xxx have silicon and magnesium as major addition elements. The Mg_2Si precipitations produced by these alloying elements result in increased hardness,

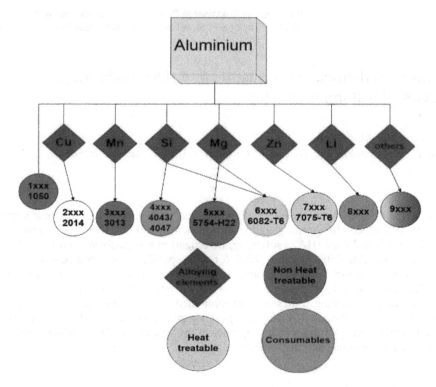

Figure 1.   Grouping of aluminium alloys.

and silicon reduces the tendency for hardness, improves corrosion resistance and the molten alloy fluidity. However, the silicon content of more than 13% already reduces the machinability of the alloys (Balogh et al., 2015). In addition to high hardness, the members of the group are also characterized by good formability and weldability as well as excellent corrosion resistance. Their tensile strength ranges from about 124-400 MPa. Generally speaking, low-alloying material qualities can be attributed to the fact that the magnesium and silicon content has an optimum because of their crack-reducing effect. The amount of both alloys' ranges from 0.3 to 1.5% by weight. Other alloying materials include Ca, Cr, Mn, Pb, Bi (Balogh et al., 2015). Magnesium has a low melting point and may be lost or oxidized during welding (Gene 2002).

There has been considerable industrial interest in these alloys because two-thirds of all extruded products are made of aluminium and 90% of those are made from 6XXX series alloys (El-Shennawy et al., 2017). Typical applications for aluminium alloys in Group 6xxx include vehicle manufacturing, construction, navy (offshore structures), etc. They are widely used for making welded products and can be incorporated into many structural components. Extruded types are often used as raw material for machined machining, adding small amounts of low melting alloy (lead, tin, bismuth) to further improve the machinability of such aluminium alloys. These alloys can be anodized, which may be necessary for products where hard, high-strength, corrosion-resistant surfaces are important. In hard anodized condition, they are ideal for braking systems, electronic valves and pistons (Balogh et al., 2015).

2   THE HFQ PROCESS

The formability of modern aluminium materials in the automotive industry is fundamentally impaired by the alloys found in them and by the brittle dispersions resulting from them (Tisza et al. 2017). The reduction in formability is such that there is already a risk of cracking during the shaping of parts with a simple geometric design, as well as a high degree of shrinkage

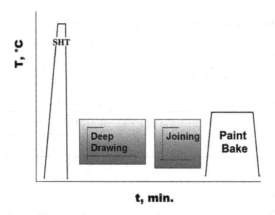

Figure 2.   The whole cycle of HFQ process.

(Pósalaky et al., 2015). In order to develop alloys of improved formability, advanced forming technologies are also being investigated to form complex shaped parts from these alloys (Hayat 2012). Solution heat treatment, forming, and in-die quenching (HFQ) is one such technology (Fakir et al., 2014). In this process, the blank is first heated up to its solution heat treatment (SHT) temperature (Gáspár et al. 2018). At this elevated temperature, the solid solubility is increased and the alloying elements, or precipitates, fully dissolve into the aluminium matrix. Consequently, the yield strength and the residual stresses are reduced, and the material becomes more ductile due to the fewer obstacles to dislocation movement, enabling more complex shapes to be formed. The blank is then transferred to a cold die, formed at a high speed and held in the cold tool to achieve a rapid cooling rate to room temperature. The fast pace of the process allows a super-saturated solid solution to be obtained (Garret 2005, Temmar 2011).

After the successful pressing process, the car-body elements should be joined to each other, usually with RSW. Finally, the assembled car-body is moved to the paint shop, where the aluminium can reach its increased mechanical properties during the ageing. However, the time of the painting process is much shorter than the conventional ageing. It means that the aluminium sheets are joined in a solution annealed (softened) and furthermore formed condition. Although it might be considered that the sheets are primarily joined at the less shaped parts of the car-body elements. Since the artificial ageing is connected to the painting of the car-body, therefore the joint connections (RSW, FSW, clinching, adhesive bonding etc.) should be prepared before. It means that joints are heat treated as well, which might influence the mechanical properties of the weld metal and the heat-affected zone. The whole process can be seen in Figure 2.

If we consider the typical strength distribution in the cross section of the fusion welded joint of different kind of aluminium alloys, we can see that a significant strength reduction happens in the weld and HAZ of heat treatable aluminium alloys. However, in case of solid-state pressure welding, the softening is lower compared to fusion welding (Oyedemi et al. 2017, Meilinger et al. 2015). The application of HFQ process, including the past ageing thermal cycle, may positively affect the strength level of the whole welded joint.

3   EXPERIMENTAL METHODOLOGY

To implement the new technological process into a manufacturing environment, it must ensure that the process is practical for its intended application. Aluminium is a particularly delicate material quality, so it is important to analyse the typical properties of aluminium alloys for weldability. As a next step, the appropriate welding procedure (Balogh et al., 2015) should be selected taking into account these properties and suitability for the purpose.

Table 1. Chemical composition of AA6082-T6 aluminium alloy (wt%).

Mg	Si	Mn	Cu	Cr	Zn	Ti	Fe	Al
0.70	0.90	0.40	0.09	0.02	0.08	0.03	0.46	Balance

Table 2. Mechanical properties of AA6082-T6 aluminium alloy.

Density (g/cm^3)	Melting point (liquidus) (°C)	R_m (MPa)	$R_{p0.2}$ (MPa)	A_{50} (%)
2.7	600	280	315	12

3.1 Investigated material

The alloy type 6082, like its family, is characterized by good corrosion resistance, while it can be welded, but in the thermal zone, it can experience significant softening. Aluminium alloys 6082 can be well machined, cold-formed and heat-treated. Their resistance to fatigue is moderate, not suitable for composite profile structures. However, using heat treatment, significant changes in mechanical properties can be achieved. The T6 condition means, that this aluminium base material has been homogenizing solution annealed at 535°C for 30 min, then quenched, and finally aged at 190°C for 8 hours. The base material chemical composition is shown in Table 1, and the mechanical properties are shown in Table 2.

3.2 Physical simulation

HAZ properties can be limitedly analysed by conventional material tests, therefore physical simulators (i.e. Gleeble) is used to produce different kind of HAZ. (Gáspár et al. 2015, Heikkilä et al. 2013).

The Gleeble recreates specific sections of the HAZ based on the programmed thermal cycle (Lukács et al. 2015). The Gleeble can be used to create large enough sections of HAZ material for subsequent mechanical properties testing. The direct resistance heating system of the Gleeble 3500 can heat specimens at rates of up to 10,000°C/s or can hold steady-state equilibrium temperatures (Adonyi 2006). High thermal conductivity grips hold the specimen, making the Gleeble 3500 capable of high cooling rates also optional quench system can provide cooling rates in excess of 10,000°C/second at the specimen surface. During the experimental work, HAZ tests have been performed on a new generation thermophysical simulator, called Gleeble 3500 (Figure 3), installed in the Institute of Materials Science and Technology of the University of Miskolc.

The results of Vickers hardness test, which was performed by a Mitutoyo microhardness tester with HV0.2 load, are summarized in Figure 6.

Figure 3. Test device- Gleeble 3500 simulator system.

3.3 Heat source model and HAZ thermal cycle

In the Quicksim software developed for Gleeble programming, the possible HAZ simulation welding heat cycle models are F (s, d) thermocouple measurement or FEM, Hannerz, Rykalin-2D, Rykalin-3D, Rosenthal, Exponential. In this paper, heat cycles were determined according to the Rykalin 2D model by considering the 1.5 mm sheet thickness. This model describes the temperature field generated by a moving spot-like heat source on the surface of a semi infinity body. In the sheet metals, the characteristic role of heat conduction disappears and the role of convection is getting more important due to the larger surface to volume ratio. By the application of Rykalin 2D model the time-temperature points of HAZ heat cycle can be calculated as follows (QuikSim^{TM} Software):

$$T - T_0 = \frac{a}{\sqrt{b * (t - t_0)}} exp\left(\frac{c}{t - t_0}\right) \tag{1}$$

$$a = \frac{Q}{d} \tag{2}$$

$$b = 4\pi * k * c * \rho \tag{3}$$

$$c = \frac{r^2}{4k/(c\rho)} \tag{4}$$

$$Q = \sqrt{\frac{4\pi k c \rho \Delta t}{1/(T_2 - T_0)^2 - 1/(T_1 - T_0)^2} * d} \tag{5}$$

where Q = energy input, J/cm; c = specific heat, J/g/°C; r = density; g/cm^3; k = thermal conductivity, W/cm/°C; d = plate thickness, cm; d_e = equivalent plate thickness, cm; T_1, T_2 = temperature used to define cooling time,°C; t_0 = time at the end of preheating, s; and Δt = cooling time from T_2 to T_1, s.

3.4 Experimental procedure

The aluminium sheet which is used for thermomechanical testing using Gleeble 3500 simulator is short samples. The samples used for all the HAZ simulations were 70 x 10 x 1 mm rectangular piece. The geometry of the samples is shown in Figure 4 left.

A precise preparation of HAZ specimen with required geometrical shape and good surface quality is indispensable for the successful simulation. A K(NiCr-Ni) type thermocouple was welded onto the middle of sample for temperature record as shown in Figure 4 right.

In the first part our aim was to identify the critical parts of the heat-affected zone (HAZ) in terms of softening during Gas Tungsten Arc Welding (GTAW). Four HAZ peak temperatures were selected in the function of the distance from the fusion line. Two linear heat

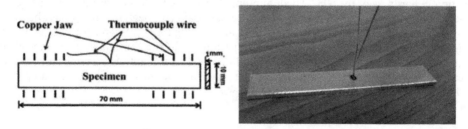

Figure 4.   left - Sample dimensions, right- Specimens for the physical simulations.

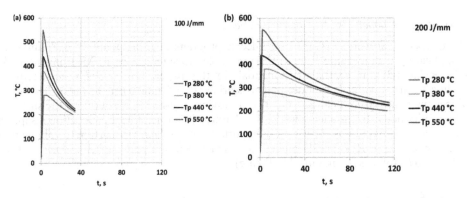

Figure 5. HAZ thermal cycles; left- low heat input (100 J/mm), right- high heat input (200 J/mm).

input values (100 and 200 J/mm) were selected in order to simulate a low and a high heat input welding at the given sheet thickness and welding technology. The desired subzones were simulated on 4 samples. This experimental part was finished with the selection of the (most) softened zone. In the second part, our main aim was to examine the effect of the characteristic heat treatments connected to HFQ process on the hardness of 6082-T6 alloy in the function of the base material T6 delivery condition. The previously determined weakest subzone was tested by three different process routes: HAZ cycle + post ageing; solution annealing + HAZ cycle + post ageing; solution annealing + pre ageing + HAZ cycle + post ageing. The programmed thermal cycles of the different HAZ subzones are illustrated in Figure 5.

4 HAZ CHARACTERIZATION

Two technological variants (Q = 100 J/mm and 200 J/mm, linear heat input energy) and four peak temperatures 280°C, 380°C, 440°C & 550°C were considered during the experiments. After the successful simulations the specimens were perpendicularly cut at thermocouples and hardness tests were presented. The examined AA6082-T6 aluminium alloy belongs to the 23.1 group of CR ISO 15608, which means that the requirement for the tensile strength of the welded joint is the 70% of the base material according to EN ISO 15614-2. If we consider the same requirement for the HAZ hardness compared to the base material, the hardness should reach 76 HV0.2 in HAZ when 6082-T6 base material has 109 HV0.2. The results of Vickers hardness test, which was performed by a Mitutoyo microhardness tester with HV0.2 load, are summarized in Figure 6. In case of 6082-T6 aluminium alloy, sub zones have always been softened by the applied linear heat input, and with the increase of linear heat input, the hardness of the heat affected zone has further decreased. However, in different zones heated to different temperatures, there was a markedly different degree of softening. In addition, the positive conclusion is that the hardness of the examined peak temperatures in three cases reached the hardness expected by the standard, and two times exceeded the 90% of the required value. The reduction in hardness of the 6082-T6 aluminium alloy is due to the deterioration in the quality of the constituents originally present in the base material. This negative change can be due to over-regeneration, which occurs in zones that are too high at peak temperatures and cause precipitation to develop. Thus, the hardness distribution of the heat affect zone depends on the interaction between solubility and recrystallization. The amount of softening was the most critical at the 440°C peak temperature subzone. Five indentations were done per sample and standard deviation coefficient was between 2 & 3%.

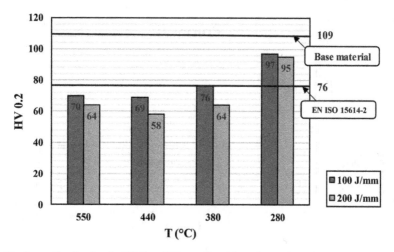

Figure 6.   Hardness distribution in HAZ at different linear heat inputs.

5   EFFECT OF HFQ PROCESS ON HAZ CHARACTERIZATION

In the second part of the investigations, the possible benefits of HFQ process on HAZ properties were analysed. Based on the performed simulations and hardness tests the most critical subzone in terms of softening has been identified (T_{max} = 440°C) for further investigations.

5.1   *Process routes*

Three possible technological routes were examined. It is important to note that the experiments did not consider the effect of metal forming (plastic deformation), only the heat-treating option was investigated. The first and the second routes can be the possible implementations of HFQ process. In case of the first one, the process starts with a solution annealing (535°C x 2min) and fast cooling, followed by a pre-ageing (220°C x 5 min), the HAZ thermal cycle with 440°C peak temperature and a post ageing (connected to paint bake, 180°C x 30 min). In case of the second route the pre-ageing was eliminated, and finally, the third one only included a HAZ thermal cycle and a post ageing. The temperature-time diagrams of the different processing routes are presented in Figure 7–9. The application of 150°C preheating was needed due to the cement curing related to the joining of thermocouples.

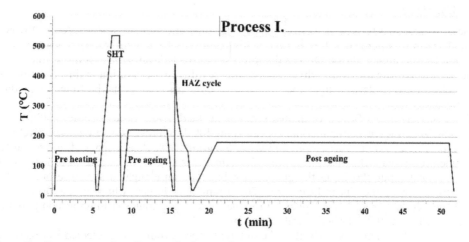

Figure 7.   Temperature-time diagram for Process I.

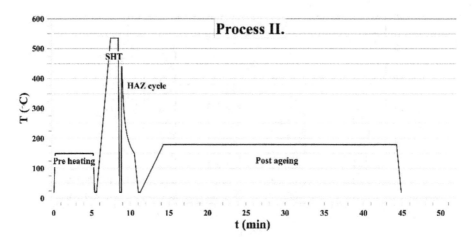

Figure 8. Temperature-time diagram for Process II.

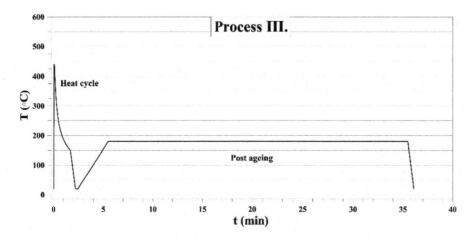

Figure 9. Temperature-time diagram for Process III.

5.2 *Materials tests*

Optical microscopic tests were performed in 200x magnification by a Zeiss Axio Observer D1m. The samples were etched by Barker-etching (5 g HBF4 + 200 ml water) which is recommended for aluminium alloys. During this process, an oxide layer forms on the surface. This optically active oxide forms with diverse speed on the different orientation grains, therefore by the application of polarized light the grains have different colours in the function of their orientation.

The grain structure of the investigated subzone at the different process routes is illustrated in Figure 10 & Figure 11 for the linear heat input 100 J/mm and 200 J/mm respectively. In case of process I for linear heat input 100 J/mm the grains have densely packed in size but for the same process with the linear heat input 200 J/mm it is larger. In the case of process II for the linear heat input 100 J/mm the grains have basically an elongated structure but for the same process with the linear heat input 200 J/mm it is larger, while in the process III the grains were larger and more spherical for the linear heat input 100 J/mm as compared to the same process with linear heat input 200 J/mm.

The results of the hardness tests after the different process routes are presented in Figure 12. From the diagram it can be concluded that none of the process variants increases the hardness

354

Figure 10.   left- Process I, middle- Process II, right- Process III, linear heat input 100 J/mm.

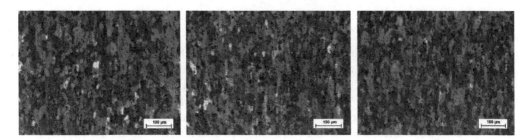

Figure 11.   left- Process I, middle- Process II, right- Process III, linear heat input 200 J/mm.

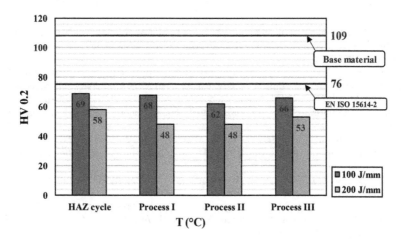

Figure 12.   Hardness of 440°C subzone at different process routes.

of the critically softened subzone, however, the measured values are still under the derived requirement. It can be supposed that the further increase of the artificial ageing time can be beneficial on the HAZ hardness, however, it results a longer production time, and thus the increase of production costs.

6   SUMMARY AND CONCLUSION

The reproduction of heat affected zone areas during the GTAW welding of 6082-T6 alloy were successfully performed, using the Rykalin 2D model in the Gleeble 3500 physical simulator. Two technological variants (Q = 100 J/mm and 200 J/mm, linear heat input) and four peak temperatures 550°C, 440°C, 380°C and 280°C were selected. Based on the performed simulations and hardness tests the most critical subzone in terms of softening has been identified (440°C). We can conclude that the hardness was under the base material hardness, for two sub zones i.e. 380°C and 280°C, Q = 100 J/mm, hardness could reach the requirement. By the modification of ageing, heat cycle including peak temperature and holding time, there may be parameters which are more advantageous for mechanical properties of the HAZ. Using both 100 J/mm and 200 J/mm of linear heat input, received very similar hardness values for the basic process, regardless of the combination of additional heat treatments. However, it is verified that it is more appropriate to select 100 J/mm of linear heat input when selecting the welding parameters. As compared to the 200 J/mm linear heat input, 100 J/mm linear heat input has given the better hardness values for each specimen.

ACKNOWLEDGEMENTS

The described article was carried out as part of the EFOP-3.6.1-16-2016-00011 "Younger and Renewing University – Innovative Knowledge City – institutional development of the University of Miskolc aiming at intelligent specialisation" project implemented in the framework of the Szechenyi 2020 program. The realization of this project is supported by the European Union, co-financed by the European Social Fund. HFQ is a registered trademark of Impression Technologies Limited, who has exclusive commercialization rights to HFQ technology.

REFERENCES

Adonyi, A. 2006. Heat-Affected Zone characterization by physical simulations: An overview on the use of the gleeble discusses the advantages and disadvantages of thermo mechanical simulation. *Welding journal* 85(10): 42–47.

Balogh, A., Lukács, J., Török, I. 2015. Weldability and the properties of welded joints: Researches on automotive steel and aluminium alloys (in Hungarian), University of Miskolc, 324. El-Shennawy, M., Abdel-Aziz, Kh. & Omar, A.A. 2017.Metallurgical and Mechanical Properties of Heat Treatable Aluminium Alloy AA6082 Welds. *International Journal of Applied Engineering Research*.12(11): 2832–2839.

Fakir, O.E., Wang, L., Balint, D., Dear, J.P., Lin, J. & Dean, T.A. 2014. Numerical study of the solution heat treatment, forming, and in-die quenching (HFQ) process on AA5754. *International Journal of Machine Tools & Manufacture* 87: 039–048.

Gáspár, M., Dobosy, A. & Török, I. 2018. Resistance spot welding of 7075 aluminium alloy. *Lecture notes in mechanical engineering* 49: 679–693.

Gáspár, M., Balogh, A. & Sas, I. 2015. Physical simulation aided process optimisation aimed sufficient HAZ toughness for quenched and tempered AHSS. *IIW International Conference, High-Strength Materials - Challenges and Applications*.

Gáspár, M., Tervo, H., Kaijalainen, A., Dobosy, A. & Török, I. 2018. The effect of solution annealing and ageing during the RSW of 6082 aluminium alloy. *Lecture notes in mechanical engineering* 49: 694–708.

Garret, R.P. Lin, J. & Dean, T.A. 2005. An investigation of the effects of solution heat treatment on mechanical properties for AA 6xxx alloys: experimentation and modelling. *International Journal of Plasticity* 21 (8): 1640–1657.

Gene, M. (1st edition) 2002. The Welding of Aluminium and Its Alloys.UK: Woodhead Publishing.

Hayat, F. 2012. Effect of aging treatment on the microstructure and mechanical properties of the similar and dissimilar 6061-T6/7075-T651 RSW joints. *Materials Science and Engineering: A* 556: 834–843.

Heikkilä, S.J., Porter, D.A., Karjalainen, L.P., Laitinen, R.O., Thinen, S.A. & Suikkanen, P.P. 2013. Hardness profiles of quenched steel heat affected zones, *Mater. Sci. Forum* 762: 722–727.

Lukács, J., Kuzsella, L., Koncsik, Z.,Gáspár, M. & Meilinger, Á. 2015. Role of the physical simulation for the estimation of the weldability of high strength steels and aluminium alloys. *Materials science forum* 812: 149–154.

Temmar, M., Hadij, M. & Sahraoui, T. 2011. Effect of post-weld aging treatment on mechanical properties of Tungsten Inert Gas welded low thickness 7075 aluminium alloy joints. *Materials and Design* 32: 3532–3536.

Meilinger, Á. & Lukács, J. 2015. Characteristics of fatigue cracks propagating in different directions of FSW joints made of 5754-H22 and 6082-T6 alloys, *Materials Science Forum*. 794–796: 371–376.

Oyedemi, K., McGrath, P., Lombard, H., Varbai, B. 2017. A Comparative Study of Tool-Pin Profile on Process Response of Friction Stir Welding of AA6082-T6 Aluminium Alloy. *Periodica Polytechnica – Mechanical Engineering* 61 (4): 296–302.

Pósalaky, D. & Lukács, J. 2015. The Properties of Welded Joints Made by 6082-T6 Aluminium Alloy and their Behaviour under Cyclic Loading Conditions, *Materials Science Forum* 812: 375–380.

QuikSim^{TM} Software, Heat Affected Zone Programming Manual: Heat Affected Zone (HAZ) Programming.

Tisza, M., Lukács, Zs., Kovács, P.Z., Budai, D. 2017. Research developments in sheet metal forming for production of lightweight automotive parts, *Journal of Physics Conference series* 896: 1–10.

Solutions for Sustainable Development – Szita, Jármai & Voith (eds.)
© 2020 Taylor & Francis Group, London, ISBN 978-0-367-42425-1

Redundancy analysis of the railway network of Hungary

B.G. Tóth

National University of Public Service, Budapest, Hungary

ABSTRACT: Available alternative routes on which traffic can be rerouted in the case of disruptions are vital for transportation networks. Line sections with less traffic under normal operating conditions but with increased importance in the case of disruptions are identified in the railway network of Hungary by using a weighted directed graph. To describe the goodness of the individual alternative routes the so-called redundancy index is used. The results show that the structure of the network is good, but the lines with the highest redundancy (lines No. 80, 2, 4 and 77 according to the numbering of the national railway operator, MÁV) are mostly single tracked and in many cases the line speed is low. The building of additional tracks and electrifying these lines while still maintaining the existing diesel locomotives for the case of disruptions of the electric support are the keys to make the performance of the rather dense railway network of Hungary sustainable.

1   INTRODUCTION

Describing a network in its fullness with a single measure is basically impossible. The choice of the measure always has to depend on what kind of property one wants to expose in the light of a specific application. If the global performance is in question, the indices of (Kansky 1989) is a good place to start as they are topology measures of the graph representing the network. In spite of their simplicity, they are effectively used even in network design (Derrible & Kennedy 2011).

To describe individual network elements based on their various roles in the network, the centrality measures are a simple but comprehensive set of indices (Lin & Ban 2013) which can also be weighted with passenger exposure (Cats et al. 2016). Recently, multi-criteria decision analysis has been also used for network element ranking (Almoghatawi et al. 2017).

But is a specific network robust against the disruptions of its most vital elements (Derrible & Kennedy 2010, Laszka et al. 2012)? Can it perform under the same supply and demand pattern even if its most important elements are disrupted (Snelder et al. 2012; Disser & Matuschke 2017)? What happens if the capacity of a link is reduced (Cats & Jenelius 2016) or even if rerouting is necessary (Oliveira et al. 2016)? Identifying these network elements is fundamental for critical infrastructure protection.

But not only the highly threatened has to be defended. The Southern Railway Bridge at Budapest is closed for pedestrians and cyclists and is guarded by armed personnel because it is the most critical element of the railway network of Hungary. Its smallest disruption affects the whole network as the total international freight transport passing through Hungary crosses the Danube via this bridge. However, the tracks and the other small bridges on the same line section are not defended in any way through any accident or other disruption can happen there and can block the line section.

One thus must be aware of the alternative routes that can be used in the case of the disruptions of the most important network elements and the line sections along these routes also have to be continuously maintained to be real alternatives for the rerouted paths. The railway network of Hungary is the fourth densest in the world (UNECE 2019), which hints that with relatively little cost good possibilities for alternative routes can be established.

In this paper, the lines that would help in keeping the system performance at sufficient level even in the case of the disruption of its most vital elements are identified by using a mathematical model. To identify these line sections, the so-called redundancy measure will be used which quantifies how important a line section in the case of the disruption of other line sections.

2 THE GRAPH MODEL OF THE RAILWAY NETWORK OF HUNGARY

To model the railway network of Hungary, a weighted graph was used. The nodes of the graph represented the stations and the edges represented the line sections between the stations. The weights of the edges were either the length of the corresponding line sections or the time needed to pass through them, the latter calculated as the ratio of the length and the line speed of the line sections. The length and allowed speed values are accessible publicly at the web page of the Hungarian Rail Capacity Allocation Office (Hungarian: Vasúti Pályaka-pacitás-elosztó Kft.) (VPE 2019). Where the line speed was different for trains with locomotives and for EMUs, the lower value was used. This means that the travel times of the model are lower bounds for the real travel times.

The model did not include the stops with no switches, i.e., where reversing of the trains is not possible. Furthermore, nodes with exactly two neighbors, the so-called "joint nodes" (Jenelius et al. 2006) were transformed out: the joint node and the two edges connecting to it were deleted and the two neighboring stations were connected with one edge weighted with the sum of the two deleted edges. The only exceptions were the border crossings and the stations preceding them not to eliminate the border effect fully.

For the reversing, an additional 15-minute increase in the travel time was assigned. When calculating trip lengths, no increase was assigned to reversing. Passing through a station did not increase the travel time or the trip length. To model the increase in the travel times, each station was represented by four nodes instead of one (Figure 1). At both "sides" of the station an arrival and a departure node were defined and the arrival and departure node on the same sides were connected with a 15-minute edge and the arrival and departure node on the opposite sides with a 0-minute edge. For neighboring stations, the departure node of a station was connected to the arrival node of the neighboring station.

However, this is still not enough: as to make the 15-minute edges not surpassable, the edges had to be directed in a proper way for the paths to arrive at a station in an arrival node and to depart from it from a departure node (Figure 1).

Wyes were represented similar to stations but because reversing on a wye is not possible without entering the respective station, the edges connecting the arrival and the departure node on the same sides were omitted.

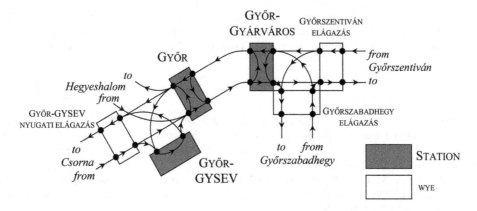

Figure 1.   An illustration of the graph model on station Győr and its neighboring stations.

In the case of termini, only one arrival and one departure node with a 15-minute edge between them are enough for this representation.

The graph on which the calculations were performed contained a total of 1136 nodes representing 291 stations and 26 wyes and 1808 edges, representing 366 line sections (and the stations and the wyes). To every edge, two weights could be assigned: one length and one travel time value. For one calculation, naturally, only one kind of weight was used. Let us denote the graph weighted with lengths by $G_\ell^0 = (V, E, W_\ell)$ and the graph weighted with travel times by $G_t^0 = (V, E, W_t)$, where V is the set of nodes, E is the set of edges and W_ℓ and W_t are the sets of length and travel time weights.

3 CALCULATION METHODS

The calculations and their visualization were carried out in the R programming language and environment (R Core Team 2012) by using the *igraph* package (Csardi & Nepusz 2006).

3.1 *Calculating the shortest path*

Since the distance and travel time values are both positive real numbers and the graph is relatively small, the easiest way to determine the shortest paths between the pairs of stations is Dijkstra's algorithm. The algorithm is implemented in the *igraph* package in the distances() function.

By calculating the shortest path in distance and in time for all $\langle a,b \rangle$ pairs of stations $(a \neq b)$ one gets two sets of 42.195 values, which number will be denoted by N°. Let us denote the duration of the fastest path between stations a and b in the G_t^0 graph by t_{ab}^0 and the length of the shortest path between them in the G_ℓ^0 graph by ℓ_{ab}^0.

3.2 *Disruption of line sections*

The term "disruption" will be used for the complete closure of a line section. Let us denote the set of edges of the disrupted line section(s) by e ($e \subseteq E$). Let the graph not containing the edges of e be denoted by $G^e = (V, E\backslash e, W^e)$ the weights being either $W_t^e \subseteq W_t$ or $W_\ell^e \subseteq W_\ell$ for time and distance weights of the G_t^e and G_ℓ^e graphs, respectively.

4 REDUNDANCY

If the shortest path passes through line section e in the undisrupted network, on the disruption of line section f there are two scenarios which are irrelevant for our calculations. First, if the shortest path passes through line section e in the disrupted network, too, then the disruption has no effect on e as it is still useable. Second, if the disrupted line section, f, makes at least one pair of stations unreachable for each other, then there is no possible alternative path (see Figure 2). Therefore, these two scenarios are left out of the calculations.

4.1 *The Network Robustness Index*

The Network Robustness Index (NRI) was introduced by (Scott et al. 2006) as a global measure to quantitatively describe the overall resilience of a network against disruptions. The NRI can be calculated for all edges of the graph based on which the importance of the individual line sections can be determined.

To calculate the NRI for line section e, the shortest paths between all pairs of stations in the undisrupted graph have to be determined. Then, the lengths or durations of these paths have to be summed, which value is denoted by c.

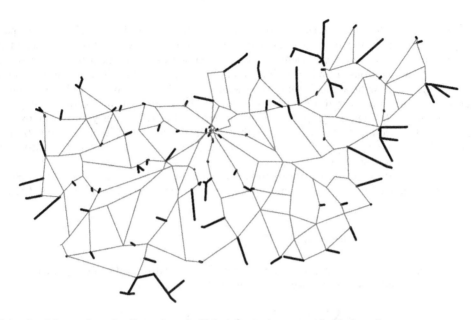

Figure 2.   Line sections the disruption of which make stations unreachable for others.

Then, the edges representing line section e are deleted from the graph. Again, the shortest paths between all pairs of stations are determined and their lengths or durations are summed. This value is denoted by c^e. The NRI is calculated as the difference of these two values and is denoted by q^e:

$$q^e = c^e - c. \tag{1}$$

The difference is made in this order for q^e to be non-negative since for most kinds of weights the deletion of a line section increases the sum of the weights of the shortest paths (or at least does not decrease it, but for a famous exception that occurs in flow models, the Braess-paradox see (Braess 1968)). This can be done for all line sections or for multiple line sections. If line sections e and f are simultaneously deleted, the NRI is calculated as

$$q^{ef} = c^{ef} - c. \tag{2}$$

The value of q^e_{ab} (the difference in the shortest path between stations a and b in the disrupted and in the undisrupted network) shows if the shortest path in the undisrupted network passes through line section e. If $q^e_{ab} = 0$, then line section e is not part of the shortest path between stations a and b neither in the network represented by G° nor in the one represented by G^e. If $q^e_{ab} > 0$, then by deleting line section e the length or duration of the shortest path between station a and b increases compared to the shortest path in undisrupted network. This means that line section e was part of the shortest path in the undisrupted network but there is still a non-infinite route between stations a and b in the disrupted network.

4.2   *The redundancy index*

The Network Robustness Index measures the increase in the total network trip length or the total network travel time in the case of the deletion of a line section. But on the disruption of line section f the exact route of the shortest path between stations a and b changes compared to the shortest path in the undisrupted network.

Let us assume that the shortest path between stations a and b in G° did not pass through line section e but in G^f it does. How much would be the additional increment in the shortest path if e would be deleted, too? In other words, how much would the shortest path between stations a and b be longer in the G^{ef} graph than it is in the G^f graph, i.e., we want to know how much total increase is caused by deleting not only line section f but also line section e for those paths that did not pass through line section e in graph G° but did pass through in graph G^f. This increase is the redundancy provided by line section e to line section f. Paths that pass through line section e neither in graph G° nor in graph G^f or pass through it in both are not relevant since they are not sensitive for the disruption of line section e.

Therefore, only those shortest paths are taken into account for which $q_{ab}^e = 0$. The r^{ef} redundancy index is defined by the sum of the increase of the shortest paths in G^{ef} compared to the sum of the shortest paths in G^f:

$$r^{ef} = q^{ef} - q^f = \left(c^{ef} - c \right) - \left(c^f - c \right) = c^{ef} - c^f.$$

(3)

By calculating r^{ef} for all f line sections that are not identical with e and summing them up one gets the total redundancy that line section e provides to line section f:

$$r^e = \sum_f r^{ef} = \sum_f \left(q^{ef} - q^f \right) = \sum_f \left(c^{ef} - c^f \right).$$

(4)

This definition was introduced by (Jenelius 2010).

4.3 *Application on 1-edge-connected graphs*

It can be seen from the definition, that if such line section(s) are deleted from the graph that makes at least one station unreachable from the others, the value of both q^e and r^e becomes infinity. The railway network of Hungary has this property, which means that the graph describing it is a so-called 1-edge-connected graph. In several cases, by deleting only one line section from the G° graph the graph will remain connected.

However, if two line sections are deleted, the number of reasonable results will rapidly decrease. If all these line sections were excluded from the calculations, only a few would remain and if only those line sections were excluded which give infinity as a result in that particular calculation, then different line section would be taken into account for each f line section, which would make the obtained r^e values incomparable to each other.

Therefore, it is practical to use the reciprocals of the travel time and trip length values of the shortest paths. By changing the order in which the difference is calculated in the summation of Equation 2, the redundancy index remains positive since longer distances mean shorter values in the reciprocal space.

By summing the values of the redundancy indices calculated in the reciprocal space for all f line sections one gets the total redundancy of a line section e:

$$\sum_f r_{\ell}^{ef'} = \sum_f \left(c_{\ell}^{f'} - c_{\ell}^{ef'} \right) = \sum_f \left(\sum_{\langle a,b \rangle} \frac{1}{\ell_{ab}^f} - \sum_{\langle a,b \rangle} \frac{1}{\ell_{ab}^{ef}} \right),$$

(5)

$$\sum_f r_t^{ef'} = \sum_f \left(c_t^{f'} - c_t^{ef'} \right) = \sum_f \left(\sum_{\langle a,b \rangle} \frac{1}{t_{ab}^f} - \sum_{\langle a,b \rangle} \frac{1}{t_{ab}^{ef}} \right).$$

(6)

However, it is more informative to normalize these values with values of the total trip length or the total travel time of the undisrupted network (which value is denoted by c_{ℓ}' and c_t', respectively):

$$r_\ell^{e'} = \frac{\sum_f r_\ell^{ef'}}{c'_\ell} = \frac{\sum_f \left(c_\ell^{f'} - c_\ell^{ef'}\right)}{c'_\ell} = \frac{\sum_f \left(\sum_{\langle a,b\rangle} \frac{1}{\ell_{ab}^f} - \sum_{\langle a,b\rangle} \frac{1}{\ell_{ab}^{ef}}\right)}{\sum_{\langle a,b\rangle} \frac{1}{\ell_{ab}^0}}, \tag{7}$$

$$r_t^{e'} = \frac{\sum_f r_t^{ef'}}{c'_t} = \frac{\sum_f \left(c_t^{f'} - c_t^{ef'}\right)}{c'_t} = \frac{\sum_f \left(\sum_{\langle a,b\rangle} \frac{1}{t_{ab}^f} - \sum_{\langle a,b\rangle} \frac{1}{t_{ab}^{ef}}\right)}{\sum_{\langle a,b\rangle} \frac{1}{t_{ab}^0}}. \tag{8}$$

The $r^{e'}$ redundancy index is the total relative decrease in the reciprocal trip length or travel time for those shortest paths that do not pass through the line section e in the undisrupted network but pass through it in the case of the disruption of line section f with line section e fixed for the calculation.

However, because of the definition, the redundancy values of line sections calculated in a specific graph cannot be used to compare with values obtained for line sections in other graphs. They have meaning only in that specific graph that they were calculated for and also only relative to the redundancy values of other line sections.

5 LINE SECTIONS PROVIDING REDUNDANCY FOR THE RAILWAY NETWORK OF HUNGARY

Calculating the redundancy values for all the line sections in the railway network of Hungary the network elements that provide an alternative route with the smallest increase in the length or duration of the shortest paths are identifiable. The numbering of railway lines according to the Hungarian State Railways (Magyar Államvasutak, MÁV) is used to refer to the lines, which accessible also at (VPE2019).

5.1 Minimal trip lengths

By calculating the value of the $r_\ell^{e'}$ index for all e line sections in graph G_ℓ, the results mapped in Figure 3 are obtained.

The results show that the electrified single tracked section between stations Görögszállás and Mezőzombor on line 80 has the highest redundancy value with 32%. The alternative routes using it have the highest relative increase in their total length were this line section unusable. The Tisza bridge at Tokaj is part of this line section which, though not thought as a particularly important crossing (compared to the Tisza bridge at Szolnok), is, in fact, vital for the network due to its high redundancy value.

The line sections with the next two highest redundancy value are also part of line 80: the Szerencs–Mezőzombor ($r_\ell' = 27\%$) and the Felsőzsolca–Miskolc-Tiszai ($r_\ell' = 22\%$) line sections. These three-line sections are part of the same Miskolc–Nyíregyháza line section, which connects the Budapest–Debrecen–Nyíregyháza line (line 100/100a) and the Budapest–Hatvan–Miskolc line (line 80a/80), both part of the Trans-European Transport Networks and handling heavy freight traffic. In the case of the disruption of these main lines, the Miskolc–Nyíregyháza line section in question is the shortest, and therefore, the most practical direction to reroute the traffic.

Lines 2 and 4 between stations Angyalföld and Almásfüzitő also have a significant 14% redundancy. This redundancy value is (or, in fact, were to) given mainly to the Southern Railway Bridge at Budapest. Though in theory paths through these lines are good alternatives being only slightly longer than the section of line 1 between Budapest and Almásfüzitő (trough Tatabánya), in reality, it is not true due to not only the poor condition of the tracks of line 4 between Dorog and Almásfüzitő but also because being single tracked and not electrified in its entire length.

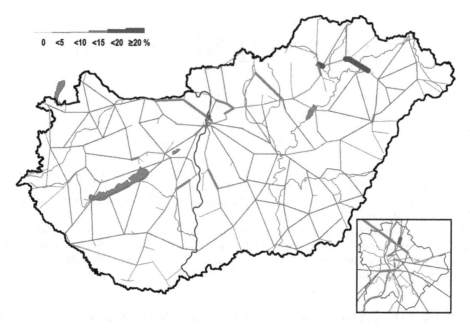

Figure 3. Redundancy values of the line section of the railway network of Hungary for paths with minimal trip length.

On the contrary, the 13% redundancy value of the single tracked but electrified line 77 between stations Galgamácsa and Vácrátót is in good agreement of the original purpose of this line: to provide a short rerouting alternative for the freight transport directed towards Western Europe through Slovakia in the case of the disruption of the main lines in and around Budapest. However, to fully utilize the advances of this route, the development of the connecting Vác–Vácrátót section of line 71 and the Galgamácsa–Aszód section of line 78 is also necessary.

At this point, the dangers in the decrease in the number of diesel locomotives have to be pointed out. As the electrification of the railway lines goes on, less and less diesel engines are needed. However, the electric network is also a critical infrastructure. Either in the case of the disruption of the overhead lines or disturbances in the electricity supply, the results are the same as if the tracks themselves were damaged if there is no sufficient number of diesel locomotives to take over. Because of this, even though using electric locomotives under normal operating conditions is much cheaper than using diesel locomotives, the lack of enough usable diesel engines can have severe economic consequences.

5.2 *Minimal travel times*

The values of index $r_t^{e'}$ for all e line sections in graph G_t, are mapped in Figure 4.

The results are similar for line 80: the line section that has the highest redundancy ($r_t' = 48\%$) is the line section between Görögszállás and Mezőzombor. The neighboring line sections also show high redundancy: the value of r_t' for the Mezőzombor–Szerencs line section is 42%, for the Felsőzsolca–Miskolc-Tiszai line section 36%, and for the Görögszállás–Nyírtelek line section 31%. Therefore, the part of line 80 between Miskolc and Nyíregyháza is the most effective alternative route between lines 80 and 100 both for trip lengths and travel times.

The line section between Debrecen and Apafa has 32% redundancy value, which indicates its bottleneck nature: if disrupted, the alternative routes run via the transversal lines between lines 80 and 100, which have lower line speeds and are therefore long detours.

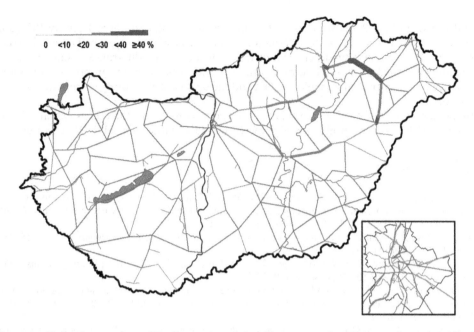

0 <10 <20 <30 <40 ≥40 %

Figure 4. Redundancy values of the line section of the railway network of Hungary for paths with minimal travel time.

The redundancy of lines 2 and 4 for travel times is much less pronounced as it was for trip lengths because of the conditions of line 4, but it is still not negligible. By developing line 77 between Galgamácsa and Vácrátót (and the connecting Vác–Vácrátót section of line 71 and the Galgamácsa–Aszód section of line 78), both ways could be used as an alternative for the Southern Railway Bridge.

The small redundancy value ($r_t' = 3\%$) of the Southern Railway Bridge means that the rerouting of those alternative routes in the case of the disruption of the bridge that does not pass through it in the undisrupted network does not have a large effect on the network as a whole. This is because the traffic of the Northern Railway Bridge in the undisrupted network is almost exclusively the traffic of line 2, which is so small that the rerouting of these paths towards Baja (in spite of the very large increase in the travel time) does not affect much the performance of the network as a whole. For the paths crossing the Danube at Baja in the undisrupted network, the detour via Budapest increases the travel time so much that crossing the Danube at either through the Southern or Northern Railway Bridge does not make a big difference.

6 CONCLUSION

The railway network of Hungary is one of the densest in the world. The benefits, however, can only be reaped if the infrastructure is maintained and developed properly. This has to include not only the lines with heavy traffic but also the lines with high redundancy, even if these lines have little traffic under normal operational conditions, in order to provide sufficient alternative routes in the case of disruptions.

The lines with high redundancy include the section of line 80 between Miskolc and Nyíregyháza, which is not only part of the RFC corridor 6 (Mediterranean) but also the connection between its two branches in Hungary. Being single-tracked in most of its length, this is the line section that has to be developed first by building a second track to be able to handle the increased rerouted traffic were lines 80 or 100 disrupted.

Paths through lines 2 and 4 are potential alternatives for the Southern Railway Bridge in Budapest, through which all of the international freight transport crosses the Danube in Hungary being the common network element of RFC corridors 6 (Mediterranean) and 7 (Orient). By building a second track and electrifying them in their whole length can provide a possible alternative route for the only double-tracked and the only electrified bridge across the Danube in the country.

By building a second track for line 77 (and its connecting lines), these lines can also be an alternative of the Southern Railway Bridge by rerouting the traffic directed to Western Europe.

REFERENCES

Almoghatawi Y. & Barker, K. & Rocco, Claudio M. & Nicholson, Charles D. 2017. A multi-criteria decision analysis approach for importance identification and ranking of network components. *Reliab. Eng. Syst. Safe.* 158: 142–151. DOI: 10.1016/j.ress.2016.10.007.

Braess, D. 1968. Über ein Paradoxon aus der Verkehrsplanung. *Unternehmensforschung* 12(1): 258–268. DOI: 10.1007/BF01918335.

Braess, D. & Nagurney, A. & Wakolbinger, T. 2005. On a paradox of traffic planning. *Transp. Sci.* 39 (4): 446–450. DOI: 10.1287/trsc.1050.0127.

Cats, O. & Jenelius, E. 2016. Beyond a complete failure: the impact of partial capacity degradation on public transport network vulnerability. *Transportmetrica B* 6(2): 77–96. DOI: 10.1080/21680566.2016.1267596.

Cats, O. & Yap, M & van Oort, N. 2016. Exposing the role of exposure: Public transport network risk analysis. *Transp. Res. A* 88: 1–14. DOI: 10.1016/j.tra.2016.03.015.

Csardi, G. & Nepusz, T. 2006 The igraph software package for complex network research. *InterJournal,* Complex Systems 1695. http://igraph.org.

Derrible, S. & Kennedy, Ch. 2010. The complexity and robustness of metro networks. *Physica A* 389: 3678–3691. DOI: 10.1016/j.physa.2010.04.008.

Derrible, S. & Kennedy, CH. 2011. Applications of Graph Theory and Network Science to Transit Network Design. *Transp. Rev.* 31(4): 495–519. DOI: 10.1080/01441647.2010.543709.

Disser, Y. & Matuschke, J. 2017. The Complexity of Computing a Robust Flow. arXiv: 1704.08241.

Jenelius, E., Petersen, T. & Mattson, L G., 2006. Importance and exposure in road network vulnerability analysis. *Transp. Res. A* 40: 537–560. DOI: 10.1016/j.tra.2005.11.003.

Jenelius, E. 2010. Redundancy importance: Links as rerouting alternatives during road network disruptions. *Proc. Eng.* 3: 129–137. DOI: 10.1016/j.proeng.2010.07.013.

Kansky, K. & Danscoine, P. 1989. Measures of network structure. In: *Flux,* numéro spécial, 89–121; DOI: 10.3406/flux.1989.913.

Laszka, A & Szeszlér, D & Buttyán, L. 2012. Linear Loss Function for the Network Blocking Game. In: *Decision and game theory for security. Third international conference, GameSec 2012, Budapest, Hungary,* November 5–6, 2012. DOI: 10.1007/978-3-642-34266-0_9.

Latora, V. & Marchiori, M. 2001. Efficient behavior of small-world networks. *Phys. Rev. Lett.* 87: 198701. DOI: 10.1103/PhysRevLett.87.198701.

Lin, J. & Ban, Y. 2013. Complex Network Topology of Transportation Systems. *Transp. Rev.* 33(6): 658–685. DOI: 10.1080/01441647.2013.848955.

Oliveira, E.L. & Portugal, L. D. & Junior, W. P. 2016. Indicators of reliability and vulnerability: Similarities and differences in ranking links of a complex road system. *Transp. Res. A* 88: 195–208. DOI: 10.1016/j.tra.2016.04.004.

R Core Team, 2012. R: A language and environment for statistical computing. R Foundation for Statistical Computing. Vienna, Austria. ISBN 3-900051-07-0, URL http://www.R-project.org/.

Scott, D.M., Novak, D., Aultman-Hall, L. & Guo, F. 2006. Network Robustness Index: A New Method for Identifying Critical Links and Evaluating the Performance of Transportation Networks. *J.Transp. Geogr.* 14(3): 215–227. DOI: 10.1016/j.jtrangeo.2005.10.003.

Snelder, M. & van Zuylen, H.J. & Immers, L.H. 2012. A framework for robustness analysis of road networks for short term variations in supply. *Transp. Res. A* 46(5): 828–842. DOI: 10.1016/j. tra.2012.02.007.

Tóth, B. 2017. Állomások és állomásközök zavarának gráfelméleti alapú vizsgálata a magyarországi vasúthálózaton. *Hadmérnök.* 12(4): 52–66.

Tóth, B. 2018a: Menetidő- és menetvonalhossz növekedés gráfelméleti alapú vizsgálata a magyarországi vasúthálózaton állomások és állomásközök zavara esetén. *Hadmérnök* 13(1): 118–132.

Tóth, B. 2018b: A magyarországi vasúthálózat zavarainak gráfelméleti alapú vizsgálata. In: Horváth, B. & Horváth, G. & Gaál, B. (Eds.): *Közlekedéstudományi Konferencia*, Győr, Széchenyi István Egyetem Közlekedési Tanszék. 505–519. (ISBN 9786155776137).

UNECE 2019. https://w3.unece.org/PXWeb/en/CountryRanking?IndicatorCode=47.

VPE 2019. http://www.vpe.hu/takt/vonal_lista.php.

Solutions for Sustainable Development – Szita, Jármai & Voith (eds.)
© 2020 Taylor & Francis Group, London, ISBN 978-0-367-42425-1

A short discussion on some influencing factors of an artificial corrosion system and obtained metallic pipe samples

H.D. Thien, B. Kovács & T. Madarász
Institute of Environmental Management, University of Miskolc, Miskolc-Egyetemváros, Hungary

I. Czinkota
Szent István University, Gödöllő, Hungary

ABSTRACT: Twentieth century beginning witnessed the rapid emergence of steel and iron pipe transportation in process industries. At the same time, potable water reticulation metallic pipe has been at least around 100 years old and fully-grown up. There is a possible threat of stray current posing to metallic materials and structures in daily reality as well. Either salty, aqueous environments or the backfill soil in vicinity of distribution conduits usually encompasses oxygen, moisture content and is inadequately resistive to permit passage of electrical current; under these circumstances, buried steel structures are subject to corrosion. As a result, these decaying infrastructures are raising concern to consumers, producers, legislators and regulators. Severely external corrosion problem is one of the most considerable interest, often attributed to the corrosiveness of the aggressive surrounding environment, particularly saline soil, commonly in coastal areas. Taking advantage of electrolytic corrosion, an artificial system was adopted to intimate potentially hostile environment to closely scrutinize corrosion behavior of steel pipes, enabling us to estimate the effect of some important factors such as degradation rate, inspected areas under similar exposure conditions. Some of these parameters are readily available in corrosion databanks after investigating experimentally 82 relatively identical specimens yielded from corrosion tests.

Keywords: artificial corrosion, ruined pipe samples, influencing factors, metallic pipelines, stray current

1 INTRODUCTION

Pipelines nowadays with their merits of consistency, cost effectiveness, environmental friendliness in transporting fluid products over great distances are becoming more and more popular all over the world (Morgan & Bruckner 1960, Orazem 2004). Flexibility in setting and manufacturing pipes such as various diameters, wall thicknesses, length and materials etc. gives them an unarguable advantage over other transportation modes such as truck/train (Kishawy & Gabbar, 2010). Above all, iron-based metals are well recognized as the most broadly employed material compared to the others type of metals (Norsworthy 2014). However, acting in an aqueous media like moist soils with the presence of stray current from time to time renders underground metal structure vulnerable to corrosion, particularly electrochemical degradation (Xu et al. 2014). From the onset of this electrolysis corrosion process, the deterioration commences accelerating and will soon burst distribution conduits to spread out the flammable and toxic material. Its potential disastrous effects could be an environmental hazard, potential fire events and even become a human catastrophe, not to mention huge losses of transported products and the cost of replacement for this high capital investment system itself. Recent trending study has made attempts at analyzing the joint effect of stray current and stresses on buried metallic structure (Wang 2014). The persistent obstacle to obtaining enough the required number of corrosive pipes from industry is usually very time consuming. Hence, the aim of current study is to propose a controlled degradation system acting on metallic pipe under the principle of electrochemical corrosion mechanism. The corrosion mechanism complies with Faraday's Law, and thus it is viable to measure the current density and judge the leakage current corrosion hazard. Examination the influencing factors are also of interest to this work.

2 RELATED ELECTROCHEMICAL THEORY

The underlying cause of corrosion on steel objects' surface in corrosive environments (e.g. water, wet saline soils, etc.) is the establishment of voltaic cells prone to iron dissolution at some positions. Whilst, at other points, oxygen existing in moist air produces hydroxide ions. It is known that the low solubility of oxygen in water droplets minimizes oxygen diffusion rate onto the metal surface; as progressing, a development of passive film on the metal surface will somehow insulate and impede the corrosion. As a result, the aforesaid routine is consequently slowed down as well. The iron isolation process can be hastened and controlled by coupling iron objects with an insoluble cathode and put both into an electrically conductive liquid (e.g. seawater or gypsum suspension). Essentially, the ions of the soluble salts do not take part in the reaction; however, acting as a noble catalyst, nascent chlorine aids in fleetingly speeding up the segregated iron process at anode. Sacrificial iron anode is dissolved while at the cathode (particularly made of copper) hydrogen gas evolves. These two processes are exhibited in the two forthcoming equations, concurrently take place to sustain charge neutrality; otherwise, a large negative charge would swiftly develop between the metal and the electrolyte and the corrosion process would stop:

Anodic process: $Fe = Fe^{2+} + 2e^-$; Cathodic process: $2H^+ + 2e^- \rightarrow 2H \rightarrow H_2 \uparrow$

If we know the current passing through the network, which is computed easily with measured voltage drop U across a given resistor, as a function of elapse time, the amount of ionized iron can be calculated in accordance with the Faraday constant. As we know that Fe^{2+} cation formed in this process liberates two electrons to the electrode during this oxidation reaction, the formula applied is as follows:

$$m = \frac{M}{F \cdot Z} \cdot \int_0^T I \cdot dt \qquad (1)$$

Where:

m: iron mass entering the solution/suspension [mg]
I: the electrical current [mA] flowing through a 1 Ω resistor derived from the voltage drop U
T: total aggregated time [sec]
M: the relative atomic mass of iron, 56 [g.mol^{-1}]
F: Faraday constant, 96500 [C.mol^{-1}]
Z: ion charge, 2 [-]

On the opposite side, two other electrons must be consumed (reduction reaction). Note that these freed electrons move only in an external path (electrical cords), not in the electrolyte.

The removed iron mass could be computed using the integral of the continuous function of corresponding electrical current and time during the experiment (Figure 1.). As the hydrogen ion concentration of the solution declines during the cathodic process, the constant value of the water ionization brings about the enrichment of hydroxide ions in the solution. Portion of the iron ions in the aqueous mixture reacts directly according to the following equation to form a grayish-green precipitate of iron hydroxide:

$$Fe^{2+} + 2\,OH^- = Fe(OH)_2$$

With the oxygen diffused from the atmosphere, the other part reacts and deposits at the bottom of the chamber, conforming to the following equation:

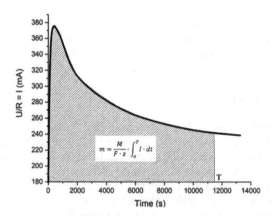

Figure 1. General concept of removed iron mass calculation.

$$4\,Fe^{2+} + O_2 + 2\,H_2O + 8\,OH^- = 4\,Fe(OH)_3$$

On account of the aforementioned chemical reactions, a major portion of the generated ions is settled in the form of precipitates, the concentration resulted from the amount iron ions left could hardly affect the ongoing process. Literally interpreting, the iron-ion concentration in the solution inside the corrosion chamber could not be computed by the removed iron amount.

3 MATERIAL AND WORKING PRINCIPLE

With the goal of examining the corrosion occurring to a small area on the outer surface of a common pipe in process industries, coated metallic pipes were chosen to be the samples in these test series. Metallic pipes in conformity with European Standard EN10220 (MSZ EN 10220:2003) are favorably utilized in fluid transportation piping network. Its information on dimensions and weights are: 21.3 mm of outside diameter, 1.8 mm of wall thickness, 0.866 kg.m^{-1} specific mass. The mechanical properties of this sort of electrically welded non-alloy steel tube with specified low temperature properties acquired from European Standard EN 10217-4:2002.

Considerable caution has been given for all steps of sample preparation. Tube samples were carefully brushed by making use of acetone-absorbed cleaning tissue to remove industrial oil on their surfaces left from manufacture process. Filling up the bottom end of each tube with silicon in order to preclude the incursion of coating liquid and electrolyte from entering the tube in the coating and experimental process, respectively. With the exception of some plain surface sites of circular shape sized either 6.3 mm, 8.45 or 10 mm, the entire exteriors of these specimens then were thoroughly coated. These perennial processes were done with 3 alternatives for number of sites (1, 2 or 3), 2 possibilities for directions of spots (either longitudinal or circumferential direction) and 3 choices of sizes (small, medium, large) (Figure 2 (a) – before & (b) after corrosion test). Thus, an aggregation of 82 samples were provided, prepared to carry out the corrosion experiment.

Either artificial seawater solution with exact 35 g salt in 1 dm^3 of distilled water or saturated gypsum solution as suspension fluid was made. Although the solubility of gypsum in water at 20°C is 2.531 g·l^{-1}, approximately 140 times lower than the solubility of table salt (360 g·l^{-1}) (Klimchouk, 1996), thorough stir in distilled water assure as much gypsum and oxygen as

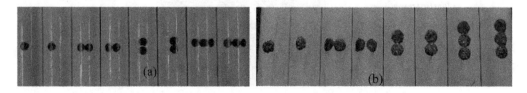

Figure 2. Pipe samples with spots on before (a) and after (b) corrosion test.

possible must be dissolved into solvent. Accordingly, we have the simulated seawater and gypsum solution with electrical conductivity values of 52 mS·cm^{-1} and 2.32 mS·cm^{-1} respectively, converting them into reasonable electrolytes of corrosion test.

4 EXPERIMENTAL SETTING

A set of 12 electrical cords correspondingly pass from both sides of an adaptable direct current power supply. On the negative side, there will be four leads connect to four small copper tubes acting as four cathodes. On the reverse, eight other electrical connections enable currents to flow through eight interchangeable resistors (1 Ω) before being affixed to eight anodes, which are carbon steel tube samples for examining (Figure 3). All of these eight daubed anodes are then installed evenly, upright and symmetrically alongside the four existing cathodes in a plastic frame (Figure 4). Imposed potential can be changed by an adjustable constant current power supply. However, within these investigating series, anodic-hold potential was always set at value 4.1 V.

Chronologically collected potential data from data logger was employed to calculate eight corresponding currents passing through eight pairs of anode-cathodes. These primary figures require one further simple calculation to obtain the iron released mass from each anode.

5 RESULTS

Several determining factors of the corrosion might be referred to incorporating incubating time, aqueous media type, number and position of sites per anode, temperature, together with sample's chemical components. Quantitative analysis was conducted into a group of comparable data, fairly similar in values, which are the removed iron masses from the identical pipe samples at a given time of testing (Figure 5.). In the same manner, another characteristic parameter is specific removed iron mass (Figure 6). This parameter eliminates the effect of spot size by showing the virtually overlapped falling range in value on this figure. The slope derived from best-fit line of the removed iron mass versus time curve reveals information about the

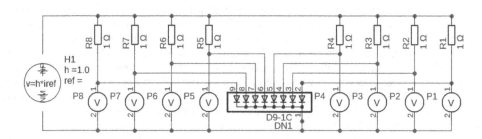

Figure 3.   Electric scheme of the measurement system.

Figure 4.   The artificial corrosion device set-up in reality.

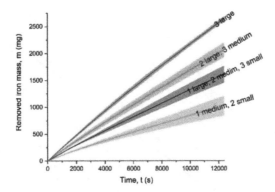

Figure 5. Common range of removed iron mass versus time for various spot sizes and numbers in the experiments.

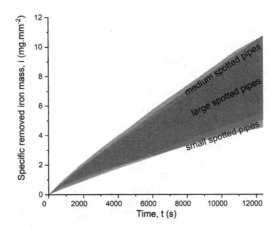

Figure 6. Common range of specific removed iron mass versus time for various spot sizes in the experiments.

isolation rate of iron in the course of the electrochemical corrosion process, which is already a time-independent parameter (Figure 7).

$$i = m_t/S \tag{2}$$

Where:
    i: corrosion intensity or specific removed iron mass [mg·mm^{-2}]
    m_t: time-lapse mass of iron entering the solution [mg]
    S: the exposing surface area [mm^2]

$$v = \partial m_n/\partial t \tag{3}$$

Where:
    v: the slope value, indicating dissolution rate of iron-56 [mg·s^{-1}]
    m_n: net removed iron mass [mg]
    t: time [sec]

$$a = i/\partial t \tag{4}$$

Where:
    a: Corrosion rate [mg·mm^{-2}·s^{-1}]

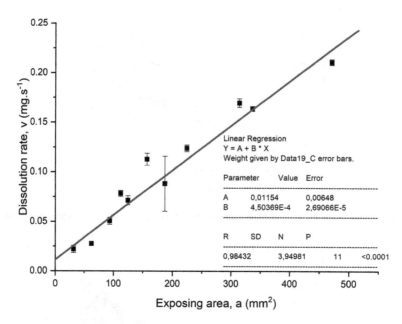

Figure 7. Removed iron mass on different size spotted pipe in a unit of time.

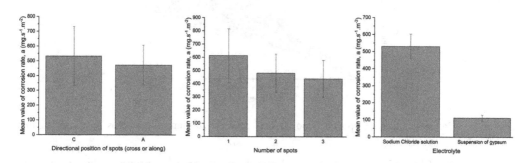

Figure 8. Difference of corrosion rate referring to different driving factors.

ANOVA quantitative analysis was taken advantage to examine the impact of directional scattering of spots, spot sizes, electrolyte varieties on corrosion rate of comparable samples, which scientifically proved the peripheral and significant factors and their interdependence (Figure 8).

Comparative analyses of the corrosion test result indicated that the corrosion intensity (specific iron mass removal) is numerically unconnected from either the number, size or direction of the sites, along with the position of the examined tube in the corrosion chamber. However, it hinges critically upon the electrolyte type (i.e. sodium chloride solution or gypsum suspension) and potentially on the pipe composition which was not investigated in this work.

6 CONCLUSIONS

The derived result of this research has disclosed no crucial distinction in rate of the corrosion intensity between pipes with divergence of number of spots, spot direction, and incubation time in a constant temperature condition of the experimental chamber. The relatively unanimous outcome of corrosion intensity obtained from almost a hundred corrosion tests have implied that there was nothing but random errors from this synthetic corrosion system measured in each working liquid. The experimental examinations are replicable and logical; hence, the trustworthiness of this artificial corrosion installation is testified.

Further evaluations of the size, shape and depth of corroded sites could be carried out with high resolution microscopy techniques (Souto et al. 2010) to attain more reliable data from ruined tubes. Diverse scenarios of hostile working environments (i.e. different kinds of soils, different electrolyte strengths, etc.) might be imitated to bring other relevant risk factors of pipes to light. Remaining strength examination on corroded metallic pipes will presumably yield a favourable outcome regarding their mechanical degradation.

ACKNOWLEDGEMENTS

The described study was carried out as part of the EFOP-3.6.1-16-2016-00011 "Younger and Renewing University – Innovative Knowledge City – institutional development of the University of Miskolc aiming at intelligent specialisation" project implemented in the framework of the Szechenyi 2020 program. The realization of this project is supported by the European Union, co-financed by the European Social Fund.

The research was carried out within the GINOP-2.3.2-15-2016- 00031 "Innovative solutions for sustainable groundwater resource management" project of the Faculty of Earth Science and Engineering of the University of Miskolc in the framework of the Széchenyi 2020 Plan, funded by the European Union, co-financed by the European Structural and Investment Funds.

We would like to thank to our colleagues Mr. András Sebők, Mr. György Czinkota, Mr. Kövesi Viktor, who worked as our supporting team, for their help and professional advice in preparing samples and collecting data. Special thanks to the Institute of Environmental Management, the Faculty of Earth Sciences and Engineering, University of Miskolc for letting us using tools and instruments.

REFERENCES

Guillerme, A. 1988. The Genesis of Water Supply, Distribution, and Sewerage Systems in France, 1800-1850. In Technology and the Rise of the Networked City in Europe and America, edited by Joel Tarr and Gabriel DuPuy, 89–115, Philadelphia: Temple University Press.

Hassan, J.A. 1985. "The growth and impact of the British water industry in the nineteenth century." Economic History Review: 531–547.

Hu, Y. & D. Hubble 2007. Factors contributing to the failure of asbestos cement water mains. *Canadian Journal of Civil Engineering* 34(5): 608–621.

Kishawy, H.A. & H.A. Gabbar 2010. Review of pipeline integrity management practices. International *Journal of Pressure Vessels and Piping* 87(7): 373–380.

Klimchouk, A. 1996. The dissolution and conversion of gypsum and anhydrite. International Journal of Speleology 25(3): 2.

Morgan, J.H. & Bruckner, W. 1960. Cathodic protection. Journal of The Electrochemical Society, 107, 138C.

Norsworthy, R. 2014. Understanding corrosion in underground pipelines: basic principles. Underground Pipeline Corrosion, Woodhead Publishing: 3–34.

Orazem, M.E., Reimer, D.P., Qiu, C., & Allahar, K. 2004. Computer simulations for cathodic protection of pipelines. Proceedings of the Corrosion/2004 Research topical Symposium: Corrosion Modeling for Assessing the Condition of Oil and Gas Pipelines, F. King and J. Beavers (eds.), Houston, TX: NACE/2004.

Souto, R.M., S.V. Lamaca & S. González 2010. Uses of scanning electrochemical microscopy in corrosion research. Microscopy: Science, technology, applications and education 3: 1769–1780.

Wang, X., et al. 2014. Synergistic effect of stray current and stress on corrosion of API X65 steel. *Journal of Natural Gas Science and Engineering* 21: 474–480.

Shao-yi Xu, Wei Li, Fang-fang Xing & Yu-qiao Wang 2014. Novel predictive model for metallic structure corrosion status in presence of stray current in DC mass transit systems. *Journal of Central South University* 21 (3): 956–962.

Author Index

Abdessemed Foufa, A. 187, 227
Agárdi, A. 151
Aguir, H. 25
Akkad, M.Z. 160
Al-Fatlawi, A. 3
Amara, M. 101, 115, 123
Arrar, H.F. 187

Babcsán, N. 205
Bartha, Z. 307
Belgacem, M.E. 12
Bihari, J. 20
Bányai, Á. 169
Bányai, T. 151, 160
Bodnár, I. 108
Bodnárné, R.S. 211, 267
Bognár, G. 129
Boros, R.R. 108
Bouaicha, A. 67, 73
Boudali Errebai, F. 115, 123
Brara, A. 12
Bézi, Z. 86

Chahboub, Y. 25
Chikh, S. 115
Chován, P. 235
Czinkota, I. 371

Derradji, L. 115, 123

Erdősy, D. 108

Fehér, Zs. 245
Ferdjani, A 12
Filali, A. 101

Ghafil, H.N. 33
Gál, B.S. 211
Gáspár, M. 347
Gubik, A.S. 307

Hamid, A. 337
Hardai, I. 169
Hornyák, O. 49, 287
Horváth, Á.K. 218

Illés, B. 169
Ismail, S.A.A. 294
István, Zs. 211, 267

Jármai, K. 3, 33, 56

Kaoula, D. 187, 227
Kis-Orloczki, M. 218
Klazly, M.M. 129
Kállai, V. 41
Koncsik, Zs. 235, 316
Kovács, B. 371
Kovács, G. 3
Kovács, L. 33, 151

Limam, A. 123
Lukács, J. 235, 325

Madarász, T. 371
Mannheim, V. 245, 254
Mileff, P. 49, 94
Mizsey, P. 41
Mobark, H.F.H. 325

Naimi, A. 337
Nehéz, K. 49, 94

Oukaci, S. 337

Petrik, M. 56

Rafa, S.A. 67, 73
Rouaz, I. 67, 73

Sami, S. 337
Sarka, F. 81
Semmar, D. 337
Semmari, H. 101
Siménfalvi, Z. 86, 139, 245
Simon, A. 254
Singh, B. 139
Sisodia, R.P.S. 347
Somogyiné Molnár, J. 108
Spisák, B. 86
Szabó, F.J. 178
Szabó, M. 94
Szabó, N. 108
Szabolcs, S. 25
Szalay, Zs. 294
Szamosi, Z. 139
Szepesi, L.G. 41, 56
Szita, K.T. 267
Szávai, Sz. 86

Takács, Á. 196
Takácsné Papp, A. 218
Terjék, A. 115, 123
Thien, H.D. 371
Tóth, B.G. 358
Tóth, L.T. 108

Velősy, A. 275
Voith, K. 254

Zajáros, A. 267